A RACE TO THE TOP

A RACE TO THE TOP
ACHIEVING THREE MILLION MORE APPRENTICESHIPS BY 2020

Edited
by
David Way

WINCHESTER UNIVERSITY PRESS

Published by the Winchester University Press 2016

First Published in Great Britain in 2016 by
The Winchester University Press
University of Winchester
Winchester SO22 4NR

British Library Cataloguing-in-Publication Data
A CIP catalogue record for this book is available
from the British Library.

ISBN: 978-1-906113-19-3

Printed and Bound in Great Britain

CONTENTS

PART FOUR
NAVIGATING THE APPRENTICESHIP SYSTEM

PART FIVE
DELIVERING APPRENTICESHIPS

ACKNOWLEDGEMENTS

I would like to thank the University of Winchester for the opportunity to produce this book on apprenticeships and especially Vice Chancellor, Joy Carter, and Director of the Winchester Business School, David Birks.

I also appreciate the continuing support of those many people with whom I worked as Chief Executive of the National Apprenticeship Service, who have contributed so enthusiastically to the book; and to more recent colleagues in learndirect and in Industrial Partnership for the energy and utilities sector.

Finally, I am most grateful to Stuart Youngs and Harry Smith at Purpose Ltd for the jacket and cover design.

David Way
May 2016

LIST OF CONTRIBUTORS

CONTRIBUTOR AND EDITOR

David Way was awarded a CBE for services to apprenticeships in 2011. He has been a leading player in the world of skills for more than a decade, including designing, establishing and running the National Apprenticeship Service from 2009 to 2013. This organisation was the government's principal vehicle for driving the changes needed to deliver an expansion in apprenticeships after they had been largely neglected for many years. David's involvement and leadership helped achieve the biggest growth in uptake of apprenticeships in a generation. He led the introduction of many innovations, including central online services to make it easier for young people to obtain apprenticeships and the introduction of higher apprenticeships at degree level. David has visited many countries to learn from international practice, including China, Australia, Germany and Switzerland. Since leaving his post as Chief Executive of NAS, David has been appointed as a Visiting Professor at the University of Winchester. He advises the energy sector on skills and is an Apprenticeship Ambassador for learndirect.

CONTRIBUTORS

Adrian Anderson is Chief Executive of the University Vocational Awards Council (UVAC). Established in 1999, UVAC is the HE and FE membership organisation with a mission to champion and mainstream innovation in vocational and work-based learning at all higher-education levels.

Emily Austin is Group Lead - Apprenticeship & School to Work Programmes at Lloyds Banking Group. She joined the Group in 2005 and has held a number of specialist HR roles, most recently in Early Careers and Emerging Talent Management. She has developed and led the Group's apprenticeship offering since its launch in 2012. Apprenticeships have transformed the way the bank sees emerging talent and the programme has become a core part of its talent proposition with commitment to create over 5,000 apprenticeship opportunities by 2017. Emily has led the design and delivery of a higher apprenticeship programme in a number of disciplines including finance, digital, marketing, IT and project management. She is a member of the Financial Services Employer Group responsible for the design of new industry apprenticeships. Emily is passionate about providing opportunities for young people to gain insight and access to business.

Sue Betts is the Director of Linking London. She is an educationalist with a varied career in and around education. She has worked as a Vice Principal with responsibility for curriculum, staff development and higher education in the further education sector. She has worked for a national awarding body, and two nationally-distributed e-learning organisations. She is an experienced project manager and developer of partnerships. Her professional interests include organisational culture and change management. She has worked as a consultant developing educational solutions for organisations. As Director of Linking London since 2006, Sue is responsible for the strategic direction and management of Linking London, a collaborative network of partners and associates, based at Birkbeck College, University of London.

David F. Birks is a Professor of Marketing, Dean of the Faculty of Business, Law and Sport and Director of Winchester Business School at the University of Winchester. Prior to working at Winchester, David worked at the universities of Southampton, Bath and Strathclyde. He has over thirty years' experience in universities, primarily working on postgraduate research, marketing, management and design programmes. David is co-author of Europe's leading Marketing Research text: Malhotra, N.K., Birks, D.F. and Wills, P., *Marketing Research, An Applied Orientation*, 4th European Edition, Pearson, 2012. David has continued to practise marketing research throughout his university career, managing projects in financial institutions, retailers, local authorities and charities.

Martyn Butlin is the External Communications Manager at EDF Energy, specifically looking after Heysham 1, Heysham 2 and Hartlepool nuclear power stations. EDF Energy is committed to encouraging young people to pursue careers in science and engineering, whether at apprentice or graduate entry level. At a company level, Martyn leads on the external communications regarding EDF Energy's Early Careers and Recruitment programme, linked particularly to apprentice recruitment.

Sara Caplan, as a Partner in PwC UK's Consulting Practice, led the Education and Skills business and has been a professional consulting leader since 2001. Sara joined PwC Australia's Management Consulting Practice, leading on Vocational Education and Skills, on secondment from PwC UK in January 2015. Her work in recent years in both the UK and Australia has focused on supporting government in delivering policy outcomes in skills and education, including by creating new access routes to professional careers. Sara was appointed by BIS to be an employer ambassador for higher apprenticeships in 2012. She was a founding member of the UK All Party Parliamentary Group on Education and Skills and of the Higher Education Commission. Sara chaired BIS's Professional and Business Services Council Skills taskforce and led the development of two new 'Trailblazer' apprenticeships in accounting and management consulting.

Mandy Crawford-Lee is a Senior Associate of UVAC and a freelance consultant specialising in higher and degree apprenticeships and higher-level skills. With over twenty years' experience of influencing skills policy, strategic planning and performance management in economic development, vocational education and training reform, Mandy is committed to engaging key partners and stakeholders in the development of apprenticeships and practical work-based learning solutions.

Matthew Creagh is the TUC Apprenticeships Policy Officer. He supports union projects, which are funded via the union learning fund, to deliver high quality apprenticeship opportunities. He project manages two European Commission funded projects which look to develop workplace learning strategies which will draw upon best practice from across Europe. One of the projects will set up a quality framework for apprenticeship programmes. He is the TUC representative on the European Trade Union Confederation (ETUC) Education and Training committee and on the board of the European Centre for the Development of Vocational Training (CEDEFOP).

John Cridland, CBE, joined the CBI as a policy adviser in 1982 and became its youngest ever director in 1991, moving on to human resources policy in 1995, where he helped negotiate the UK's first national minimum wage and entry into the European Union's 'social chapter' on employment conditions. He was promoted to the post of Deputy Director-General in 2000 and became the organisation's tenth Director-General in 2011. John stepped down as Director-General in November 2015 and was appointed Chair of Transport for the north. Beyond his work with the CBI, John served on the Low Pay Commission from its formation in 1997 until 2007. He was Vice-Chair of the National Learning and Skills Council between 2007 and 2010 and spent ten years on the council of the conciliation service, ACAS. He was also a member of the Women and Work Commission; a Board member of Business in the Community; a UK Commissioner for Employment and Skills; and a member of the council of Cranfield University.

Craig Crowther is the Network Development Manager at GTA England where he has responsibility for delivering a Gatsby funded project to increase the number of STEM apprenticeships being delivered by the Group Training Association network. Craig spent the first fifteen years of his career in senior roles including with the TTE Technical Training Group who train over 400 apprentices a year. He has led a number of high profile national sector projects including the establishment of two National Skills Academies. Since 2014 he has worked for GTA England.

Kirstie Donnelly MBE is the Managing Director of City & Guilds. Kirstie is a passionate advocate of the role of digital and social media in transforming how people live, learn and work. She was awarded an MBE for her services to e-learning in the FE/Adult sector in 2011. Kirstie's career highlights include nine years with learndirect where she created the first large-scale brand and online offer in adult learning. She also transformed Video Arts into an online digital business, produced a range of award-winning DVD and e-learning training products and created a first of its kind 'virtual chamber of commerce'. Kirstie has helped to shape policy around e-learning in the UK and was part of the Task Force Group that established the Green paper: 'Learning Age'. In 2014, Kirstie was appointed as a commissioner on the future of apprenticeships.

Dr Dominik Furgler is Swiss Ambassador to the United Kingdom. He obtained his degree and his PhD in Law at the University of Fribourg (Switzerland) before joining the Swiss Federal Department of Foreign Affairs (FDFA) in 1985. After a series of postings in Paris and London and several appointments within the FDFA in Berne, Ambassador Furgler was appointed Head of Mission at the Swiss Embassy in Cairo in 2009. He returned to London to take up his current post in July 2013.

Dr Jacqueline Hall has been instrumental in establishing a new Energy and Efficiency Independent Assessment Service at Energy and Utilities Skills, working on pedagogy, funding, and pre-implementation planning/specification design. She is responsible for the strategy and senior project management of the new apprenticeship standards and assessments for the power, water, waste and gas industries. Jacqueline has worked at a senior strategic level with FE and within a UK plc., including on strategic recovery projects for both SMEs and public sector education institutions. From 2006, she has undertaken expert evaluation within the EU Research Framework, Horizon 2020, Framework 7, ICT Policy Support Programme and Erasmus. She has also worked as an associate inspector for Ofsted.

David Harbourne is Director of Policy and Research at the Edge Foundation, which champions all forms of technical, practical and vocational learning. He was involved in the design of Modern Apprenticeships at the National Association of Master Bakers and the Hospitality Training Foundation; apprenticeship delivery by the Hotel and Catering Training Company; apprenticeship funding by North Yorkshire Training and Enterprise Council and the Learning and Skills Council; and apprenticeship inspection as a non-executive.

Steve Holliday is the Chief Executive of National Grid plc. and was a non-executive director of Marks & Spencer plc. from 2004 to 2014. He is chairman of the board of trustees at Crisis, the homeless charity. Steve joined National Grid Group as the board director responsible for the UK and Europe in March 2001, becoming Chief Executive for the group in 2007. Prior to joining National Grid he was on the board of British Borneo Oil and Gas and was responsible for the successful development of its international businesses in Brazil, Australia and West Africa. Steve spent much of his early career with Exxon, where he held senior roles in refining, shipping and international gas. Steve is Vice-Chair of Business in the Community and Vice-Chair of The Careers and Enterprise Company. He is a fellow of the Royal Academy of Engineering and volunteers his time to lead a number of skills and STEM activities. Steve also chairs the Young Offender Programme led by National Grid and has been awarded honorary doctorates from Nottingham and Strathclyde universities.

Jason Holt CBE is CEO of the Holts Group and founder of Holts Academy. He employs around one hundred people. Jason holds the Queen's Award for Enterprise Promotion and is a Freeman and Liveryman of the City of London and Goldsmiths Company. In October 2014, he received the Enterprise Policy Influencer of the Year Award and in January 2015 was awarded the CBE for his services to the jewellery sector and apprenticeships. Following his authorship of the Government Review on *'Making Apprenticeships More Accessible to SMEs'*, his recommendations have changed policies so it is now easier for employers to recruit apprentices. He was subsequently appointed as the government Apprenticeship Ambassador for Small Business. Jason chaired the Craft Trailblazer which represents all areas of craft in the UK. Jason was also appointed Chair of the Apprenticeship Stakeholder Board, which has been set up to hold the government to account on delivering three million apprentices in the next five years.

Martin Hottass is the Manager, Skills and Siemens Professional Education UK (SPE), at Siemens. SPE is the brand name for all Siemens' activities in vocational education. He is part of Siemens' global expert team for vocational education and has been one of the main drivers of the approach that the company has pioneered to keep talented young people who apply for apprenticeships within the industry. He has been with Siemens since 2006 and responsible for the UK apprenticeship programme since 2008.

Dr Deirdre Hughes OBE is a Principal Research Fellow at the University of Warwick, Institute for Employment Research (IER). She was Chair of the National Careers Council in England, from May 2012 to September 2014, reporting to three skills ministers. She is co-editor of the British Journal for Guidance and Counselling: International Symposia Series. Deirdre is also Director of DMH & Associates Ltd., specialising in global, national and local careers policies, research and practice.

Andrew Jones is Deputy Director at Linking London. Andrew previously worked as a Higher Education Coordinator in further education colleges in Brighton and London, as well as taking on the role of Aimhigher Project Officer. He has a great deal of experience of both supporting vocational learners into higher education and of identifying and helping to address the barriers those learners need to overcome. As Deputy Director, Andrew leads on the IAG work-strand and leads on the development of a wide range of innovative resources and activities to support college learners and advisers. Andrew chairs Linking London's BTEC and Access to HE Diploma practitioner groups.

Chris Jones is Group Chief Executive of the City & Guilds Group, a global leader in skills development. The Group includes City & Guilds, the Institute of Leadership and Management, The Oxford Group and City & Guilds Kineo. Chris plays a prominent role in driving the national and international skills agenda – something of which he has personal experience as he followed the vocational education path himself. He is a member of the Business in the Community Talent and Skills Leadership team and a governor at the Activate Learning group of colleges, formerly the Oxford and Cherwell Valley College Group. Before joining the City & Guilds Group, Chris held several senior management positions in Pearson and Reed Elsevier.

John Longworth was Director General of the British Chambers of Commerce from September 2011 to March 2016. Over the course of his career, John has done business on every continent, and has held a number of executive and non-executive board positions, including Chief Executive and Chairman. In 2009 he embarked on a portfolio career, co-founded and became Chairman of a professional services, IT and laboratory testing business. SVA Ltd was a 'gazelle business', which John guided to a successful exit in May 2015. John is a regular contributor to broadcast, print and online media and has had frequent engagement with government and politicians in Westminster and Brussels at the highest level, now and over the last thirty years. Originally trained as a scientist, John is also a Chartered Company Secretary.

Bill Lucas is Professor of Learning and Director of the Centre for Real-World Learning at the University of Winchester. With colleagues at Winchester, he has undertaken acclaimed research into vocational pedagogy and apprenticeships including *Remaking Apprenticeships: powerful learning for work and life*. Bill is the author of more than forty books about learning, most recently, with Guy Claxton, *Educating Ruby: what our children need to learn*. Bill is the co-creator of the Expansive Education Network – one of the world's largest communities of teacher researchers.

Dr Anthony Mann is Director of Policy and Research at Education and Employers, a UK charity devoted to improving understanding and delivery of employer engagement in education. He is the author of more than thirty works on the subject and was lead editor of the 2014 Routledge collection, *Understanding Employer Engagement in Education: Theories and Evidence*. He is a member of the Editorial Board of the *Journal of Education and Work* and regularly advises government and its agencies on effective approaches to school engagement with local economic communities.

Stella McKnight, as Director for Employer Partnerships, is responsible for developing and managing the University of Winchester's Employer Engagement Strategy. In her capacity as Programme Leader for the University's employer-led degrees she is active in leading a number of university-wide initiatives and groups, including the highly successful CGI Sponsored Degree Programme and the University of Winchester's Degree Apprenticeship Programmes. Stella has a track record of involvement in professional bodies locally, nationally and regionally, including the Chartered Management Institute (CMI), Young Enterprise, and Enterprise Business Partnerships (EBPs). Stella works with the EnterpriseM3 Local Enterprise Partnership (EM3LEP) on the Employability and Skills agenda.

David Meller has been Chair of the Apprenticeship Ambassadors Network since 2014. David is Chair of the Meller Group, one of the largest luxury home and beauty suppliers in the UK, a non-executive board member for the Department for Education, and sponsors five academies and two University Technical Colleges, through the Meller Educational Trust.

Judith Norrington has extensive experience of assessment and curriculum policy and practice and has been a contributor to government assessment policy reviews over many years. Formerly a Group Director of Policy,

Research and Regulation at City & Guilds and a Director of Curriculum and Quality at the Association of Colleges, she is now an education and skills consultant. She serves on the Skills Commission, which is an independent body of leading education and skills thinkers.

Frances O'Grady is the General Secretary of the TUC. She first joined the TUC as Campaigns Officer in 1994 and went on to launch the TUC's Organising Academy in 1997. Frances headed up the TUC's organisation department in 1999, reorganising local skills projects into Unionlearn which now helps a quarter of a million workers into learning every year. As Deputy General Secretary from 2003, Frances led on the environment, industrial policy, the NHS and winning an agreement covering the 2012 Olympics. She has served as a member of the Low Pay Commission, the High Pay Centre and the Resolution Foundation's Commission on Living Standards.

Roger Peace qualified as a Chartered Accountant with KPMG. After a background in industry, including with Severn Trent plc., Roger joined learndirect in 2005 as Chief Financial Officer, and was appointed CEO in November 2013. During his time at learndirect, Roger was responsible for the management buy-out with LDC and the subsequent merger with JHP Group Ltd. He led the successful bid for the driving theory test for the DSA and the acquisition of Tabs Training Ltd. Roger is a non-executive director of Birmingham Children's Hospital.

Simon Perryman is the Director of a small skills and industrial strategy advisory business. He is also chair of Barnsley College and has a number of other non-executive roles. Simon retired from his role as Executive Director at the UK Commission for Employment and Skills in April 2015. His role was to lead the Commission's industrial investment programme, support the development of Sector Skills, delivery of the government's 'Employer Ownership Pilot' and development of National Occupational Standards. Simon previously held a number of senior roles in government relating to skills, small business development and regional development. Simon is a Fellow of the Chartered Institute of Personnel and Development.

Toby Peyton-Jones started his career in the Army as an Officer in the Royal Engineers. He subsequently worked at the Centre for Leadership and Development at Brathay Hall Trust and later in management consulting. Joining Siemens in 1989, he has held a number of high level leadership roles in the UK, China and Germany ranging from general management to mergers and acquisitions. In 2008, he returned to the UK as the Siemens Director of HR for UK and Assigned Countries. He was first appointed as a Commissioner for the UK Commission for Employment and Skills in 2012 and has served on the CBI Employment and Skills Board since 2013.

Steve Radley is Director of Policy at the Construction Industry Training Board and is responsible for understanding the construction industry's changing skills needs, developing evidence for the policies required to meet them and translating these needs into outcomes that deliver the skills required. He works closely with the industry, government, other policy makers and partner organisations and has been instrumental in addressing key issues such as attracting talent into the industry and increasing diversity as part of his role with CITB.

Sir Michael Rake FCA, FCGI was knighted in 2007. He is chairman of BT Group plc. (2007) and Worldpay Group plc. (2015) as well as deputy chairman of Barclays PLC (2012). Sir Mike's business advisory roles include his membership of the Advisory Council for Business for New Europe, and Board of the TransAtlantic Business Council. He is also a Senior Adviser for Chatham House and a member of the Oxford University Centre for Corporate Reputation Global Advisory Board. Mike is a William Pitt Fellow at Pembroke College, Cambridge. He was the President of the CBI (2013-2015), member of the Prime Minister's Business Advisory Group (2013-2015), Chairman of easyJet plc. (2010-2013), Chairman of the UK Commission for Employment and Skills (2007-2010) and Chairman of Business in the Community (2004-2007) among many key senior leadership roles. He was FTSE 100 Non-Executive Director of the Year in 2013.

Laura-Jane Rawlings has spent the majority of her adult working life supporting others seeking work. Having spent years in recruitment, she understands the view many employers have about the employability skills of young people today. She gained an in-depth knowledge of the structural barriers that young people face when they leave school and begin the search for a job. Laura-Jane sought to use her commercial, employment and careers expertise to help raise the employability skills and aspirations of young people; and has been a key voice in the field since 2010. Passionate about the views and needs of young people and careers education, Laura-Jane is committed to supporting all young people to realise their potential. Since founding Youth Employment UK in 2012, she has consolidated their position as the leading youth employment specialists.

Neil Robertson joined the National Skills Academy for Rail in June 2015. He spent the previous three years as CEO of the Energy & Utility Skills Group, where he led and secured a £55m collaboration bid in energy and utilities aimed at increasing workforce skills and improving diversity. Prior to this, he was Director of Training at Babcock International, after holding senior posts in government including DfES and DIUS (now BIS) where he had responsibility for basic skills, targets, employability, ESOL and migration. Neil has also worked in the Government Office North East and with the British Institute of Innkeeping and City & Guilds and at the Scottish Qualifications Authority. His first role was as a trainee quantity surveyor.

David Russell is the Chief Executive of the Education and Training Foundation, the body which supports quality and workforce development in Further Education and Training. He was the DfE Director responsible for national apprenticeships policy for 16 to 18-year-olds in 2009-10 and again in 2012-13. David has been both a school and college governor.

Graham Schumacher MBE is now an Independent Skills Training Partner after a long career in a major aerospace engineering company and with over thirty years spent involved with training at all levels of the business. The past fifteen years were spent leading the Learning Services Function, supporting 50,000 people in sixty countries delivering over 400,000 standard development programmes per year. Throughout his career, Graham has taken a very close interest in the development of a world-class apprentice training process.

Stewart Segal was appointed to the role of Chief Executive Officer of AELP in 2013. Stewart has worked in the funded work-based learning sector for over twenty years and has supported AELP since its formation. Stewart worked as an independent consultant within the training sector, specialising in business development and funding issues. Following a background in HR and general management in the private sector, Stewart joined Hertfordshire Training and Enterprise Council in 1994 as Chief Executive until 1998 when he joined Spring Skills as Chief Executive. Spring Skills was then the largest independent training provider, specialising in the service sectors. Stewart is now Strategy Director with 3aaa and remains a Board member of AELP.

Andy Smyth is the Vocational Learning Development Manager for TUI UK and Ireland, part of the TUI Group. Prior to joining TUI in 2005, Andy spent eighteen years in the logistics industry. Andy now leads the Vocational Learning team and directly supports the Early Talent, Leadership Development and People Development functions. Andy has sat on a number of advisory boards to support policy development and the sharing of best practice. In addition to this he has been involved in the modernisation of the qualifications system including actively supporting the current round of Trailblazers and levy reforms. Andy is Chair of the City & Guilds Industry Skills Board. His ambition is creating an effective modern system that is accessible to all, with simplicity and efficiency at its heart.

Mike Thompson is the Director Early Careers at Barclays. Mike has been working in Barclays for over twenty-five years in a variety of frontline and support leadership roles across the business. His past six years have been in HR where he has built up extensive experience of implementing management and leadership development programmes, employee engagement programmes and managing the development of Barclays' Learning and Development Curriculum. His achievements include the roll out of the highly successful Leadership Excellence Programme across the UK, Africa and Europe and the implementation of the Barclays Apprenticeship Programme. Mike is passionate about developing young talent and tackling diversity and equality issues as well as providing meaningful solutions that address youth unemployment.

William Walter is an Associate Director at Kreab, a leading global political and corporate communications consultancy where he specialises in advising clients on a range of policy areas, including further and higher education. Prior to joining Kreab, Will worked for three Conservative MPs, including as research assistant to the Minister of State for Universities & Science, David Willetts. In 2014, he authored *Varsity Blues: Time for apprenticeships to graduate?'* The report was the first of its kind to investigate and compare the earnings, employment and taxpayer returns of apprenticeships to those of university degrees. Launched in Parliament, the report's findings were widely reported.

Phil White is the Technical and Health and Safety Training Manager at United Utilities. He joined UU as a mechanical engineer apprentice in 1997 and been with UU ever since. He rose through operational managerial roles at UU before moving to HR and L&D in 2012. Key responsibilities include all technical and health and safety training for the 5,500-strong workforce; all apprentices in the organisation; and running the new Technical Training Centre in Bolton. He is Vice-Chair for the Water Development Group and Chair of the Water Assessment and Assurance Panel for the new Trailblazer Standards.

Tom Wilson was Director of Unionlearn, the learning and skills arm of the TUC, from 2009 to 2015. He has worked for a number of education unions in research, policy, negotiation and management. He is currently a self-employed skills consultant involved in FE, HE and other skills organisations, in the UK and abroad. Tom is a Visiting Research Fellow at the UCL Institute of Education and at Wolverhampton University. He also chairs GTA England, the national network of Group Training Associations.

Stuart Youngs is Executive Creative Director at Purpose. Stuart has acquired a wealth of experience in creating, delivering and managing brands. Before joining Purpose as a partner, he spent much of his career working with design industry grandee, Mike Dempsey RDI. He has worked with a rich and varied number of clients including Royal Mail, Clarks Shoes, British Land, Design Council, Pearson, City & Guilds, British Heart Foundation, Goldman Sachs and National Apprenticeship Service. Stuart is a frequent speaker and commentator on branding. He has been recognised in over seventy industry awards. Stuart is currently exploring the intersection between science and creativity to help brands with a team of behavioural change experts. Stuart is also co-founder of a Silicon Valley-based tech start-up which specialises in Augmented Reality (AR) based products.

INTRODUCTION TO THE BOOK

David Way CBE

Imagine a day in 2020 when the government announces that its ambition for three million more apprenticeships has been achieved. This will be a significant milestone in the growth of apprenticeships.

Imagine as well a time when most young people of all abilities clamour for their apprenticeship and employers compete to get the very best applicants. Where the status of apprenticeships is second to none and all parents are as proud of their children for becoming apprentices as they are of those who go to university.

This book is concerned with how we achieve both of these imaginations and explores how we make them real.

It showcases many aspects of apprenticeships where we should be proud of our successes; and highlights those many aspects of current policy and practice where we are getting it right.

It also looks at what more we must do, with the benefit of insights from more than forty people who are influencing apprenticeships today either as leaders, analysts or practitioners.

A Race to the Top reflects the clear link that has been established between economic competitiveness and skills. This is a race that we need to win if we are to increase productivity and employment in the UK and therefore sustain economic and social prosperity.

1

In running this race today, it is surprising that, with all this momentum and the evidence of high returns for public, employer and individual investment in apprenticeships, this race feels uphill. Apprenticeships clearly don't sell themselves. This book therefore provides insights into why this is the case and what we can do to make our progress easier.

The book concludes with a summary of the main messages from all of the wide range of contributors. They all believe in the importance of apprenticeships and know them well. Their selection of themes for making apprenticeships even better is therefore instructive and compelling.

Their conclusions are reinforced by a clear sense of how the race to achieve the ambition for high quality apprenticeships should be judged. Achieving three million more apprenticeships will be very important but it is not the sole measure if England and the UK are to achieve the transformational change in attitudes and behaviours needed to win the race that really matters.

When I was Chief Executive of the National Apprenticeship Service (NAS), I was very fortunate to live and breathe all matters apprenticeship. The NAS had been given 'end to end responsibility' for apprenticeships and this enabled me to look into any and all aspects of apprenticeships from policy formulation to how they worked for employers and individuals.

It was this all-round view of apprenticeships that was the inspiration for this book, along with an appreciation that there are very many people and organisations that are already contributing to making apprenticeships a success in England and across the UK. I wanted to draw all of these voices and perspectives into one strong and persuasive set of messages about the current state of apprenticeships and about the future.

Over recent years, the importance of apprenticeships has been recognised and promoted by all the main political parties. The Prime Minister has raised the bar with the ambition of three million more apprenticeships by 2020. This follows previous statements to the effect that he wants all young people who leave school or college to either go to university or to take up an apprenticeship. These are notable statements of ambition that merit support and deeper consideration.

This book is not akin to previous reviews of apprenticeship programmes. It has not been commissioned by the government, nor restricted to looking at specific aspects. It evaluates apprenticeships in a holistic manner; nothing is off-limits because of political sensitivity. It addresses all the main issues that I encountered in trying to increase the number of high quality apprenticeships.

While the ambition of three million more apprenticeships is central to the book, its thinking is not constrained by it. This book is not about incremental change. Nor is it advocating 'more of the same'. This is too good an opportunity to waste. It considers how we can transform apprenticeships so that they are unarguably a great choice for all young people and they are the skills supply pipelines for all successful employers.

The book brings together leadership pieces from employers, the TUC and those delivering apprenticeships. There are also analyses from experts and academics in the skills system. Finally I have included a number of case studies giving practical insights into the challenges and successes of running an apprenticeship programme. In the UK, we have a tendency to keep reinventing rather than sharing best practice. I believe there is much in the UK of which to be proud and on which to build.

The book is intended for everyone with an interest in apprenticeships, whether employers, policy makers, advisers, individuals, parents or skills professionals. It is for everyone working to put apprenticeships at the heart of our future economic and social strategy and giving young people a great start to their working lives. It draws together the collective wisdom of people who know about apprenticeships in 2016 and have a real passion for achieving the 2020 ambition and much more.

Even in a book of this scale, it is not possible to hear from all sectors and interests or look at all aspects. There are people and issues that I have not covered but they can expect to feature in subsequent updates.

I have not set out to be critical of government policies and certainly not of its ambition for apprenticeships. This is reflected in the book, though I have not restricted opinions. Everyone who has contributed cares about apprenticeships and about enhancing future prospects for young people, skills and employment.

The intention is to feed the ideas that emerge from the contributors into the ongoing debate about how we make apprenticeships even more successful to the benefit of millions more young people and businesses. I hope that it will stimulate those running the employer-led skills infrastructure and those in policy positions to consider these perspectives and the clear signals they send as to future priorities for action.

As the slogan goes, 'There's never been a better time to be an apprentice'. In my view, 'There's never been a better time to write about them either'.

The current government has set out its plans until 2020 and is in the process of reforming funding and governance. Change is in the air, making for an exciting period in apprenticeship history and a vital time to get it right.

I have chosen five main themes for this book that I believe are crucial for understanding, analysing and addressing the growth in apprenticeships that we need if we are to see the productivity improvements to which the government and nations aspire.

1. Apprenticeship design and quality
The book begins by looking at the fundamental requirements of apprenticeships - what are they trying to achieve and who are they really for? This question is considered both from an employer perspective and for young people. How can they be designed so that more employers and young people demand them? Why aren't they as popular as everyone wishes they were? Can they attract the higher achievers and really make their peers think about apprenticeships as an option? Is the recent expansion of higher apprenticeships to degree level and beyond the opportunity to transform the image of apprenticeships and change employer recruitment practices? If so, how can employers and government speed up their extension? Will the current ambition for apprenticeships be sufficient to ensure the UK has the technical skills that it needs to compete internationally? How do we ensure that the new ambition for apprenticeships delivers on quality as well as quantity; and that standards

are maintained during a period of rapid expansion? Can this really be achieved without a serious programme to ensure higher quality teaching and training? Three case studies examine big changes in the way in which apprenticeships have been designed. They cover degree apprenticeships, and the introduction of new trailblazer standards and traineeships into UK companies.

2. Promoting apprenticeships

Apprenticeships are a success story but we still have a long way to go to convince all employers and parents that they offer a great choice for their own business and their own children. This section begins by considering why employers are attracted to apprenticeships and how the case can be further strengthened. It looks at approaches that have been tried to improve the promotion of apprenticeships, including through using committed employer ambassadors and apprentices themselves. What is the advice from marketing and communications professionals about what works best, the biggest challenges and how to ensure current positive attitudes to apprenticeships are deepened and do not regress? How can we use the power of schools to inform more young people about apprenticeships and to present them objectively alongside higher education options? How can apprenticeships lose their perception as having second-class status? Finally, this section considers the importance of presenting the UK's apprenticeship practices more positively (as so many other countries manage to achieve) and help persuade more people that they represent a system on which we can build with real pride. The case studies look at how one big financial institution has actively taken its apprenticeship offer out into communities and achieved great success; and at how a leading training provider has succeeded in engaging more employers in apprenticeships.

3. Apprenticeships and business

While government has set the ambition for apprenticeships and largely controls the system to support them, the growth in apprenticeships depends primarily on employers. This section therefore examines the appeal of apprenticeships to business and the extra help that small and medium-sized enterprises and large businesses need to take on more of them. If the benefits of more apprenticeships are to really make a difference, they need to tackle the skills barriers to productivity growth. Do Industrial Partnerships provide the opportunity to ensure sector skills shortages are addressed? Should they be given more time and support to show their true worth?

Trades unions already play a vital role in training. What advice and help can they offer about raising the status of apprenticeships? How do we ensure that apprentices are prepared for a career and not simply a job? Do the wages that apprentices receive inhibit some employers from providing more opportunities? How can the wages more closely relate to productivity? What should be done to deter those employers who take advantage of apprentices? Plans for large national infrastructure projects provide great opportunities for apprenticeships and for business. How can we ensure employers and contractors recruit apprentices? Employers need to ensure that apprenticeships work well for them and boost productivity. How can they ensure that is the case? What lessons have we learnt about how employers can get the best out of them during and after their apprenticeship? What too have we learnt from the Employer Ownership pilots that will attract more businesses to offer apprenticeships?

The case studies look at how one business has been working to ensure that the surplus apprentices that are attracted to working with them are retained in the sector and by their supply chain through new clearing-house arrangements; how one business has been tackling the challenges of recruiting more employees into STEM occupations; and lessons for the UK from the much admired Swiss approach where employers collaborate to minimise youth unemployment and maximise skills to drive their economy.

4 Navigating the apprenticeship system

When the National Apprenticeship Service was first established, its primary task was to make it easier for employers to take on apprentices. However, our skills system still seems complex and off-putting for very many employers. This section looks at how to make navigating the skills world easier for employers and young people. It first looks at why simplicity matters with reference to a selection of other countries. It then explores the reasons why so many employers call for demystification and stability; and offers ideas for simplification from employers and sector experts, including employers who face four different funding systems in operating across the UK. The biggest change in apprenticeships arrangements faced by employers for a generation or more is the introduction of an apprenticeship levy. How do we ensure that this achieves the boost to employer investment in skills that is needed rather than be simply a tax and additional bureaucracy? This chapter looks at these issues through the eyes of experienced levy practitioners.

Finally, what about young people wanting an apprenticeship? How can we provide better support to them through improved careers information, advice and guidance arrangements, including realistic information about career earnings and employment prospects? What are their perceptions of apprenticeships and how do we present them with the inspirational messages that would mean they took apprenticeships more seriously? We know that the opportunity to go to university at some time in their career is an important consideration for most young people and their families. In practice, how easy is it to achieve progression from an apprenticeship to higher education?

5. Delivering apprenticeships

This section begins by addressing the critical question of how to achieve greater investment in training by employers. This is vital for successful delivery if employers are indeed to be in the driving seat. It then considers the growth ambitions from the important perspectives of the principal deliverers of apprenticeships. How do independent training providers and higher education institutions plan to capitalise on the opportunities that the government wants to stimulate? What innovations might we see to help them to propel apprenticeships onto this new higher trajectory? Is this the time for higher education and further education to come closer together? Finally, greater employer ownership of the skills system means businesses will expect something very different from apprenticeship providers in future. Government has already signalled new expectations in respect of assessment in order to ensure the effectiveness of delivery arrangements to meet employer needs. What does international experience teach us about best practice?

The case studies look at the changed expectations of one employer that found current services falling short of their expectations and how they took greater ownership of delivery arrangements; as well as the continuing role that Group Training Associations take in addressing critical skills shortages especially in STEM occupations.

Conclusion

This final section draws together the strongest and most persuasive messages from the collection of expert articles and sets out the principal messages for government, employers, training organisations and all those who are critical to the UK winning the Race to the Top for apprenticeships.

It captures the state of apprenticeships in 2016, comments on the direction of travel, and highlights key messages to policy makers, employers and everyone working to build high quality apprenticeships in the UK.

It goes on to propose the key areas by which the successful transformation of apprenticeships needs to be judged. It argues that while three million more apprenticeships are vital, a wider set of performance considerations are needed if the UK is to win any race with our economic competitors and create a system that will drive productivity and social mobility.

PART ONE

APPRENTICESHIP DESIGN AND QUALITY

INTRODUCTION TO PART ONE

David Way CBE

Successive governments have been immensely proud of having brought back apprenticeships after years of relative neglect. The importance of apprenticeships and their successful renaissance (as it has been called) is widely proclaimed by political leaders. It has become a rare event when the Prime Minister or Chancellor does not include apprenticeships in keynote speeches about their major achievements and plans for the future. Apprenticeships are increasingly presented as an important part of the answer to a wide range of social and economic problems.

This section of the book first looks at why apprenticeships are important to young people and to employers as well as to the living standards of us all. It reflects on employer concerns that skills shortages are the biggest concern facing businesses today and that apprenticeships need to be the primary vehicle through which the UK addresses this and creates a virtuous circle in which skills and productivity increases reinforce each other.

John Cridland had taken a close interest in education and skills even before he became Director General of the Confederation of British Industry in 2011. Under his leadership, the CBI has been a regular and respected analyst and commentator on current and emerging skills issues. With the growth in apprenticeships highly dependent on more employers offering high quality apprenticeship opportunities and the introduction of the new levy, John's views on apprenticeships and insights into the future prospects for growth are of particular interest.

But every apprenticeship role needs an apprentice as well as an employer. While we regularly hear about those apprenticeships that have a higher ratio of applicants than do the top universities, there is a relatively low level of interest in apprenticeships from young people compared with other countries where apprenticeships enjoy a higher status.

This chapter looks at how we can change this and ensure that apprenticeships are not perceived to be for the less able or for 'the other 50 per cent'. It identifies the key building blocks for achieving this and the lessons from our international competitors.

David Russell is a former Director in the Department of Education with a longstanding interest in these questions. He became the first Chief Executive of the Education and Training Foundation in 2014. The Foundation prioritises excellence in teaching and learning.

When John Hayes was the Skills Minister in 2012, he was justifiably proud of extending apprenticeships up to higher levels so that young people could see a ladder of opportunity all the way to the top if they took the vocational route. It also helped transform the recruitment practices for many top businesses that had previously adopted graduate-only recruitment.

Employers recognised that they were missing out on some very talented young people who were choosing not to go to university. This new group of recruits also had a different profile and brought different strengths and perspectives to the business.

The next chapter explores why employers saw the essential business case and social benefits of introducing a high quality vocational recruitment and training route. It sets out some of the obstacles that had to be overcome and the successes they achieved. Why did PwC take on this role for the professional and business services sector? What lessons have been learnt that might help employers following a similar route? Why was gaining the recognition of professional bodies so important? How do we complete the higher apprenticeship map so that the vocational route to the top is clear across all occupations? Why was networking apprentices so important? And how can employers be encouraged to take on similar leadership roles within and across sectors?

Sara Caplan is a partner at PwC and she was at the forefront of the development and introduction of higher apprenticeships, playing a vital role in leading and coordinating work across the professional and business services sector. Sarah and I jointly launched the first level 7 apprenticeship in 2013, equivalent to a masters degree. This was an ambition considered out of reach only a few years earlier.

International reviews of the UK apprenticeship system highlight the comparative lack of apprenticeships in high skill technical occupations. Growth in recent years has mostly been in the services sector as the apprenticeship footprint has extended into all parts of the economy. The government has responded by underlining the importance of Technical Pathways.

The next chapter looks at why technical apprenticeship growth is critical to the UK and how the traditional barriers to growth can be overcome. Will the ambition of three million more apprentices even meet demand? It looks at this question from the important perspective of the engineering manufacture sector and looks at the potential of university technical colleges and creating skills hubs. It also explains why apprenticeships are important for employees of all ages.

Graham Schumacher has first-hand experience of the UK and German apprenticeship systems having been at the heart of apprenticeship policy discussions and practice for many years, including at Rolls Royce.

The next two chapters are concerned with quality.

When the Public Accounts Committee examined apprenticeships in 2012, it was considered inevitable that the period of rapid expansion over the previous five years would have also seen inconsistency in quality. There was certainly no intention to reduce quality standards over that period and the normal quality safeguards were in place. However for many months, the National Apprenticeship Service and the Skills Funding Agency had to combat quality issues on a number of fronts, which understandably undermined work on expanding apprenticeships.

Each chapter examines how we can ensure that we do not simply become fixated with a quantitative figure of three million apprenticeship starts but also take the great opportunity to raise quality. Giving employers and individuals a great, productive and rewarding experience must surely be the best way to underpin apprenticeship growth and stimulate future demand.

Kirstie Donnelly of City &Guilds addresses the need for there to be real and meaningful learning at the heart of all apprenticeships and argues that the current ambitions for apprenticeships presents the opportunity to reinvigorate the skills system and to embed processes and best practice that will create the right outcomes and desperately needed skills for employers, individuals and the UK economy. This includes establishing apprenticeships as a high-aspiration route and ensuring that the prospects for progression are clear.

When I first met Bill Lucas, he was working on his latest research report about how the UK needs to improve the ways in which it teaches vocational learning. It underlined that the opportunity to build apprenticeship numbers to unprecedented levels in the UK does not mean settling for more of the same. Indeed improving the quality of the teaching is the best way of ensuring that the increased demand is sustained.

The chapter from Bill Lucas looks specifically at how the UK might address raising standards of teaching and draws especially on research and experiences from around the world. It looks at how we must motivate all of those people involved in apprenticeships to understand and deliver an outstanding apprenticeship experience that can truly be described as world-class.

Bill Lucas is the UK's leading researcher in this area. He has been the Director of the Centre for Real-World Learning at the University of Winchester since 2008. Bill is part of an impressive group of researchers who we have in this field whose expertise we surely must deploy more effectively.

This part of the book that looks at apprenticeship design and quality is completed by considering three case studies covering relatively recent innovations in apprenticeship development: degree apprenticeships, trailblazers and traineeships. These examples of successfully implementing the new policies are valuable in themselves but also demonstrate the need for all of the moving parts of the system to work together if apprenticeships as a whole are to succeed.

The University of Winchester has been at the forefront of developing degree apprenticeships and Stella McKnight and David Birks share their learning from a ten-year association with CGI (formerly Logica) to support the company's need for higher-level skills in the digital industry. This is described as 'Growing your own graduates'. They also provide insights for the wider HE sector as more collaborations take place with degree apprenticeships as the training vehicle.

Trailblazers in England have been the clearest recent articulation of employer ownership in action. Jacqueline Hall has been at the heart of developing and introducing the earliest trailblazers. She shares her experiences and explains the progress that has been made through working with employers in the energy and utilities sector who have seized this opportunity to define new standards and to back it up with independent assessment arrangements. As Jacqueline describes it, the employers have really 'grasped the nettle'.

Finally, there is an account of E.ON's approach to apprenticeships and how the introduction of traineeships as a feeder programme transformed their recruitment and training approach. The new entry programmes are now in great demand across the organisation and have strong advocacy at all levels of the company.

1.
WHY ARE APPRENTICESHIPS SO IMPORTANT AND WHAT ARE THEY TRYING TO ACHIEVE?

John Cridland
former Director General, Confederation of British Industry

During my time at the CBI a lot changed: the decades between 1982 and 2015 saw tectonic economic and political shifts. From when I started out as a junior staff member or when I was Director General, one perennial issue that I heard from businesses of all sizes and sectors was a clamour for people with the right skills and talents. Too often government policy interventions have been variations of the same approach – with all too little emphasis placed on the future needs of a constantly transforming labour market.

Education and skills is a core business issue. As a result, it has always been territory where the CBI has developed ideas and sought to shape effective policy. It is a legacy of which I am proud – both by association and personal involvement. Whether campaigning to raise the status of vocational education and alternative routes to higher-level skills, to advocating greater business involvement in education and support for teachers from industry, these kind of policy initiatives are important not only for business, but for people who benefit from them and whose livelihoods improve as a direct result. The best way to raise living standards – for individuals and the country as a whole – is through education and skills.

As a CBI representative I was personally involved with a number of bodies directly involved with improving the UK skills situation. From the role of Vice Chair of the National Learning and Skills Council – where it was our explicit remit to increase the standards and range of learning opportunities for businesses, communities, and individuals alike – to being a UK Commissioner for employment and skills. The countless

16

conversations that I had with learners of all ages and stages — whether with apprentice bricklayers and welders in London or the West Midlands, or with those later in their careers looking for a change of direction by studying part time – provided valuable insight into the challenges faced by British businesses and the UK economy as a whole.

Apprenticeships have a vital contribution to make to the skills landscape – with widening skills gaps making high-quality provision a particular priority. The continuing negative perception from which they suffer, however, suggests that there is still work to be done. Writing this chapter having now stepped-down as Director General of the CBI, the skills landscape continues to change — with the apprenticeship levy fundamentally altering the UK's approach to skills funding. The purpose of apprenticeship policy and an apprenticeship itself should be the same: raising the skill level of an individual and the economy, setting them on a path to a good career.

The introduction of an apprenticeship levy marks a significant shift in skills policy, and its success depends on business playing a vital role. This book, with contributions from across UK plc., is part of the ongoing debate about apprenticeships – which are important not only for the impact they can have on the UK economy, but for the impact they can have on the lives of individuals for whom an apprenticeship can be the start of a great career. This chapter contributes to this narrative, by first considering their purpose within the UK economic context. I will then explore apprenticeship policy – most notably the apprenticeship levy – suggesting how this can be an intervention that encourages, not damages, apprenticeship provision, before making some suggestions as to how the wider education and skills system can contribute to closing the skills gap.

The scale of the skills challenge facing the UK is apparent when considering that by 2022, an additional two million jobs will require higher-level skills (a sobering thought, given that the cohort entering the labour market in 2022 are already at secondary school). When the CBI researched and produced a report on boosting living standards in 2014, *A Better off Britain*, it noted that skills are the essential currency to ensure that people can move between roles and progress their careers and prospects. However, there is gulf between the potential economic opportunities and the readiness of school and college leavers, graduates as well as the existing workforce, to seize these chances. Where shortages are particularly acute is at the technical level, or rather the level 4 and 5 skills required for roles in the 'new middle' of our labour market – roles such as account executives, air traffic controllers, engineering technicians, financial analysts, IT assistants, and paralegals. Last year, for the first time in the seventeen-year history of the CBI's Employment Trends Surveys, businesses' concerns about skills shortages emerged as the biggest perceived threat to UK competitiveness.

If the UK is to reshape its economy around high-value, high-skill activities that will enable us to excel in an intensely competitive global economy then we need a talented workforce equipped with the higher-level skills required to maximise future business opportunities. Without this, the UK will fall short in its attempts to rebalance the economy towards investment and exports that will be essential to achieving future growth. As domestic demand drives forward, companies are increasingly flagging the difficulty of finding staff with the right skills for their business. This anxiety is widespread – indeed half of businesses (55 per cent) surveyed in the CBI/Pearson Education and Skills Survey 2015 expressed that they are not confident of finding people with higher-level skills. Other CBI economic and business surveys have shown that shortages have become increasingly acute since the recovery began to solidify in 2013; they are now well above pre-crisis norms, and are close to their last peak in 2006. With skills shortages acting as a brake on growth and undermining the UK's economic competitiveness, it is therefore critical to address them if we are to ensure sustainable growth over the long-term.

There are many routes to higher skills, and variety is vital, but apprenticeships are an important part of the solution to the skills gap and the need for training which helps build meaningful careers. There is no mistaking however, that this is a huge challenge following perpetual changes to the skills landscape – both of process and perception – that have seen a decline in the status and up-take of technical and vocational education. In recent years, apprenticeships have undergone something of a renaissance, enjoying a resurgence of political and public interest. Historically, apprenticeships provided training for skilled manual workers in artisan trades, and later in manufacturing. Post-war, from the late 1940s until the 1970s, apprenticeships were an essential entry-route into a career for most males of school-leaving age and a played a key role in the transition from the school gates to the shop floor. A fall in apprenticeship numbers in the 1970s and 1980s was for the most part attributable to a decline in the UK's manufacturing industries − a trend that was exacerbated by the introduction of cheaper alternative training schemes (the Youth Training Scheme for one). Another critical factor in the decline of apprenticeships was the perception that this model of training was no longer relevant, as the labour market increasingly demanded flexibility and the need to change jobs during the course of one's career.

After falling into relative obscurity, an attempt in the mid-1990s to rebrand apprenticeships as 'Modern Apprenticeships' to improve their brand image and relevance to business need, was in part hindered by the Labour government's drive to get 50 per cent of young people into higher education from the late 1990s. The subsequent opening-up of apprenticeships to adults, and widening to include level 2 qualifications, did little to increase enthusiasm for them. Fast forward to the coalition government's focus on apprenticeships, and although there was an increase of more than 60 per cent in apprenticeship starts overall in the year immediately following the 2010 general election, these did not lead to a corresponding narrowing of the skill gaps as the majority increase was at level 2. It is important to point out however that one should not conflate level and quality − in fact businesses are involved in providing high quality apprenticeships at every level. The strategic focus on apprenticeships under the coalition government, and the push for greater employer ownership − in tandem with the creation of other routes such as university technical colleges (UTCs) and studio schools − was welcomed by business.

Determined to better the 2.3 million apprenticeship starts during the 2010 − 2015 parliament the current Conservative government made a manifesto pledge of creating 3 million new apprenticeship starts, which alongside policies on fiscal tightening led to the apprenticeship levy.

The apprenticeship levy is a significant shift in skills funding policy − not one which the CBI supported in my time, or since, because of the risk that it increases business costs without delivering a greater degree of high-quality training. As the levy policy shifts to design and delivery further critical details on the operational side are yet to be determined. But to achieve success for the economy and for would-be apprentices, it is imperative that the system is led by business and economic need – and not by policymakers – especially as businesses are now paying the bill. For years there have been promises of less bureaucracy and more business input to workplace training, but too often the reality has been very different. Therefore, the new system must mark a genuine transition from a government-led system to a truly business-led one, with the Institute for Apprenticeships having a critical role to play. As one of the founding members of the Low Pay Commission, I can testify to the expertise, rigour and commitment that genuinely independent advisers can bring to complicated policy issues – which can be best addressed out of the hands of politicians.

The ambition to increase the number of apprenticeship starts is a positive one – but increases are only meaningful if they are targeted to close skills gaps and set people on their way to a good career. Whilst level 2 is an important entry point and stepping stone – for which there are many positive apprenticeship examples across industry – it is growth in apprenticeships at levels 3 and 4 that is of critical importance. Apprenticeships are valuable to both the individual apprentice and the businesses too when filling genuine gaps within the workforce – indeed it is true that many employers who recruit apprentices report that they are more likely to remain with the company for longer – enabling firms to realise a greater return on investment in training.

Addressing the skills crisis is complex so there is no one solution or route that will lead to success.

Apprenticeships sit within a wider education and skills ecosystem, making the success of any policy or individual apprentice dependent on many factors. The development and delivery of the levy will be critical, and alongside that I'd highlight two other priorities.

Firstly, increasing business-school engagement, where there has been great progress in recent years but there is still a way to go yet. Business can bring so much in the way of support for teachers and inspiration for students, and during my time as CBI director-general I was particularly proud of the contribution the organisation made to the education debate – through working with business, educators, trade unions and academics to develop policy on education with the intention of improving opportunity for all. Beyond policy and into the practical, the positive impact of engagement with employers has been expertly profiled by the Gatsby Foundation, demonstrating how business can promote social mobility by providing experiences of the workplace and increasing exposure to business. In the area of careers, business input can be particularly helpful – signposting routes into good careers, including through apprenticeships.

Secondly, if we're serious about setting young people up for success the education and skills system needs to reflect the diversity of talents, aspirations and careers people have. Too often policy has swung in favour of one training route, only to subsequently shift in correction at the unintended consequences when they become apparent. Whether university or apprenticeship, a one-size-fits-all approach to education and skills will never suit people looking for training and opportunities, or the economic needs of the country. Again, one of the privileges of representing businesses of all shapes, sizes and sectors I always enjoyed was the chance to see the different approaches which companies took to developing people – in the existing workforce and for the future – and hearing success stories. From companies running UTCs, with their blend of the school and working environment, to those who partnered with universities on cutting edge innovation, to others who ran 'earn while you learn' routes at all skill levels – this diversity needs to become more widely known and championed amongst teachers, parents and employers as well as supported by policymakers.

Given that the skills landscape is being re-shaped as I write, the result of this 'race to the top' will have to wait for another book. What is clear at this stage is the scale of the challenge, but also the size of the prize if we get it right. A highly-skilled workforce is the key component in creating a virtuous circle with productivity improvement stimulating increased investment in skills, which in turn drives UK productivity still further. For individuals, education and skills give them currency in a fast-changing world. Apprenticeships can, and should, be one of the premier means to address skills needs, although success will be determined by the design and implementation of the levy. While the solutions are not easy, getting the variety of skills routes right and ensuring that they are well signposted will help ensure that not only are people aiming for the top but that everyone can get there.

2.

WHY AREN'T APPRENTICESHIPS AS POPULAR AS EVERYONE WISHES THEY WERE? A PERSONAL VIEW

David Russell
Chief Executive of the Education and Training Foundation.

———————

This chapter looks at the fundamental building blocks of an apprenticeship system that will not simply attract more apprentices but will transform the way in which apprenticeships are perceived by people and families of all backgrounds, ambitions and achievements. It addresses what we need to do in order to develop a shared narrative of social value so that everyone views apprenticeships as a good thing for their own children as well as for other people's.

Before we can understand what is right and wrong with apprenticeships in England today, we must know the true history of how they came to this point.

In 2007, the Labour government declared that it cared about apprenticeships. Up until that point, they had been a low-key aspect of an ambitious over-arching reform programme called The Skills Strategy. The Skills Strategy had been given considerable political impetus from inside No 11. The Chancellor, Gordon Brown, strongly believed that by breaking out of the so-called 'low skills equilibrium' Britain could reinvent itself as a high-wage, high-value-added knowledge economy, and that this would create a virtuous circle whereby higher employment and higher tax-take would generate sufficient return on the huge public 'investment' the government had made in education, skills, welfare and the state since 1997.

Aside from the Chancellor, political drive behind the Skills Strategy was relatively low-powered; not for want of talented or committed ministers, but simply because of the very rapid turnover of education secretaries and ministers during those times, which has been chronicled elsewhere. (Hodgson, 2015) The strategy was logical, coherent, ambitious and deliverable. It was underpinned by economic analysis – albeit contested, not least by economists in rival government departments – and it combined national vision with detailed propositions for funding rules, systems and structures.

However, it paid little attention to the teaching profession – the teachers and trainers who would actually 'deliver' these new 'skills'. It paid even less attention to the learners, be they in college, adult education or in the workplace. And this was ultimately to prove a fatal weakness – a programme for education reform not based on insight into what makes great teaching is as effective as a new set of furniture in a classroom: not entirely irrelevant, but rather tangential to the core business of the place.

Apprenticeships had for several years up until 2007 been managed mainly from a backwater of the Learning and Skills Council: a technocratic concern with loyal supporters and an arcane body of knowledge which was disconnected from the dynamism of the Skills Strategy.

So what happened in 2007 to cause them to re-emerge into the political spotlight? The answer was typical of the Blair/Brown years: machinery of government changes. The Department for Education and Skills was split – almost literally overnight – into the Department for Children, Schools and Families under Ed Balls, with a remit for education up to age 19, and the Department for Innovation, Universities and Skills (19-plus) under John Denham. Both men cared about skills policy. Denham wanted apprenticeships in his brief, a jewel in the modest crown of his new Department; Balls wanted it equally.

An internal struggle ensued over the governance of the Apprenticeships Programme. The details are of interest only to Whitehall-watchers of the first order, but the uneasy outcome was the National Apprenticeship Service with a convoluted set of reporting lines and organisational relationships; a bureaucratic disaster averted only by its direct oversight by a joint DCSF/DIUS Skills Minister to whom all the various departmental officials were accountable.

Why does any of this matter, in 2007 or in 2016? How can it possibly affect the choice of a 16-year-old in Gateshead who is leaving school and wondering what do with her life, an apprenticeship or otherwise? She has never heard of any of these people, or organisations, nor does she need to.

It matters because at the heart of England's apprenticeships programme is a set of rules. Laws. Funding rules. Policy intentions. Incentives for employers. Qualification rules.

If the political leadership of an area of public life is contested, or if the management and governance of that area is fuzzy, convoluted, compromised, incoherent, then the result is inevitable. The people on the end of the chain – be they learners, employers or teachers – will get a sub-optimal service. Messages will be unclear. Rules will be obscure. Simplicity will be absent. Conflicts will be passed down the line instead of resolved at the policy stage. Above all, institutions will be weak.

Quite simply, if two departments of state are each trying to define apprenticeships as something which furthers their respective agendas, then they are starting policy design at the wrong end of the street. 'How do we make apprenticeships best support our overall policy ambitions?' is the question each department asks, and it is entirely the wrong one.

So what is the right question? It is this: 'What makes excellence in an education or training system?'

There are clearly identifiable features that every excellent education system has. The precise forms these features take will vary – from country to country, from early years to secondary, from higher education to work-based training. But my contention is that they can be discovered and either confirmed or disproved by research. More of that research, I hope, will be done in future; for now I set out a theory to be tested, based on the incomplete but privileged experiences I have had over fifteen years working in education and skills policy at the 'heart of the machine'.

Excellence in an education system – a definition
The first place to start is the most obvious. In a high quality education or training system, learners have high quality teaching and learning experiences. External signs can measure these - high learner satisfaction,

strong value-added in attainment measures, and, in the case of vocational education, increased productivity and employment. (Note that each of these is a result of high quality, but does not in itself guarantee high quality; for example, high learner satisfaction could occur in a poor quality system if learners' expectations are low).

There are also certain inputs which will bring about high quality teaching and learning experiences, and these are comparatively well researched. Prominent amongst them are expert teaching skills (aka pedagogical knowledge); strong subject knowledge (aka industrial expertise in the vocational sphere); and good institutional utilisation of teachers' knowledge and skills (aka good management).

The second characteristic of an excellent education system (or sub-system) is that it has strong institutions. As with high quality teaching and learning, strong institutions can be identified by external signs. They have clear purpose and values; longevity and recognition; strong professional standards; autonomy (though not isolation); and sufficient assets. Interestingly, as well as being signs or signals of strong institutions, these are also inputs, which in themselves add to the strength of institutions in a circular way. To these inputs we may add good governance.

Both high quality teaching and learning and strong institutions contribute over time to the third, emergent characteristic of a strong education system situated within a society and an economy, namely a shared narrative of social value. The 'over time' part is very important. A shared narrative of public value – or, more simply, a widely held good opinion by all and sundry – will only build up over time if institutions remain strong and teaching and learning remains high quality.

Again, we can identify characteristics of this shared narrative of social value. These are: high status of those who work in that sector; good pay of those who work in that sector; public investment in it; and well-respected qualifications of those who enter and advance within that sector. Shared good opinion of an education system (or sub-system) can be identified by a further feature it leads to - private investment by individuals.

How good is our apprenticeships system and how about our current
reforms?

If this theoretical framework is useful and correct we should be able to
'plug in' any education system to it and test it. How would our school
system stack up against each of the characteristics? How would our
university sector? Early years education? Vocational education and work-
based training?

My contention is that our vocational education system – including and
especially our apprenticeships system – would not fare well against this
model of excellence. The area where it would score best is the first. We do
indeed see high satisfaction from apprentices (especially younger ones),
we do see increased productivity and employment, and the crude
measures we have of value-added in attainment have moved very
positively.

So good news then – the fundamentals, while far from perfect, are sound.
So why are apprenticeships not as popular with young people as we would
all like them to be? And to be brutally honest about the subtext to this
ubiquitous question, why are they not more popular with the middle
classes and with our most able young people (two different but
intersecting groups), as they seem to be in some other OECD countries?

I believe the answer starts to emerge when we look at the second test:
strong institutions. With a very limited number of exceptions, our
apprenticeships system simply does not have strong institutions at its
heart. Think of the institutions that would feature in an assessment of
state secondary education; or higher education; or private education. And
now look for the institutions at the heart of the work-based education
system. Can we even identify them with confidence? Many colleges have a
proud history and would pass the test, but increasingly few look strong by
all the measures postulated, and in any case the majority of
apprenticeships followed in England do not have much to do with
colleges.

In some other OECD countries, powerful institutions beat at the heart of
their apprenticeships system, for example, Chambers of Commerce, or
partnerships between employers, unions and educators. In a way, the
precise shape or identity of the institution matters less than its strength.

This is because without stable, well-respected, autonomous, well-run institutions we can never build the one thing which will lead to many more people 'buying in' to apprenticeships universally as a good thing – namely the shared narrative of social value.

The government has set a clear target of three million apprenticeship starts over this parliament. This is excellent. It has not set any target for how many must be young people; nor at higher levels; nor in growth sectors; nor any targets on quality.

Many young people already opt for apprenticeships – there were 126,000 apprenticeship starts by people aged 16 to 18 in 2014/15. However over recent years this level has only represented around 6 per cent of the cohort. Many apprenticeship opportunities are heavily oversubscribed, about eleven applications to each vacancy in 2013/14. So it might seem that employer supply is the fundamental problem, not demand from young people. But this is a superficial analysis. In fact many employers do not fill their apprenticeship places, and many more do not offer them at all because of a complex set of factors about the candidates which can be boiled down to one word: quality.

When policymakers or employers ask 'why don't more people apply for apprenticeships?' what they usually mean is 'why don't more people apply who are well-educated, well-qualified, with good attitude and sufficient social capital?' All the sideways talk of 'promoting the apprenticeships brand' really means 'getting more high achievers to apply'. It is not accidental that the Prime Minister, David Cameron, uses the formula 'university or an apprenticeship'. It is not thoughtlessness that has led officials to create the concept of a 'degree apprenticeship'.

Before addressing the question of how to get more high-achievers taking the apprenticeships route, we should first challenge it. Why does this matter? High achievers are already the group best served by our education system, why are we focusing on them again? Isn't that missing the whole point of the apprenticeship route, namely that it gives a positive, structured pathway to success in work for those who cannot (or will not) follow the tried and tested academic-based pathways? Perhaps.

It is certainly depressing to have watched for fifteen years the predictability with which decision-makers and influencers use their own opinions and life experiences as a yard-stick for all, like so many well-meaning but solipsistic Protagorases.

But if we are concerned with creating an excellent apprenticeships system for all, it may be that we need to watch the choices of the high achievers closely. For the choices made by those who have the most options are a very strong indicator of the extent to which an educational sub-sector has achieved a shared narrative of strong social value.

Furthermore, it is their choices that do most to create this shared opinion and shape the culture. From this shared social perception flow all the positive influences which will lead to apprenticeships for young people reaching 'escape velocity' and entering the upper atmosphere in a way they never have in recent years, despite concerted efforts to grow them. When high achievers choose apprenticeships then the middle classes will do so en-masse (note of course that 'high achievers' and 'middle classes' are not synonymous, though there is in England still a strong correlation between them, despite steady progress made in recent years to 'close the gap' between educational attainment of the poorest and the rest). When this happens, so much follows: high status of the chosen sector, support for public investment, and the rest.

The paradox is that this virtuous circle cannot be achieved without the fundamentals first being in place: high quality teaching and learning, and strong institutions. Unfortunately, while the current government's reform programme may achieve very useful things (employer ownership of standards, greater employer investment via a levy), at the time of writing it will largely be good luck if it results in higher quality teaching and learning experiences. And it will be frankly miraculous if it results in strong institutions.

Without a joined-up, long-term, sustained plan to create these two conditions, it seems dubious that a strong, shared narrative of positive value will emerge around apprenticeships. And that, in turn, means it will not be seen as a career path of choice for most who have other attractive options. And that, in turn, means it will never take its place in the public consciousness nor the public policy sphere as a route with equal status to university, however much politicians honestly and fervently wish it to be so.

So what should we do?

If the ideas postulated here are even 50 per cent correct, we can get on track with a new revised national ambition for three million high quality apprenticeships quite easily, and that will just be for starters:

1   The current reforms can continue. They don't address all necessary aspects and therefore won't guarantee quality, but they may well bring about other benefits, especially in terms of industrial relevance and wider employer participation.

2   A serious, large-scale programme should be put in place to support high quality teaching and learning. We know how to do this. It is about supporting strong dual professionalism in teachers and trainers; ensuring business relevance and excellent employer engagement in a 'two-way street'; and supplying high quality CPD to teaching staff.

3   Government should think again – with industry and vocational educators – about how to establish and nurture strong institutions at the heart of the system. This should include colleges, but not solely colleges – the key is that they are autonomous, respected, long-lived, well governed, and with clear purpose and professional standards.

With excellent outcomes and strong institutions, government would no longer need to think about how to promote apprenticeships to young people. It would become no more of a problem than promoting A-levels, university degrees or professional qualifications. And perhaps three million starts over a parliament will one day be looked back on as a low ambition, and a reminder of how far we have travelled under the power of investment in teaching quality and strong institutions.

Reference
Hodgson, Ann. (2015) *Coming of Age for FE?* London: Institute of Education Press.

3.
HIGHER APPRENTICESHIPS:
COMPLETING THE MAP

Sara Caplan
Partner, PwC

_____

Why do we need higher level apprenticeships in the UK? And why should employers be driving their development? How can we help small employers to take on apprentices? How do we change negative perceptions that still surround the apprenticeship agenda? To achieve three million apprenticeships in the UK we need to answer all of these questions and put in place new practices. This article explores some of the ways PwC tackled these questions from 2010 to 2015 and the lessons we learnt through doing so.

In 2010, we at PwC were deciding how to improve our school leaver programme. We had a thriving and quite structured graduate programme, and we wanted to do something more formal for our school leavers. The driving force behind this was twofold: we were looking at new ways of diversifying our talent pool, and we wanted to do something more about social mobility.

On the first, we had successively widened our recruitment net for graduates, to the point that we recruited from most of the universities in the UK, plus from overseas, but we knew we were missing out on recruiting talented people who hadn't, for whatever reason, gone to university, wanted to go, or succeeded on that university route.

On the second, we were proud members of the Social Mobility Compact, and were working on removing barriers to entry, but we knew there was more we could do to attract people who may never have aspired to work in professional services. The majority of accountants still come from a family where one or more parent has been in professional services – that's a small gene pool! We wanted to widen it.

So what did we do?

We started to look at apprenticeships, to see if they would provide that more formal route for non-university entry. There was nothing there that we felt would meet our needs as a business. The programmes seemed outdated, inflexible and not aligned to what people were doing in our workplaces.

At that point we were lucky that the UK government announced the launch of the Higher Apprenticeships Development Fund. The fund was there to create new routes to higher-level qualifications (level 4 and above), provide progression from lower level and where appropriate, to articulate with university degrees.

On the basis of this we decided that, instead of creating something that was just for PwC, we would try to bring together a range of employers and create a programme for our whole industry. For me, this was the critical moment that turned something that we had been thinking of as an individual employer into an iconic programme that would become a whole new way of entering professional services, and revolutionise recruitment to our sector.

We put together a bid to the Fund that was based on creating a higher apprenticeship at level 4 in Professional Services with three pathways: Audit, Tax, and Management Consulting. Our proposal focused on four areas: engaging employers; engaging schools; designing and developing the apprenticeship framework; and creating progression routes.

The first challenge was to bring together employers in our sector. This was challenging, as we work in a highly regulated environment, where competition, not collaboration is the dominant feature. We overcame this by looking at what we were trying to achieve. We all wanted more talented people to enter professional services; to create qualifications that were fit for our businesses to use; to spread the word about the great careers we can offer; and to help people to get on the first step on the ladder of employment.

So we decided collectively that creating a thriving, high quality talent pool is an area on which we can collaborate. It is a win–win for each employer on our sector, large or small. We compete on getting people out of that talent pool into our own organisation. By working in that way we overcame the natural hesitance in working together and avoided any hint of anti-competitive behaviour.

Once that principle was established, we created three working groups – one for each work stream. The groups were comprised solely of employers and each had a relevant professional body representative: The Institute of Chartered Accountants for England and Wales (ICAEW) for audit; the Association of Taxation Technicians for tax; and the Management Consultancy Association for consulting. We had a mix of size of employers and geographies, so that we made sure that what we created was fit for all.

We started with blank sheets of paper and wrote down what tasks we would want new trainees to undertake. We then converted these into learning outcomes. We thought about new trends in our sector, how jobs might change, and above all how we could make sure that trainees would learn most of their skills in the workplace. We wanted everything to be relevant to what they did in practice. What we found was that there was a common core of skills and abilities that no matter which pathway was being studied, the trainees needed to learn. These were competencies such as working in teams, presentation skills, communication, data analysis, and building relationships.

We used the same competency-based part for each pathway and created a technical part that was specific to the pathway. An important feature of this was that we worked, from the beginning, with the various professional bodies that 'license' people to practise as accountants, auditors, tax advisers and, although not licensed, management consultants. This was critical in making sure that our programmes would give the apprentices the professional qualifications they needed to be able to work in the sector and offer exemptions from the professional exams.

Once we had established this, we brought in some FE colleges, universities and private training providers to advise us on whether what we had created was deliverable, check that we had used consistent terminology and standards, and help us design the assessment regime.

It was difficult to get the providers to engage with us. We had turned the design process on its head and they were used to being in the driving seat.

However there were some dynamic and forward-thinking organisations that responded positively to our request for help and gave us excellent support and advice.

Our next challenge was to have our framework signed off as a recognised apprenticeship. This required various layers of scrutiny - from OfQual, the National Apprenticeship Service, and the Learning and Skills Council. Additionally, of course, the professional bodies had to give recognition so that the trainees would gain their professional qualifications as a result of completing the apprenticeship. Every single one of these organisations pulled out all the stops and undertook their part of the process in record time. We were overwhelmed by the quality and speed of their support, challenge, discussion and decision-making.

Our framework took eight months from start to finish to be ready for delivery!

Alongside this activity on the design and development, we were working hard on the schools engagement side of things. This proved very difficult. Most schools were very resistant to anyone talking to their pupils about apprenticeships, even if they were higher-level opportunities leading to great careers. In addition, there was no centralised careers support for pupils or schools, so it was incredibly difficult to get a message out about the programme – we knew we couldn't tackle 27,000 schools individually!

Interestingly, the independent schools sector was much more proactive – we were approached by several of them to come and talk to parents and pupils and to present at their events. We often did this together, with several of the employers from our working groups going to the schools. No state schools approached us.

We decided that we would overcome this in two ways. First, we found opportunities to speak at major conferences – teaching union conferences, events that were specifically about raising awareness of apprenticeships, chambers of commerce, party conferences, anywhere that we would have an audience that could influence uptake of higher apprenticeships.

Second, collaboration. At this stage we weren't even wedded to talking about our own sector. There were several other sectors developing higher apprenticeships and we all decided that to crack this we would go out as ambassadors for the whole suite of subjects. A great camaraderie and network built up across employers from different sectors at this stage – we were all trying to engage with pupils, parents, teachers and careers advisors and it made no sense to go it alone.

I was privileged to be appointed as an employer ambassador for higher apprenticeships, which gave me an extra mandate to go out and talk to people about this new way of creating jobs and careers.

Putting it into practice
In 2011, having had our framework approved, the first apprentices started on the framework. At PwC, we recruited apprentices to the tax and consulting work streams. Audit came a year later as we had just made a change to the previous school leaver programme and needed to run that out.

Our aim of widening the talent pool came true – our apprentices came from very different social backgrounds, with different characters, experiences and expectations. Not all had come straight from school – some had tried other jobs, some had started university but decided it wasn't for them.

There were some interesting similarities: most of the apprentices had needed to overcome huge pressure from parents and peers not to go down the apprenticeship route; many of the school leavers had been taunted about not being good enough for university. This made them doubly determined to prove that they had made a good choice and they were motivated, committed and enthusiastic. They knew they were in a new programme, and we sought their feedback throughout on what was working and what wasn't. This worked really well and led to us changing some of the structure. They created their own network and met regularly to share experiences and support each other, in addition to the buddies and coaches we had assigned to them when they started.

It was this creation of the network that led us to start thinking about small employers, and the challenges they face in talking on apprentices. Small employers often take on a huge risk employing an apprentice – they have to take time out from the day-to-day work to train them and support them and if it doesn't work out there is big cost to their business. They also tend to take on one or two apprentices, who can then feel isolated. We were lucky in that we took on seventy, but how could we help apprentices in small employers to come together?

The UK government had just announced a new initiative called Employer Ownership of Skills and we decided to bid for some of the grant funding to help smaller employers overcome these challenges. Our programme focused on apprenticeships in professional services occupations in their broad definition, including areas such as management, HR, marketing, finance, accounting, and recruitment, as well as the audit, tax and consulting that we had developed.

We focused on level 3 and above. Our aim was to help the employers to identify where they could take on new apprentices, help them find the right training provider, streamline the process of getting the funding as we would handle that for them if they wanted us to, and support them in keeping the apprentice on the programme and complete it successfully.

We created a team within PwC who visited employers, discussed their needs, and helped them to get the right training for their apprentices. We found that many employers had been put off apprenticeships in the past for a number of reasons: their poor perception of the quality of training providers; being inundated with offers of 'free' training that was substandard; being embroiled in bureaucracy surrounding funding; and not wanting to take the risk.

The business-to-business approach that we took seemed to work better for the employers we engaged. There was a level of understanding of the business and trust in the support that meant they felt more comfortable with it. Having said that, it took a long time to overcome some of these issues and persuade each employer that this would be good for their business. We worked with employers from every sector – manufacturing, hospitality, motor vehicle, and construction as well as our traditional stamping ground. Many employers had only thought about apprenticeships in their front line activities, and not about taking an apprentice who could do, say, finance or HR (or both).

This was great, as it meant that we were actually increasing the number of apprentices in the UK workforce, not replacing existing ones. We also had rules: every single apprenticeship had to be either a new job created or someone progressing to a new role who had to develop significant new skills; and every employer had to make a contribution to the training – if it wasn't a cash contribution then they had to evidence what it was. We wanted to make sure that we were supporting real, new training and that every employer was investing in it.

It was hard work, but very rewarding. Our model didn't always fit with what the Skills Funding Agency, who were administering this, had planned – we had to keep changing our delivery profile as we were totally focused on what employers needed and that didn't fit a set profile – it had to be dynamic. But we worked through it and reached a good way of working with the Agency. Through this we tackled the issues of small employers taking on apprentices more easily, but not those concerning networking.

We came up with an idea which we called the 'London Professional Apprenticeship'. What we wanted to do was to create a cohort of apprentices across London (as a pilot area), from all different employers, who would come together to learn about, for example, the London economy and its place in the global economy, how business works, how to build relationships and networks and do a project together. We felt this would create a network of professionals that would grow as they grew, would connect their employers and help overcome any issues of isolation of apprentices in smaller employers.

We got together with the SFA, the Mayor of London's office, and BIS and found a way to fund the development of the programme together. We designed a set of training modules, worked together to get employers to nominate apprentices in their workforce to be part of the programme, and the BIS Secretary of State, The Rt Hon Dr Vince Cable, MP, launched it in December 2013.

It was a huge success! Apprentices came from local authorities, social media companies, accountancy firms, PwC, shoe designers – you name it, they were there. We ran three cohorts. We talked to them about being the future leaders of business. We brought in guest speakers. They did business simulations. They learnt about how to use social media to grow their business. They got to know each other and about each other's work and after each session they went back to their employer with new knowledge to help the business to thrive.

For each one, the transformation of the apprentices from quiet mice into confident, outgoing networkers was astounding! I spoke at the first and the third sessions of the first cohort and at the latter, which was a networking session with their employers. I could hardly make myself heard. The employers loved it too – they said they had never been part of something like this before and had also learnt a lot about how to build relationships across businesses. The first cohort of London Professional Apprentices graduated in September 2015. We can only hope they keep in touch and help business to grow and connect in London!

So, what lessons did we learn from these three programmes?

1. Apprenticeships have to start with what the employers want for their workforce. By that, I don't mean bodies that are formed to represent the voice of employers, I mean actual employers, large and small
2. There is a huge benefit in employers collaborating to achieve change in the apprenticeship world. Anything that government can do to initiate and support this would be welcome.
3. There needs to be a step change in the way careers advice is given and available in the UK. School, parents, teachers and children don't know what they don't know. I was constantly amazed by the number of parents who didn't know there were any options for their children other than university. This has to change. We need to remove the postcode lottery on careers advice and make sure every child has access to informed people who can help them with choices.
4. Small businesses need extra help to take on apprentices and having larger businesses take the lead to do this can be an effective way of doing this
5. Apprenticeship networks work – they help businesses to connect, apprentices to flourish and they spread the word about how apprenticeships grow business.

6. Existing government funding structures don't support employer led/controlled approaches as they are not responsive to actual need.
7. We need more higher-level apprenticeships. The trailblazers are doing a great job, but there are still many employment sectors where there is no or not enough progression to levels 4 and above – another higher apprenticeship development funding round would be a great investment to fix this
8. Employers need to step up and take responsibility.

I worked with some great people across all different organisations. Each of us was passionate about apprenticeships, about changing perceptions, about bringing our own organisation into the apprenticeship world and about spreading the word to others. In each organisation there will be someone who can do this – business and industry leaders need to find them and support them in taking this forward so there is a deafening employer voice on why apprenticeships are the right thing for the UK economy. By showing that we as employers are committed to the apprenticeship agenda, by offering high quality jobs, with great training, good career prospects and fair pay we will change the culture and persuade parents, teachers and pupils that apprenticeships are a great choice.

Interestingly, these are lessons that I have also been able to apply in Australia. Here the nation is facing some very similar issues – a poor perception of vocational training and apprenticeships, a lack of higher-level vocational routes, skills shortages and few ways for employers to lead on the skills agenda. But things in Australia are changing too. A programme of Vocational Education and Training (VET) reform is gaining momentum and there is a desire to focus on more, high quality apprenticeship opportunities including those at higher level.

New Skills Service Organisations (SSOs) will work with industry to make sure that the employer voice leads the development of new qualifications, based on the current and future needs of business and industry in terms of skills to support a growing and thriving Australian economy. At PwC Australia we are lucky to have been given the opportunity to operate one of these SSOs and help six key industries to create the training and qualifications they need to attract a great talent pool and cement their future.

There are challenges, not least the huge geography, remoteness of some communities and industries and the interplay between states and territories and the Commonwealth. But these challenges create opportunities for innovation and fresh approaches, and the willingness of employers, training organisations, industry bodies and government to come together to back Australian skills and economic growth will overcome anything that is thrown at us.

4.

INCREASING TECHNICAL APPRENTICESHIPS: WILL THREE MILLION APPRENTICESHIPS BE ENOUGH?

Graham Schumacher MBE

———————

This chapter looks at the ambition of three million apprentices by 2020. Is it enough to meet the real need? What more needs to be done? It provides an expert point of view from an engineering manufacturing perspective and looks at improved careers advice, more higher apprenticeships, a clearer focus by trainers on the apprentice's first job opportunity and the development of skills hubs.

The problem with judging if three million more apprentices over the time of this parliament are going to be enough is that it depends. It depends on the performance of the skills system in delivering them and more critically can we move back to a position where the majority of companies become involved again in early career entry programmes such as apprenticeships.

So what are the areas of the system that need attention if we are to meet this target and understand the number of apprentices the country really needs? What do we need to do to help people understand the apprentice offer to meet these needs?

The Apprenticeship Offer

The apprentice training route has become better understood in spite of the school and university route still being the focus of much careers advice and guidance in schools. Apprentice training activity in the UK has made great progress over the past few years with traditional sectors like engineering seeing apprenticeships as a key part of their early career entry pipeline. Why? Because they continue to see work-based learning, in the environment and facilities of work, as well as the underpinning knowledge gained from FE colleges and universities as a proven way of developing people who understand and are committed to their organisations.

The growth of the higher apprenticeship has made a big difference for many young people who enjoy working and learning and applying that combined experience to real work. Managers and work colleagues also enjoy sharing their experience and teaching these young people, developing their 'workplace behaviours' of self-management, problem-solving, communications, working to standard processes, and being able to offer ideas for improvements and then being able to implement them.

The higher apprenticeship is starting to broaden the range of programmes traditional apprentice training organisations have offered, adding to the skilled and technician jobs with programmes for roles such as programme management, purchasing, supply chain planning & control, engineering professional roles and management.

Nearly two-thirds of the overall employment growth in the European Union is expected to be in the technician and associated professional roles.

Higher apprenticeships have also seen companies such as PwC offering professional services apprenticeships and taking a lead in the development of these programmes and selling the benefits to both young people and other companies in their sector.

The benefit for the young person is that the apprenticeship offer extends from level 3 through to levels 6 and 7 with a clear alignment between training and education and real work with the added attraction, from the apprentice's point of view, of the employer paying the FE or HE fees and the apprentice being paid.

The apprenticeship offer seems to be growing in alignment with needs and future activity.

Employers and opportunities
However not enough companies are involved and offering entry level opportunities to young people to enter the workplace for work experience from age 14 through to apprenticeships at age 16 and 18-plus.

The opportunity to rebalance the economy in the engineering manufacturing sector by 'on shoring' work from the low cost employment countries is a real possibility, only threatened by skills shortages in technician and professional technical areas.

On one hand, the technology is available to companies to produce high-value, high-volume products to the very highest quality and at a competitive cost, enabling the UK to be able to compete with the rest of the world for this work.

For companies in this sector to take advantage of this opportunity there is a need for a range of skills from manufacturing systems engineers, able to plan lean manufacturing operations; manufacturing operations managers, able to run these systems to deliver high quality right first time and on time production; through to operators of these systems who are technically capable and able to intervene to ensure production standards are maintained and improvements are identified and implemented.

The skills required for these roles, and others in a modern context, need to be based on standardised, repeatable ways of working and focused on continuous improvement. In this technically demanding real workplace the roles required need to have three aspects:

1. Knowledge, from level 3 technical through to level 7 management and professional engineering qualifications.
2. Skill, the application and deployment of that knowledge in the workplace.
3. Workplace behaviour, the ability to work and understand processes, work in teams, communicate, solve problems, and lead on projects, self-management and many others. This element of an apprenticeship is often missed but is key to a successful programme.

Key to these roles working well are the three aspects coming together and deployed with the attitude required to make the maximum contribution in the workplace. Given that the minimum educational standard is met, attitude is the most important quality needed at selection and recruitment and is the quality that will land the offer of an apprenticeship.

Apprenticeships are the perfect vehicle for developing and honing this attitude and developing young people into well-rounded individuals capable of delivering in the workplace. The knowledge, skills and workplace behaviour can be then learnt through the apprenticeship.

On the other hand, the UK is short of the skills required to take advantage of the opportunity that technology gives to be able to automate high-value, high-volume manufacturing. With 'baby boomers' retiring, further pressure will be put on companies who may find the opportunities for 'on shoring' work impossible to take advantage of because of the shortage of skills.

Whilst some larger companies have entry-level programmes to the workplace that deliver the skills, knowledge and behaviours required, many don't. A high proportion of the UK's enterprises have no entry level programmes and rely on experienced hires to fill their vacancies.

Apprenticeships give companies the chance to build capability in foundation roles. Young people will need time to develop and grow before filling some of the shortages we have today at project and team leadership levels in particular.

To help take advantage of the technology-based opportunity to grow our engineering manufacturing sector, the often missed other skill resource is in the current workforce.

People who have been in the workplace for some time and know how things work are in a strong position to grow into the developing roles by undertaking apprenticeship or bite-sized programmes which may be focused on knowledge or skill.

Apprenticeships can help turn late developers into employees who can thrive in the changing workplace and where the production leaders, technical specialists and process improvement capability can be grown. Age should not be an issue in deciding who can take advantage of an apprenticeship. Attitude of those who wish to take part should.

The skills system

We have many of the programmes and the delivery organisations to support companies to develop people to fill these skills shortages and a growing interest from young people and parents who are taking an interest in apprenticeships again.

In Derby we have opened a manufacturing university technical college (UTC) with full intake of year 10 and year 12 students keen to be involved in a technical education, as well as a wide range of companies keen to help with work placements, projects and material.

We have a range of educational and training programmes to address the needs of the engineering manufacturing sector. The two routes, both are needed and valuable, from foundation education into work are:

1. The university route that principally delivers deep knowledge and (to be truly effective) is combined with work experience or a training programme to develop the skills and workplace behaviour.
2. The apprenticeship route that teaches the three aspects of roles in the workplace in context during a programme delivered at levels 3 through to 7.

Group Training Associations (GTAs), large companies, colleges, universities and many other training providers work with companies to make programmes that fit the needs of the employers. Experienced workers can need similar apprenticeship programmes or bite-sized elements of them to meet the needs of their roles.

The university sector has moved to be involved in the development and delivery of apprentices, particularly the higher apprenticeships, however the number of companies involved in these programmes has not grown enough.

We can't just rely on large companies like Rolls-Royce and BAE Systems training all the apprentices for the sector. Typically these companies receive 20,000 plus applications for 700 to 800 apprenticeships and graduate training places per year. But what about the 19,000 applicants who fail to get a place? The UK clearly has a supply of young people keen to be apprentices. The demand from businesses remains low even with growth opportunities and the growing threat of the baby boomer leaving the workforce. Companies, big and small, need help to offer more entry-level programmes for those young people who want to get a job and start a career, and for companies to thrive.

To help companies grow and take the opportunity to 'on shore' work the UK, the skills systems needs to step up to the challenge of working with employers and help meet the skills shortages.

To get more young people into jobs and to address skills shortages, companies need to be helped by the 'skills system' (colleges, group training associations, universities and large companies working with supply chains) on how to start offering entry-level programmes and how to hire and train young people.

We need to help the organisations in the skills system to think and act in a manner that is focused on helping the needs of individuals and the needs of employers. To deliver to the needs of individuals and employers (and the country) training and educational organisations need to see themselves as part of a bigger system with partners up and down the system and not as stand-alone organisations.

The skills system should start from year 9 or year 10 when young people should have access to information on career options with high quality guidance and help to start understanding their own learning styles. Employers can play a part in helping schools with current understanding of roles in the work place, with visits to look and see as well as employees going into schools to talk about their jobs and the skills and knowledge required.

As an apprentice ambassador, I have often heard employers talk about being refused access to talk to young people about work. This is seen as driven by schools focused on keeping 'their' students until they are 18 and have finished A-levels so as to maximise funding.

Some young people are ready to move to an environment that is closer to a work situation with UTCs offering exactly that type of mixed technical and academic experience with sight of the local university programmes. The most successful UTCs have strong links with local employers who can offer a clear line of sight to work whether this is directly into apprenticeships at 16, or at 18 after A-levels, or via a university course. The students often select UTCs for a number of reasons:

- Technical specialism
- Opportunities for work experience
- Links with business
- Facilities and equipment
- Better placed to get a job

At year 12, for some young people, the next part of the skills system would be joining an employer as an apprentice or moving to their local college. The apprenticeship programme can be delivered by the employer, a GTA, the college or by a training company. Whichever it is, they should be joined up, working together and having a clear view of the first role after the completion of the apprenticeship.

The system should be open to employers to shape the curriculum, have their employees deliver within the college or GTA, and offer training for the students on work experience back in the company.

The organisation within the system should look to implement the ten recommendations of the Commission on Adult Vocational Teaching and Learning (CAVTL, 2013) Four of the key recommendations are:

1. Adopt the two-way street so that colleges, training providers and employers are directly involved in shaping programmes that reflect the up-to-date needs of occupations and workplace. Employees should deliver in the college and at the training provider, while trainers should spend time in the workplace experiencing current practices and challenges.
2. Develop a core and tailored approach to vocational qualifications.
3. Revise and strengthen the education and training arrangements for vocational teachers and trainers including introducing Teach Too, a scheme to encourage experienced professionals to pass on their expertise.
4. Reinstate employers' presence and influence across providers of VET.

Apprenticeship Design and Quality

Many companies who in the past trained apprentices have lost the capability to become involved in the apprenticeship programme today, feeling the process has become too complicated and that they have difficulty attracting and selecting young people capable of succeeding in the work environment.

The skills system should not only be more joined up but should be seen as the 'go to' organisation and the local skills hubs for employers, capable of supporting employers in all activities, from attracting and selecting young people through to training and developing them.

The skills hubs should also be seen as the place to go for up-skilling their existing workforce. The delivery of skills and of apprenticeship programmes needs to become the norm again. These skills hubs should also be close enough to the employers that they understand their needs and are able to support them with their workforce development throughout a person life at work.

The key players in developing the skills hubs are FE colleges, GTAs, training providers and large companies with area collaboration and links to the local university.

Large employers should be encouraged to establish programmes for the delivery of apprenticeships to smaller or supply chain companies. This helps young people embark on a career while learning skills that employers need for the workplace with the 'kite mark' of quality associated with the large company. The larger company can benefit by helping build training capability in these companies, building trust within the supply chain and which can be the vehicle to work on process improvement activity together.

However, just merely boosting apprenticeship targets isn't enough by itself. We need to run widespread campaigns in schools and colleges to generate interest in various sectors and the opportunities to train for available roles and the careers. The routes and options to these careers also need to be part of developing up to date advice and guidance process.

The other important factor with entry-level programmes is changing mind sets of people about traditional industries. For example: why is

47

manufacturing still seen as a sector for boys, while other sectors are for girls? Engineering manufacturing companies should make efforts to attract female talent by showcasing the roles in high value manufacturing and by offering higher apprenticeships programmes for girls who may be more attracted to technical roles requiring teamwork, problem solving, communication skills and planning. Core manufacturing functions need support from other critical parts of a business and a career in purchasing, supply chain management, purchasing or management, may be more appealing to young people who hitherto haven't seen an apprenticeship as an option for them.

Is achieving three million apprentices by 2020 enough?
This question is difficult. If the retiring baby boomers are to be replaced and the economy stays at the same level as today then the answer is maybe. However with an aging population and life expectancy increasing there will be a need for new jobs in the care and health sectors then maybe not.

If we can make the skills system work to support companies who can grow and those that can on-shore work by use of technology probably not.

So what should we be doing to hit the first target of three million apprentices?

- We must focus on making the skills system work to deliver the needs of the individual looking to develop skills that will provide them with an income for life and for needs of business both at an entry level and the through career development of their workforces.

- At the 14 to18-year-old phase, high quality careers advice and guidance is needed, with close relationships with employers, so that accurate information on roles and career paths, along with earning potential, is available to young people. Clear information on the pathway options available to be successful in the next phase of young person's education and training supported with help in understanding the student's preferred learning styles. To support this advice and guidance, the young people should have the opportunity to 'go, look, see' other young people taking the pathways and have the opportunity to speak to them.

- Colleges, GTAs, large employers and other training companies should work at bringing employers into their organisations to influence the curriculum offer and support the organisations in becoming 'local skills hubs'. These skills hubs should focus on understanding the local employment profile and reach out to build relationships with these businesses. Over time, these skills hubs could work to understand the business needs, which may range from bite-sized skills training, short courses, or help to develop their capability to training apprentices. The aim would be to become the training and development preferred 'go to' option for both entry-level programmes and through career development of the workforce.

- Higher apprentices are well thought of by employers and young people involved with the programme today. More work should be done by the skills system, including universities, to ensure this option is more broadly available and understood by employers and young people.

- The higher apprenticeship programme must have a focus to be developed across other sectors and roles. If two-thirds of employment growth is in the areas of technician and professional roles this programme is ideally suited to meet the demand.

Reference

Commission on Adult Vocational Teaching and Learning, 2013. *It's about work... excellent adult vocational teaching and learning (summary report of CAVTL)* (www.excellencegateway.org)

5.
PUTTING QUALITY AT THE HEART OF APPRENTICESHIPS

Kirstie Donnelly MBE
Managing Director, City & Guilds

Introduction

The government's commitment to apprenticeships shows policymakers are starting to realise the myriad benefits of professional and technical education. The challenge now is to ensure that apprenticeships are recognised for what they are - an aspirational and proven route through to a great career. To achieve this, learning needs to be at the heart of apprenticeships and we need to avoid a narrow focus on reaching an arbitrary target figure at the expense of quality. Now is the time to reinvigorate the system and make it one of the best and most varied in the world, giving our young people real choice, and providing British industry with the workforce it desperately needs.

This chapter will explore the means to achieving this. There has already been a considerable amount of policy change in the apprenticeship system, with reforms shifting the balance of apprenticeship design and delivery towards employers. What we need to focus on now is implementing the policy changes. Employers must be embedded in the apprenticeship process from start to finish for the best outcomes for business and young people. Only then will we see the positive effect apprenticeships have on overall productivity.

Many of the points we raise in this chapter are explored in our report *Making Apprenticeships Work* which was produced by City & Guilds and its Industry Skills Board (ISB) – a group of leading employers who are all involved in apprenticeships delivery. The report itself set out a number of key actions for politicians, business, employer groups and the education sector to ensure apprenticeships become as sought after and accessible as securing a place in higher education.

Professional and technical education and apprenticeships are a proven way of improving UK productivity.

The current government rightly believes that apprenticeships can result in productivity gains and reduce skills shortages and youth unemployment. Despite rising employment and increased hiring levels from big businesses in 2015, youth unemployment and productivity levels still show negative trends. Output per worker in the UK is now 21 per cent lower than the average for other G7 countries, while youth unemployment figures are still far too high at over 15 per cent, with 21 per cent of those out of work for over a year. (ONS, 2015)

Good quality professional and technical education will play a vital role in raising employment levels and improving productivity. Recent research from the Centre for Economics & Business Research (Cebr) (Cebr/City & Guilds, 2015) highlights the effect vocational education has on UK productivity:

- 1 per cent increase in vocational skills would uplift UK GDP by £163bn within ten years
- 10 per cent increase in vocational education enrolment by 16 to 18-year-olds would lead to a 1.5 per cent drop in youth unemployment
- Businesses see average productivity gain of £10,000 per annum per hired apprentice
- Apprenticeships have a return on government investment of £16 to £21 for every £1 invested

The return on investment is there for all to see, but to guarantee consistent and healthy returns we must ensure the system is set up to create quality apprenticeships.

Quality should be the key driver not quantity
With apprenticeships now enjoying their highest profile amongst policymakers in recent times, we need to ensure this agenda isn't just driven by numbers. Nor should success be framed in this way. To stimulate real growth, it is essential that the key driver for the

apprenticeship agenda is focused around the quality of the provision on offer.

If the quality of the product can be achieved, apprenticeships will begin to be seen as a valuable alternative to other education pathways and a genuine choice for young people to consider alongside A-levels and the higher education route and, ultimately, a sustainable route into a great career.

To achieve this quality, we need to focus on four key areas within the apprenticeship system, these include:

- Recruitment into apprenticeships that are intrinsically demanding and worthwhile. This will result in a broader range of apprenticeship occupations offered at all levels as new jobs or roles, substantially improving the offer both in perception and reality.
- Training and learning programmes that use a range of effective methods and are built on the support of highly-skilled adults in the workplace. This needs to be led by employers to ensure the skills most desirable in the workplace are provided.
- High standards built into a demanding assessment at the end of the apprenticeship. One recommendation in relation to this would be to issue a certificate for the end assessment, an idea that already has interest from employers.
- Progression opportunities that display the potential career routes beyond the initial apprenticeship. It is vital that young people understand and are aware of the career opportunities available to them when embarking on an apprenticeship. (City & Guilds, 2015)

This isn't just an argument the education sector has been making. Responding to the announcement on the introduction of an apprenticeship levy, Neil Carberry, CBI Director for Employment and Skills, stated 'We must not sacrifice quality for quantity...The best way to drive up quality is to give employers real control.' (CBI, 2015)

Similarly, commenting on the quality over quantity debate, the British Chamber of Commerce stated: 'The government is right to turn the spotlight on apprenticeships, but wrong in focusing on counting numbers rather than the quality of apprenticeships'. (Mason, 2015)

Putting learning at the heart of apprenticeships

As we have established, quality should be the key driver behind the apprenticeship agenda. One of the main elements to ensuring this quality is the provision of effective training and learning. There is currently limited debate around ensuring learning is firmly seen as the core element of apprenticeships. Meanwhile learning methods remain a neglected area of both national action and government policy. At present, there is no overall learning framework for apprenticeships or overarching policy on a number of imperative aspects, including workplace mentors or the training of trainers, as was previously the case. This has had the unintended consequence of a move towards a system where the college or training provider is providing the apprenticeship, rather than the employer, who in reality provides a large amount of the apprenticeship content. (City & Guilds, 2015)

Apprenticeships need to be seen as a valuable step towards progression into higher skilled jobs. It is therefore vital to develop an offer which makes it easier for young people to progress through technical and professional education. They must be founded on high quality learning, so enabling those young people to develop the skills most desirable to employers, including those around autonomy and the ability to work productively without extensive supervision, knowing that employers trust them to carry out the needed work.

The move to an independent end assessment for all apprenticeships is an important step to ensuring the quality of learning from start to finish and signals a move away from the existing models of continuous assessment that are arguably no longer fit for purpose. The change would mean that apprentices would be assessed at the end of their apprenticeship, once they have gained the maximum experience during their training. This allows for greater flexibility over how training is carried out, giving employers more control over the structure of the course and freeing up providers to focus more on the quality of the learning and less on assessment. (Ibid)

In *Making Apprenticeships Work*, we outlined our recommended quality model which we believe achieves the aim of putting learning at the heart of apprenticeships. The model includes a combination and range of learning methods which will help improve the overall learning experience for young people and also recognises the flexibility needed in the learning process as different employers will prefer different types of methods. Proposed methods include:

- On-the-job training and learning from and with experts and peers
- Off-the-job education, training and online learning
- Coaching, mentoring, formative assessments, review and feedback
- A nurturing, supportive and visible learning environment where apprentices have a voice. (Ibid)

Quality Apprenticeship Model

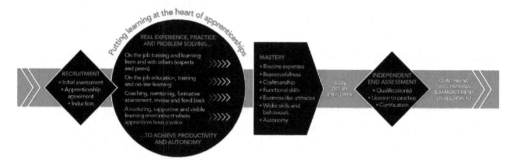

Giving employers control

There is a challenge in both securing the long-term stability of the sector and driving up the quality of the apprenticeship offering. We believe that full employer commitment and involvement in apprenticeships is necessary to enable the system to grow sustainably and in a way that meets the needs of business.

The government agrees and in November 2015 announced the creation of an independent apprenticeship body. This chimes with our recommendation in *Making Apprenticeships Work* that the majority of apprenticeship policy should be governed by an employer-led body, autonomous from the government, while giving government the control and power to focus on the political direction of the system as well as maintaining responsibility for pay levels, funding allocation, levy collection and stakeholder representation. (Ibid)

The UK Commission for Employment and Skills promotes the case for increased employer control over training and apprenticeships in their 2011 report. They explore solutions to 'encourage greater employer ownership of skills, working to secure a partnership for the long term'. This includes allowing employers increased freedom over the 'design and delivery' of training, including apprenticeships and work experience as well as 'funding employers directly for apprenticeships, for example through the tax system, and incentivising work experience'. (UKCES, 2011)

Willingness from employers to be involved in the apprenticeship process is relatively varied. We need to ensure that as many employers as possible are motivated to play a key role in shaping the sector and extending the offer currently available for young people. We have identified two key levers which employers should have responsibility for which we think would help increase apprenticeship employment, the apprenticeship levy and licence to practise. (City & Guilds, 2015)

The full detail of the levy is still being worked out and there is currently much more information needed about how it will work in practice. What's important is that employers don't rush to develop large-scale apprenticeship programmes purely to ensure they are spending their levy allocation. As we've said before, apprenticeships will only be successful if they are allowed to grow sustainably and with the best interests of the apprentices at heart.

Turning to licence to practise, rather than the introduction of any type of compulsory extension, we recommend in *Making Apprenticeships Work* that a national employer-led governing body should take forward a voluntary approach to extension, sector-by-sector and work with relevant employer groups.

There also needs to be flexibility in the system to enable it to work for all employers regardless of size or sector whilst still maintaining quality. From our experience, there is huge variety in the way apprenticeship programmes are managed and we can support greater employer take up by providing best practice examples that showcase different types of learning styles and outcomes.

To support *Making Apprenticeships Work*, we produced a case study supplement[13] that demonstrates how each member of the Industry Skills Board uses their apprenticeship programme to support their business growth, The more opportunities we can create to move the apprenticeship discussion away from one of employer duty towards a focus on the benefits apprenticeships bring to employers the greater we believe the enthusiasm and employer take up will be.

Securing and incentivising employer involvement in the apprenticeship agenda is key if the sector is to continue evolving and providing effective opportunities for young people, allowing them to progress into the careers of their choosing. Increased control from industry would also address the skills shortages currently facing numerous sectors as employers would have greater oversight of the makeup of apprenticeship programmes, and the ability to ensure training was targeted at the right regions and skill areas.

Access to apprenticeships
Some of the main challenges for the apprenticeship agenda are the current barriers around access for young people. It should be as simple and clear how and where young people can access apprenticeships, as it is for those wishing to go into higher education. Yet at present, a majority of people aren't even aware this route is available to them or have the first idea how to go about finding a suitable apprenticeship. The knock-on effect of this for industry is that businesses will miss out on some of the highest quality apprentices purely because potential candidates don't know how to access the programme.

There is also a continued stigma around apprenticeships, and the notion persists that they are 'for other people's children'. I sat on the Demos commission on apprenticeships and as part of that work Demos commissioned a poll of 1,000 parents of 15 to 16-year-olds to understand their views of apprenticeships. Worryingly, but perhaps not unsurprisingly given the historical issues of quality in the apprenticeship system, while nearly all (92 per cent) parents thought apprenticeships were a good thing, only a third (32 per cent) think they are the best option for their own children. Now that the system has improved so much and is so focused on delivering a quality experience for apprentices that can equip them with a degree-level education as well as a great career, it's time for this attitude to change.

This access issue is particularly acute when attempting to recruit 16 and 17-year-olds for some of the highest quality apprenticeships, as the majority of schools and sixth-form colleges are incentivised to keep students in full time education, with the intention of supporting them in securing a place at a university. As a result of this, in most regions in the UK, employers find it difficult to recruit within this age group, even though this could be the best option for many young people who are looking to gain practical experience in the workplace.

Inaccurate careers guidance unsurprisingly has an impact on access. With so many young people encouraged to focus on gaining a place at university, this is limiting the opportunities they feel are open to them. Currently, the advice offered in schools is far too focused on the steps which need to be taken to get to the next stage in their education, rather than beginning by addressing what a young person's preferred career path is and then from that, working out the best way get to their desired end point. This comes as no surprise when teachers are still responsible for the majority of careers advice in schools. Our research shows us how valuable interactions with employers are for young people when they are making decisions about their future. It should become the norm for employers to visit schools and provide role models for young people, inspiring them to consider careers they might never have thought about and alternative routes to get there.

We saw this lack of bias towards higher education first-hand in our recent research, *Great Expectations,* (City & Guilds, 2015) which surveyed over 3,000 young people about their career aspirations and then worked with economic modellers EMSI to match aspiration against the realities of the UK jobs market. The vast majority (70 per cent) of young people want to go to university despite EMSI telling us that under a third (30 per cent) of all job roles will be at graduate level in 2022. University will always be seen as the best, and indeed only, route through to a good career unless we can change perceptions of young people, their parents and their teachers.

We've already highlighted the issues around access to apprenticeships and the knock-on effect this is having on business. To address this, it needs to be much simpler for young people to know how and where to apply for an apprenticeship, and employers should seek to increase the proportion of apprenticeships currently offered as vacancies, particularly between July and October, with the government also getting behind this drive for better access. We need a process that can improve the connection between development stages for young people and correlates to some extent with the timetable of the expected final year of full-time education.

This will increase visibility and ensure engagement on apprenticeship options at a time when young people are thinking about what the next step is. Flexibility is also essential in promoting access and any system should provide the opportunity for people to leave some full-time education courses part way through a year if they discover an apprenticeship that suits them, while continuing to complete the course. (City & Guilds, 2015)

We propose a UCAS-style system for apprenticeships for young people looking to leave full time education in years 11, 12 and 13. This would involve a process whereby employers, providers and colleges post vacancies earlier in the year and carry out a selection process similar to that of the current university UCAS system. Such an approach would ensure the alignment with the process currently most familiar to young people looking for their next opportunity, and will simplify the process for them by putting all options on a level playing field, in turn addressing one of the biggest challenges currently facing apprenticeships. (Ibid) It would also allow apprenticeships to be focused at a higher level, taking a young person right through to a degree. This raises the ambition and means there is no limit to the progression of an apprentice.

Conclusion
The benefits of apprenticeships, and more broadly, professional and technical education, have been clearly set out in terms of the contribution the sector makes to raising employment levels and improving productivity. This is reflected in the government's recent commitments to the apprenticeship agenda. With the spotlight firmly set on apprenticeships, it is now time to take advantage of this rise in profile.

This focus from policymakers has been seen as a positive step in the right direction but we need to be more ambitious in terms of driving the agenda forward. Quality needs to be placed firmly at the heart of all apprenticeships and should be the key consideration when developing any apprenticeship programme. If we are to achieve this level of quality, apprenticeships must be built around the core foundation of learning, to ensure people are confident that the skills they will develop will take them a step closer to the career of their choice.

To enable this, we need collaboration from all parties involved, employers in particular. The key to ensuring employer involvement lies in providing them with the opportunity to control and shape the apprenticeship offering, giving them the space and support to make informed decisions over quality, access and process which will in turn mean they are in the position to provide an increased number of high quality apprenticeships.

City & Guilds has extensive experience in working with employers across the UK in developing and providing quality apprenticeships, and we will continue to do this to ensure progression on the key challenges we have identified here, and to further the aims of both the government and industry on raising the profile of professional and technical education. It is time for policymakers, providers and industry to work together to ensure that quality becomes the driving force behind the apprenticeship agenda.

References

Carberry, Neil. (21 August 2015) 'CBI comments on apprenticeship levy', *CBI*: news.cbi.org.uk.

City & Guilds Group. (2015) *Making Apprenticeships Work: Case Studies*, London: City & Guilds.

City & Guilds Group. (2015) *Making Apprenticeships Work: The Employers' Perspective*, London: City & Guilds.

City & Guilds Group. (6 July 2015) *Chancellor must support vocational education to boost productivity*, London: City & Guilds.

City & Guilds Group. (November 2015) *'Great Expectations research'*, City & Guilds: www.cityandguilds.com.

GOV.UK. (1 December 2011) *Employer Ownership of Skills, UKCES vision*, GOV.UK: www.gov.uk.

Guy, Richard. (31 October 2015) 'Independent end assessment', *FE Week*: feweek.co.uk.

Maason, Marcus. (3 September 2015) Quality over quantity with apprenticeships', *British Chambers of Commerce*: www.britishchambers.org.uk.

Office for National Statistics. (September2015) *UK Labour Market: September 2015*, Office for National Statistics: www.ons.gov.uk.

6.
NEVER MIND THE QUANTITY, FEEL THE DEPTH

Professor Bill Lucas
Director, Centre for Real-World Learning, University of Winchester

———

Introduction

There's nothing like a large numeric target to motivate ministers and their civil servants. And the idea of three million apprenticeships is just that. An audacious goal for England. If it motivates all those involved – employers, colleges, training providers, curriculum designers, teachers, coaches, mentors and apprentices – really to understand what an outstanding apprenticeship experience is, then the target will have served its purpose. But if in 2020 we look back and see it as mere work-based learning flag-waving it may have ended up damaging the brand of apprenticeships.

This chapter takes us into the engine room of apprenticeship. It explores the distinctive features of the apprentice's journey in terms of its pedagogy, the teaching she or he receives, and the learning experiences which are offered. To do so, it draws on research (Lucas, Claxton and Spencer, 2012; Lucas and Spencer, 2015) which, to continue the metaphor of a ship, looks in more detail at the components of the engine – the crank shaft bearing, pistons and drive shafts as it were - to be able to really understand what these are, and what anyone delivering apprenticeships needs to understand if the outcomes are going to be, to use that sometimes lazily used word, 'world-class'.

A theory of change

Unless we change the way we approach apprenticeships it is foolish to think for a moment that we will dramatically increase their numbers and significantly improve their quality. We can't simply assume that by doing more of the same, with the exception of the new employer role in specifying standards and contributing to a levy, we will actually make a step change in what we are doing. If we are going both to expand the number and dramatically increase the quality of apprenticeships we need a theory of change, a cunning plan.

Let's start by reminding ourselves about what apprenticeships are. Our definition is:

> An apprenticeship is a job with significant inbuilt learning designed to prepare the apprentice for future employment, employability and active citizenship of a high quality. (Lucas and Spencer, 2015)

This compares to the relatively encouraging one offered by government:

> An apprenticeship is a job that requires substantial and sustained training, leading to the achievement of an apprenticeship standard and the development of transferable skills. (Department of Business, Innovation and Skills).

The problem with the BIS definition is that, despite the interesting development of the phrase 'transferable skills', critically important if apprentices are going to function effectively in the real world, this subtlety all too easily gets lost and the definition becomes simplified as here in, at the time of writing, the most recent iteration from the House of Commons Library:

> Apprenticeships are full-time paid jobs which incorporate on and off the job training. (Delebarre, 2015)

Apprenticeships, like vocational education more generally, can either be positioned as a second class alternative to academic pathways, with a set of largely instrumental concerns about funding structures and delivery systems dominating discussions, or they can be offered as an ambitious, expansive and powerful alternative to academic routes, suitable for a wide range of learners and with a well-articulated pedagogy of their own. I argue that it is the second of these two routes that will enable England's system to be truly world-class. The focus now and over the coming years should be on really understanding how different kinds of apprenticeship learning can lead to the accomplishments we need in twenty-first-century apprentices. Figure 1 seeks to articulate a theory of change that might bring this about:

If we:
- clearly articulate a set of expansive and contemporary outcomes for apprenticeships, as well as
- the knowledge and skills which are required, and

if we:
- really understand the distinctive features of apprenticeships

then:
- it will be much easier to select the best possible teaching and learning methods
- as well as the most effective assessment methods

so that:
- the quality of apprenticeships increases, and
- there are more higher-level apprenticeships

so that:
- more employers want to employ and support apprentices, and
- more young people choose the apprenticeship route rather than university or as part of an FE or HE pathway, and
- the UK has more and higher-quality apprentices.

Figure 1. Theory of change for improving quality of apprenticeships

Being clearer about outcomes
It is all too easy to get lost in the details of which skills are needed, what knowledge is helpful, and why this or that competence is more or less important. Each of these is a legitimate concern, but a preoccupation with each of them can all too easily distract those designing apprenticeships from the real task of being clear about what the bigger picture is.

There are six essential desirable learning outcomes of apprenticeship if we want to create a genuinely high quality offer:

1. Routine expertise – the essential competences required for a particular occupation.

2. Resourcefulness – the capacity to think and act in situations not previously encountered.

3. Craftsmanship – pride in a job well done and an ethic of excellence.

4. Functional literacies – literacy, numeracy, digital and graphical.

5. Business-like attitudes – customer and client-focused, entrepreneurial and aware of value for money, whether in for-profit, public sector or third sector roles.

6. Wider skills for growth – the dispositions and wider skills for a lifetime of learning and change.

All too often the focus defaults to the first and the fourth of these outcomes and the case for the effectiveness of apprenticeship is consequently diminished. For higher-level apprentices and the new degree apprenticeships, while the language of these six outcomes may need refining, we believe that the concepts are equally valid.

In the plethora of English government documents which exist about apprenticeships, there is a signal lack of ambition in terms of the point of apprenticeships. It is one (admirable) thing to seek to develop transferable skills (as per the definition cited on page 62); it is another to then ignore the implications of how you do this in almost every document produced.

But there is something else too. Apprenticeships have distinctive features which means that they are different from any other educational pathway:

1. Both on-the-job and off-the-job learning. This has now been agreed as 280 hours of guided learning with 100 hours (or 30 per cent, whichever is greater) delivered off the job as well as training to level 2 in maths and English.

2. The essentially social nature of apprenticeship learning. Even in a small enterprise, apprentices join a group of co-workers within a specific vocational group. They have real opportunities to learn from those around them who are more skilled.

3. The partnership arrangements between those providing the employment and learning. Employers, colleges, training providers, higher education institutions, professional bodies and others collaborate to ensure they provide learning for apprentices. This calls for high levels of coordination and, we believe, drawing from educational research, visibility of learning processes.

Figure 2 summarises the argument so far and seeks to redress this balance by introducing a range of teaching and learning methods which might be used.

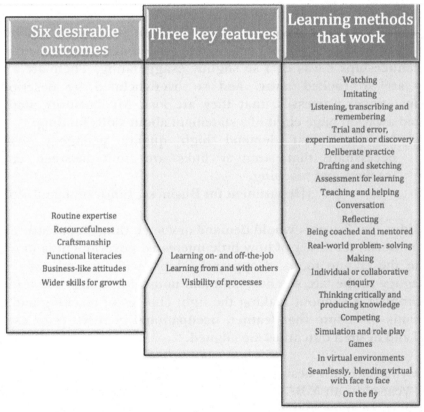

Figure 2. Desirable outcomes, key features and learning methods for a pedagogy of apprenticeship

Once you describe such a breadth of outcomes, along with the three core features of apprenticeships, it becomes clear that certain kinds of teaching and learning methods are likely to work best to deliver truly outstanding apprenticeship experiences.

But here is where even strong supporters of apprenticeships like me realise that in government documents there is a stunning omission – almost any detailed description of the learning and pedagogy which will be required to ensure that, as new standards are designed, the three million new starts planned will be anything other than low quality.

A pedagogy of apprenticeship

In the right hand column of Figure 2 is a list of tried and tested learning methods. But before we explore these, let's take a step back. When I said that there is no mention of pedagogy in government documentation about apprenticeships I was ever so slightly exaggerating. There are a few, but they are so tucked away, and so overwhelmed by descriptions of organizational processes, that they are lost. My favourite statement is tucked away on page eight of a statement about skills funding:

> Learners must demand high quality pedagogy which will necessitate that stronger links are built between employers, teachers and teaching.
> (Department for Business, Innovation and Skills, 2014)

The idea that learners would demand any such thing is patently absurd. It is a powerful reminder of how little interested government is in what goes on in the engine room of apprenticeships, with its pedagogy. I define pedagogy as the 'art, science, craft and nous of teaching and learning'. At its simplest it is about making the right choices of teaching and learning methods to ensure that learner, occupational context, prior experience, level and desired outcomes are aligned.

How you select the best pedagogy is different if you are teaching a class of nine-year-olds, an MBA or, in the case of this chapter, an apprenticeship. There is an idea that may be useful here: 'signature pedagogy'. First coined by Lee Shulman in 2005, it refers to the types of teaching and learning which most suit or match the way a specific profession or vocation operates:

> Signature pedagogies make a difference. They form habits of the mind, habits of the heart and habits of the hand [...] signature pedagogies prefigure the cultures of professional work and provide

66

the early socialization into the practices and values of a field. Whether in a lecture hall or a lab, in a design studio or a clinical setting, the way we teach will shape how professionals behave.

(Shulman, 2005)

So, for example, a signature pedagogy for developing apprentice engineers might involve problem solving, inquiry-based learning, project work, and real-time simulations – all using the engineering design process of prototyping and testing. It would not sit engineers in rows and ask them to learn things by rote or undertake mathematical tests in isolation or write essays.

I am not alone in seeing the gap between quality and rhetoric in the debate in England about apprenticeships. In 2013 the Commission on Adult and Vocational Teaching and Learning (CAVTL) raised expectations with regard to really understanding pedagogy in vocational education. It argued that:

> We need to strengthen and make more visible the distinctive pedagogies of vocational teaching and learning.

Let's go back to our six desired outcomes for a moment and think about implications for pedagogy.

Routine expertise

Learning how to become routinely expert at something is at the heart of apprenticeship. Apprentices clearly need to be able to do the job for which they have been trained, and demonstrate their routine expertise. Challenges to developing routine expertise in apprentices include:

- adequate opportunities to practise and the motivation of the learner to keep practising until the skill has become routine;
- the availability of an 'expert' teacher/coach;
- the skill of the 'expert' as a 'teacher', especially their ability to articulate the important steps of a process and make tacit processes 'visible'.

Resourcefulness

This desirable outcome is about being able to deal with the non-routine and unexpected, about transferable skills. While reliable skill is essential,

67

in most workplace situations things happen which are beyond the routine. As apprentices progress, so they will need to be able to stop to think and draw on resources other than their own knowledge of the routine. Challenges to developing resourceful apprentices include:

- having adequate opportunities and time to practise skills in unfamiliar settings;
- being allowed to develop such a higher-order capability when there is an emphasis on productivity;
- having an expert teacher to suggest resourcefulness strategies.

Craftsmanship

Craftsmanship involves an unambiguous aspiration in a worker or learner for excellence. With it comes the sense of pride in a job well done. The idea of craftsmanship is central to apprenticeship and was at the heart of the medieval forms of apprenticeship in England. In Germany the term 'meister' carries this meaning. For Richard Sennett, author of *The Craftsman* (2009), the desire to do a job well for its own sake is a basic human impulse. Everyone, he argues – the computer programmer, the doctor, the parent and the citizen – can put into practice the values of craftsmanship. Matthew Crawford (2009) and Mike Rose (2004) similarly explore the satisfaction and pleasure individuals derive from the cognitive aspects of a job well done. The challenges to developing craftsmanship in apprentices include:

- the perceived opportunity / cost trade-off of doing a job fit for purpose, and doing a job perfectly with deadlines looming;
- peer pressure from other apprentices for learners not to appear too 'pedantic', and the availability of good role models;
- employers who may be more concerned with profit than with quality and potential wastage of material if something isn't 'quite right'.

Functional literacies

As well as being functionally literate in numeracy, literacy and ICT, apprentices also need a level of graphical and digital literacy. In England in 2012, Functional Skills became a mandatory part of all apprenticeship frameworks, replacing what were known as Key Skills. The role of Functional Skills has been further complicated by the latest requirement for all 16- to 19-year-old learners to study towards GCSE/level 2 maths and English if they have not already achieved a GCSE A*-C in these subjects. Developing apprentices' maths and English to level 2 brings some real challenges including:

- the small amount of day release time and guided learning hours available to apprentices;
- the availability of appropriately skilled staff in the workplace and off-the-job trainers to support the authentic development of maths and English;
- the ability of workplace staff to identify and articulate the Functional Skills they are using.

Business-like attitudes

This desirable outcome is about dealing with clients, suppliers and customers appropriately. Another word for it might be 'professionalism', a way of behaving, whatever your occupation. To act appropriately in the world of business, whether for- or not-for-profit, is an essential requirement of any apprenticeship. Developing business-like skills in apprentices brings some real challenges including:

- ensuring that the definition of business is set expansively to include both basic self-organisation and higher-level communication and work skills;
- deciding which sorts of business-like attitudes and skills count as important;
- providing opportunities for learners to develop if they are rarely client-facing.

Wider skills for growth

Across the world 'wider' skills are known by different names such as 'dispositions', 'attributes', 'capabilities' and so on, each one of which comes with a slightly different emphasis. As well as at work, apprentices need to be able to thrive at home and in the community. Within apprenticeships in England the Personal, Learning and Thinking Skills (PLTS) required at levels 2 and 3 are helpful. The Confederation of British Industry (CBI, 2012) has recommended that we should go further – clarifying what these wider skills are and measuring them.

Developing wider skills for growth in the workplace brings some real challenges including:

- balancing the needs of the apprentice as a learner and the requirement for him/her to be productive as a worker with reference to particular sectors and organisations;
- finding ways of assessing skills that go beyond the more routinely assessed tests of knowledge, physical control, manual dexterity or mental facility;
- ensuring that qualified and experienced employees and teachers model the skills required.

Each of these six desirable outcomes, in other words, brings with it questions about teaching and learning for any designer of an apprenticeship. The same is true with regard to its three distinctive features (as summarized in Figure 2).

On- and off-the-job learning

Apprentices learn both on-the-job through their employer and off-the-job at a college or with a learning provider, with the bulk of their learning importantly located within a workplace. Ofsted has also summarised some on-the-job learning activities which are valuable:

formal and informal training; placements within and outside the company to obtain experience and assessment evidence; mentoring by colleagues; attendance at trade shows; visits, participation in competitions, and manufacturer training. May include learning support visits.

(Ofsted, 2010)

The critical issue in terms of ensuring effective learning within an apprenticeship is the degree to which those responsible for the on-the-job elements talk to and understand those who provide the off-the-job learning. One mechanism which might provide a focus for such conversations is the Individual Learning Plan (ILP), the record of all the elements of an apprentice's planned learning. An ILP can be a very dull document full of little more than previous exam results and dates of specific courses. But it could be much more powerful as a prompt for desired activities (such as mentoring and coaching) and a reflective space for learners to plan and review on their experiences.

Learning from and with others
Communities of practices and of learning are an essential element of all apprenticeships (Lave and Wenger, 1991; Fuller and Unwin, 2008). In some larger workplaces there will be a number of apprentices, where in a small business there might be just one. But even if the apprentice is alone she or he will necessarily be learning and working with others. Such collaborative learning comes naturally to apprentices. Colleges, training providers and employers can all facilitate the process of learning from and with others.

Visibility of learning processes
In the last decade we have begun to understand the relationship between quality of outcomes and learning in which the processes are clearly visible to learners and teachers. John Hattie highlights four features of high quality learning in *Visible Learning* (Hattie, 2009):

- The learning arising from any learning experience is given explicit attention in the moment.
- Learners have specific, challenging, and practical, goals in mind and learning tasks are constructed with those goals in mind so that they are beneficial.
- Feedback is clear and plentiful. Learners recognise the need to welcome and listen to feedback.
- Teachers recognise learners' self-concepts and are fully able to coach them to develop improved learning dispositions and strategies.

Each of the features above requires all those involved to be actively involved in making the processes of learning visible, all being able to give a precise name to what is happening in terms of an apprentice's learning.

Learning methods that work for apprenticeships
In Figure 2, I listed methods which are appropriate to apprenticeship. Some methods will be more likely to lead to one or more our desired six outcomes. Different apprentices in varying contexts learning contrasting occupations will require those orchestrating their learning to make considered choices to ensure that the experience is of a high quality. Here the methods are grouped into nine clusters for ease of comparison.

Learning from experts - by watching and imitating, and by listening, transcribing and remembering

We learn by watching and trying first to work out what someone is doing and then to try it out ourselves. Such learning is at the heart of the medieval apprentice model, where novices watch experts, just as it is also at the heart of learning that takes place within family groups.

Practising - through trial and error, experimentation or discovery, and deliberate practice

Human beings have always learned well by experiencing things at first hand. Generally referred to as experiential learning (Kolb, 1984) this cluster of methods assumes that we learn well when we can combine both theory and practice. Trial and error responds to our natural human motivation to be curious. Deliberate practice is a particular kind of practising involving a focus on improving particular tasks, (Ericsson, 2008).

Hands-on - by making, by drafting and by sketching

If writing about and talking about skills and knowledge is the default way of operating in general 'academic' subjects, hands-on learning is its parallel default setting in vocational education. Of course not all apprenticeships involve physically making things, but 'hands-on' has a more general meaning here implying that the learning is 'first-hand' wherever possible.

Feedback for learning - using assessment for learning approaches, through conversation, by reflecting, and by teaching and helping others

Feedback for learning is any communication whose emphasis is on understanding and improving the processes of learning. Feedback is information provided by someone to a learner on an aspect of their performance and is essential to all kinds of learning. Feedback is seldom neutral, providing, as it does, information about the values and attitudes of the feedback giver as well as about the person or task. John Hattie and Helen Timperley (2007) have helped us to see that there are essentially three core questions informing effective feedback: Where am I going? How am I doing? Where to next?

One-to-one - by being coached and by being mentored

One-to-one interactions enable apprentices to develop the right attitudes, knowledge and skills in the context of a trusted relationship and where the focus is on them as an individual learner. Coaching is where two individuals meet regularly to reflect on progress and work on aspects of performance. Mentoring, also a relationship between two people, tends to focus on career transitions and progression and is normally provided by a more experienced and expert worker.

Real-world learning - by real-world problem solving, through personal or collaborative enquiry, and by thinking critically and producing knowledge

Real-world learning recognizes that the workplace is more like the real world than a classroom. In the real world you encounter challenges, ask and answer questions, engage your critical faculties and use your creativity or nous to solve problems. Many apprentices choose the route precisely because it appears to offer the prospect of real-world learning. Real-world problem solving requires apprentices to be able to identify problems and have a range of strategies to find solutions, both working as an individual and in a team, thinking critically as they do so.

Against the clock - by competing, through simulation and role play, and through games

Learning against the clock is real and important in the sense that, in any workplace, deadlines will be important. But, by contrast, really deep learning transcends time, inviting engagement which is more than timetabled lessons or sessions. Apprenticeship learning is, by definition, a kind of learning against the clock as the apprenticeship has a specified overall time and within that, specified learning hours. Simulations and games provide opportunities for apprentices to explore contexts which otherwise would not be available to them. Constructive competition is increasingly being seen as a way of developing the skills of apprentices, especially those most skilled.

Online - through virtual environments and, seamlessly, blending virtual with face to face.

Online learning, unleashing the power of the internet is growing in importance and sophistication. As a cluster of methods, 'online' sits apart from the others we have listed so far, being a means of delivering many of the other methods. For example, learners do not learn simply by 'being

73

online'. They learn through 'watching' while online, or 'thinking critically' while online.

Anytime - on the fly

This last category is a simple reminder that much of what apprentices learn is not planned, stressing instead the need for them to be ready to learn. On the fly learning is unplanned and informal, the result of an unexpected occurrence from which something can be gleaned. Sometimes it is these on the fly moments – an unexpected conversation with a visitor, equipment which does not work and forces a rethink, a chance encounter, an exchange on social media – which provide apprentices with useful know-how.

A yawning gap between rhetoric and reality – some conclusions

Having a real understanding of the learning methods described above is critically important for all designers of apprenticeships. The blend of methods which is chosen is, de facto, the pedagogy of a particular apprenticeship. If we are really wanting to develop world-class apprenticeships in England we need to develop a rich apprenticeship pedagogy which unequivocally aims to deliver the six desired outcomes with which this chapter began. This, in turn, requires a debate about the pedagogy of apprenticeship between employers, providers and researchers within and across sectors to identify best practices. As a consequence we can develop accessible guidance for employers and providers about the pedagogy of apprenticeship. If government is serious about developing high-quality apprenticeships, it needs to ensure that its documents about apprenticeship include explicit reference to teaching and learning.

References

Apprenticeship Frameworks Online. (2014) *Frameworks Library:* www.afo.sscalliance.org.

BIS. (2013) *The future of apprenticeships in England: implementation plan.* Department for Business, Innovation and Skills. London www.gov.uk .

BIS. (2014) *Skills Funding Statement 2013-2016.* Department for Business, Innovation and Skills. London www.gov.uk.

CBI. (2012) *First Steps: A new approach for our schools.* www.cbi.org.uk.

Crawford, M. (2009) *The Case for Working With Your Hands: or Why Office Work is Bad for Us and Fixing Things Feels Good.* London: The Penguin Group.

Delebarre, J. (2015) *Apprenticeships Policy, England 2015.* London: House of Commons Library.

Ericsson, K. A. (2008) *Deliberate Practice and Acquisition of Expert Performance: A general overview.* Academic Emergency Medicine 15(11): 988-994 .

Fuller, A, and Unwin, L. (2008) *Towards Expansive Apprenticeships: A commentary by the Teaching and Learning Research Programme.* London: TLRP.

Hattie, J. (2009) *Visible Learning: A synthesis of over 800 meta-analyses relating to achievement.* Abingdon: Routledge.

Hattie, J. and Timperley, H. (2007) 'The Power of Feedback', *Review of Educational Research* 77(1): 81-112 .

Kolb, D. (1984) *Experiential Learning: Experience as the source of learning and development.* Englewood Cliffs, NJ: Prentice Hall.

Lave, J. and Wenger, E. (1991) *Situated Learning: Legitimate peripheral participation.* Cambridge: Cambridge University Press.

Lucas, B., Spencer, E., and Claxton, G. (2012) *How to Teach Vocational Education: A theory of vocational pedagogy.* London: City & Guilds.

Lucas, B. and Spencer, E. (2015) *Remaking Apprenticeships: powerful learning for work and life.* London: City & Guilds.

McLoughlin, F. (2013) *It's About Work: Excellent adult vocational educational teaching and learning.* London: CAVTL .

Ofsted. (2010) *Overview: The successful training of apprentices: key steps.* www.ofsted.gov.uk.

Richard, D. (2012) *The Richard Review of Apprenticeships.* London: Department for Business, Innovation and Skills .

Rose, M. (2004) *The Mind at Work: Valuing the intelligence of the American worker.* London: Penguin Books Ltd.

Sennett, R. (2009) *The Craftsman.* London: Penguin Books.

Shulman, L. (2005) 'Signature pedagogies in the professions', *Daedalus* 134 (3): 52-59.

7.

GROWING YOUR OWN GRADUATES THROUGH DEGREE APPRENTICESHIPS: A CASE STUDY OF COLLABORATION BETWEEN THE UNIVERSITY OF WINCHESTER AND CGI

Stella McKnight
Director for Employer Partnerships, University of Winchester
and
Professor David F. Birks
Dean of Faculty of Business, Law and Sport, University of Winchester

―――――

This chapter evaluates how the University of Winchester successfully implemented a business-facing model of apprenticeship development and engagement. It describes a ten-year relationship between Winchester Business School and CGI (formerly Logica) and discusses how the relationship grew from an initial single programme development into jointly pioneering degree apprenticeships. The relationship initially succeeded through recognising and being highly responsive to CGI's needs. Over the years for both parties, success has grown as a result of mutual commitment in turning strategic vision into reality. The chapter concludes by agreeing with the Tudor and Mendez (2014) contention that achieving mutual value is essential for both university and employer as each seeks to maximise the benefits from collaboration.

Introduction
UK universities have many students who already have some form of industry engagement as part of their higher-level studies. The higher education (HE) sector now has a new type of student with significant and focused levels of industry engagement: the Degree Apprentice. For these students learning mostly takes place away from the university and is categorised as business-led, work-based learning. The government has

pledged to create three million more apprenticeships by 2020; approximately 60,000 of these will be degree apprenticeships.

The University of Winchester (UoW) has been at the forefront of degree apprenticeship development and in September 2014 launched two of the first degree apprenticeship programmes in the country, becoming a Tech Industry Gold endorsed training provider for the digital economy.

Being one of the smallest universities in the country, one might be prompted to ask: How has the University of Winchester managed to achieve this?

In responding to this employer-driven, government-sponsored initiative of degree apprenticeships, the University of Winchester does not anticipate that being at the forefront of change will be without its challenges.

This chapter evaluates the challenges encountered by the University of Winchester in developing a business-led curriculum which has work-based learning at its heart. It is a story that began ten years ago when the university was approached by two forward-thinking individuals working within one of the leading businesses in the IT sector, CGI (formerly Logica). It is these ten years of collaborative employer partnership experience and a continued analysis of lessons learned that has led to the University of Winchester being at the forefront of degree apprenticeship development.

This chapter will begin by discussing some of the challenges in designing and delivering an employer-led curriculum. It will then seek to capture and share what is perceived to be the benefit to the university of 'good practice' in building an effective employer partnership which has led to the design and delivery of a number of successful sponsored degree programmes and more recently the design of two degree apprenticeship programmes. First, a little bit about the University of Winchester.

The University of Winchester
Established in 1840 as King Alfred's College, the University of Winchester is a values-driven institution with around 7,000 students with faculties that cover the Arts; Business, Law and Sport; Education and Social Studies and Humanities. University status was awarded in 2005 and Research degree awarding powers granted in 2008.

The location of the university is within a growing and vibrant area of employment. With the right attitude to employer engagement, it is well placed to embrace and address the need for a demand-led approach. The Director for Employer Partnerships has responsibility for delivering the employer engagement strategy. This includes working with employers to identify their higher-level skills needs, and in turn, to work with faculties to develop responsive provision with the aim to build the long-term capability and the capacity of the university to engage with employers.

The School delivered a number of business and management related modules and programmes but did not establish a fully-fledged Business School until 2008. It has grown rapidly in size and reputation, with over forty academic staff supporting over 900 undergraduate and postgraduate students.

The Business School has four departments: Accounting and Investment; Applied Management Studies; Global Issues and Responsible Management; and Marketing and Innovation. Developments have been built around three key themes of future-thinking, creativity, and responsibility, with curriculum and Research and Knowledge Exchange activities reflecting these values. Creativity is manifest through an emphasis on innovation and entrepreneurship, both in business start-ups and established enterprises, and more recently through the relationship between business and creative aspects of fashion, digital media and journalism.

The holistic learning environment created by combining workplace training within the business and information technology and information systems degrees is a distinctive and successful part of teaching and learning practice. The theme of future-thinking is represented through a focus on leadership development, insight management, and business improvement in a number of new industry sectors. The Business School is one of thirty Global Champion Business Schools in the United Nations' Principles for Responsible Management (PRME) initiative. It hosts the UK Chapter for PRME, resulting in a focus on business ethics, corporate social responsibility and sustainability across all programmes.

Winchester Business School, where degree apprenticeship programmes reside, has developed strong links with the business community from its inception. A key part of the establishment of the Business School was foundation degrees, which included significant elements of work-based study. Of particular relevance in these development came in 2006 with the development and nurturing of collaboration with CGI (formerly Logica). This involved students studying part-time at the university whilst working for CGI in various project and service delivery roles. Day-release teaching sessions at the university are complemented by additional work-based studies, guided by workplace mentors. This relationship means that the Business School has experience in the design, development and delivery of employer-sponsored higher-education programmes at the heart of its formation and growth which has ensured that it is well-placed to respond to the government's apprenticeship growth agenda.

Working with CGI (formerly Logica)
Founded in 1976, CGI Group Inc. is the fifth largest independent information technology and business process services firm in the world. Approximately 68,000 professionals serve thousands of global clients from offices and delivery centres across the Americas, Europe and Asia Pacific, leveraging a comprehensive portfolio of services including high-end business and IT consulting, systems integration, application development and maintenance, infrastructure management as well as a wide range of proprietary solutions.

In July 2006, CGI (then Logica) approached a number of universities, voicing a sense of frustration in the preparedness of graduates for the workplace. Logica wanted a new recruitment initiative to supplement their UK graduate recruitment scheme, which although successful, did not entirely fulfil their need for well-rounded candidates with a broad understanding of IT and management skills. To help overcome this, Logica began investigating apprenticeship schemes and other ways of attracting non-graduates to facilitate training tailored to their needs, offering a mixture of academic study and work-based learning.

The approach from Logica coincided with publication of the Leitch Review (2006) which recommended that the UK should commit to becoming a world leader in skills by 2020. This created opportunities for academic institutions, such as the University of Winchester, to redefine their strategic priorities and gave impetus to developments of relationships with industry and commerce as highlighted by Lambert's (2003) *Review of Business-University Collaboration*. This set the scene for the University of Winchester and CGI partnership which has endured despite a changing economic climate and the challenges of merger and acquisition activity.

Following the approach by Logica in July 2006, the University of Winchester began working with Logica's telecommunications and media arm, based in Reading. The brief was to design a degree programme in Information Technology Management that was flexible and responsive to their training needs.

The University of Winchester proposed an alternative to the more traditional university Bachelor Honours programme. The solution was a tailor-made, flexible foundation degree in Information Technology Management, with an option to top up to a Bachelor Honours degree. The target market was school leavers and those who had not previously applied to higher education, nor necessarily aspired to attend university.

The first sponsored degree was launched in September 2007 and further jointly developed degrees followed. CGI fully funded the degrees and paid their new employees a competitive salary whilst they undertake work-based learning at the company and academic study at the University of Winchester's campus.

Early challenges for the university
Having signed a contract in 2006, the partners were immediately faced with a number of formidable targets, including designing and validating the programme, marketing, recruitment of students, and staff development for those teaching on the programme, many of whom were unfamiliar with the concept of work-based learning. All of this had to be achieved within short timescales, with limited budget and scarce human resources. This was not for the faint-hearted and could only be achieved through a shared vision and with enormous commitment and collaboration on both sides.

Each partner soon recognised the need to be open in terms of strengths and limitations. For example, CGI were unfamiliar with undergraduate

81

recruitment and had no relationship with their local schools and colleges. These of course are fundamental to a university and could be shared with CGI. In return, CGI were able and willing to share with the university contemporary developments in designing and using communication tools and technology.

Another early challenge the university encountered was the need to be more agile in their processes, systems and decision-making in order to be responsive to the employer's needs. In failing to be agile, the university could lose out to more nimble commercial providers. However, the university had to balance this risk against its own need to maintain quality and to uphold academic standards.

There were, and continue to be, on-going challenges in achieving effective work-based learning that is seen to make an impact not only on the learner but also on the employer. Most students find the concepts of the work-based learning situation difficult to grasp. Some also lack motivation to invest effort in what they perceive as a less well-defined area of their study. This lack of interest and engagement are not the only challenges. Much time and effort have been invested, by both the university and CGI, in identifying a diversity of learning situations and range of assessment methods that engage student interest and build business confidence.

Initially, the university anticipated that much of the responsibility for the effectiveness of work-based learning would lie with the employer. This was found not to be the case. CGI had limited resource and expertise to support students engaged in work-based learning. The university therefore needed to train managers in the workplace to take on mentoring roles. Nevertheless, support was achieved through a consultative process and has since led to a strong model of mentoring in which shared responsibility for work-based learning has been achieved and an enhanced student support mechanism established.

The challenges encountered in the university's early engagement with CGI have supported curriculum thinking in the Business School, leading to more flexible modes of delivery with the potential to offer degree programmes which better meet employer needs.

Work-based provision is now delivered to CGI on an extended basis throughout the year to meet their work-based degree apprenticeship requirements and allow sufficient time for academic provision. Academic tuition takes place one day per week, with students employed in real world roles. Much of the assessment is work-based, designed around academic study with support from workplace managers.

The degree students, who are also apprentice employees, benefit from focused training that prepares them for their role in the business. Each student benefits from having both a workplace mentor and a personal tutor at the university. This is crucial as students do not have as much contact as they might if they were studying full-time at the university. The success of our partnership in preparing students for real world roles is captured by Lucy Waterman, who has progressed her career from student to manager:

> From being one of the first students to start and complete this programme, to becoming the programme manager for CGI's Sponsored Degree, I have seen the evolution, development and the success of our programme and could not be more proud to have been part of it and to have helped shape it. The calibre of students, and the results they obtain, we have seen over the past 10 years, makes this our flagship UK Student Programme and a model of best practice which we share across the world.
> Lucy Waterman School Leaver Programmes Manager, CGI 2016.

The university has a dedicated person who serves as the main point of contact in the partnership. This person plays a vital role in building, managing and sustaining the partnership.

Regular communication has taken place between the university and CGI to monitor quality, jointly deliver and assess learning and, where appropriate, to identify where further customisation of the programme can enhance the student learning experience. The university regularly visits CGI and vice-versa to enable a continued and enhanced understanding of their businesses; and senior managers meet to ensure that the strategic vision continues to be shared. This approach helps maintain responsive delivery and build relationships at all levels of the partnership.

There is a shared commitment to marketing and communications. Although it is the business that employs the student, joint recruitment has been found to lead to sound decisions in terms of recruiting employees who not only fit into the work-place environment but are also academically able to cope with the challenges of balancing the demands of work and study.

Progression of the partnership: positive impacts for the university and CGI
Almost ten years on, and having worked through a number of programme developments, the university and CGI harnessed their combined expertise in the design and development of two new degree apprenticeships that were validated in May 2015.

As with our sponsored degrees that were designed with and for CGI, these degree apprenticeship programmes were also tailored in order to continue to meet CGI's strategic needs:

> UK-based companies need to move into higher-level aspects of IT. We need employees of a high calibre who are able to help drive UK PLC to be a global technology leader. The combination of work-based learning and teaching during the programmes is designed to ensure graduates of the university will be fully equipped to rise to this challenge.
>
> Tim Gregory, President, CGI in the UK.

Successes of the University of Winchester/CGI programme
It goes without saying that employers will wish to see impact, not only in terms of student results, but also in terms of how their students contribute to the business and contribute to their economic bottom-line. This partnership provided strong evidence of the value of the sponsored degree programme which is an early version of the degree apprenticeship in all but name. It is therefore a good indication of the potential success and benefits to employers and the students/apprentices who participate in these work-based learning programmes.

Strong student retention
Student retention rates for this programme are strong, regularly averaging over 85 per cent in each intake year:

Year	%
2010	95
2011	79
2012	86
2013	95
2014	100

Table 1. Student Retention Rates 2010 - 2014

Excellent academic achievement
Academic achievement on the sponsored degree programme has strengthened over the years as identified below.

Year	1st	2.1	2.2
2010	40%	50%	10%
2011	17%	50%	33%
2012	29%	57%	14%
2013	61%	28%	11%
2014	64%	36%	
2015	72%	28%	

Table 2. Academic Achievement Record 2010 - 2015

Strong return on investment
For CGI, the driver behind the development of their sponsored programme was to address a recruitment problem which previously saw them investing in additional training in order to prepare their graduate intake for work. CGI has been extremely pleased with the results achieved in recruiting undergraduates into the business as it has reduced investment in additional training, producing graduates who clearly meet their business needs. Upon graduation, the majority of their sponsored degree students achieve at least graduate level plus one, which means they are adding value back into the business even before graduating, which adds testimony to CGI's 'grow your own' approach.

According to Tudor and Mendez (2014) achieving mutual value is of concern for both universities and employers as each seeks to maximise the benefits from collaboration. Covey (1989) argued that mutual value requires interactions which are mutually beneficial and satisfying, describing it as 'win-win'. The changing landscape of higher education points to the need for universities to collaborate effectively with employers to address the skills shortage, benefit the economy, better meet the needs of students, and gain mutually beneficial value.

The following from CGI's Director of Recruitment highlights why CGI believes this long-standing partnership has not only survived but flourished:

. Our programme at Winchester has been such a success due to the University of Winchester's open mind, forward-thinking and collaborative approach. As a partner they have listened to our needs as an employer and worked with us to help us develop graduates that are right for our business and clients' needs, now and for the future.

Sarah McKinlay, CGI 2015

Summary
The benefits of work-based learning, which underpins the degree apprenticeship model, for employers are compelling. Meaningful work-based learning experiences developed through strong partnerships provide employers with the opportunity to directly recruit talented individuals and to retain them through focussed career development. However, this demands a commitment from the university, one of which is recruiting additional staff to manage the two-tier data management system i.e. in providing both Higher Education Statistics Agency (HESA) returns and Skills Funding Agency (SFA) returns.

Champion and leadership
From an educational leadership perspective, the implementation and sustainability of apprenticeship programmes needs ownership and strong and consistent support from university senior management. As pioneers in degree apprenticeships, Winchester has addressed many challenges. Support from senior management has been, and continues to be, essential to secure prioritisation and investment in the programme. It has also been

essential to receive the support of the Faculty Dean who has also championed the project in negotiations with senior management.

Agile systems and strong quality processes
In order to respond to changing business needs, there is a need for universities to develop agile systems whilst at the same time maintaining quality standards. This requires additional processes and systems such as partnership agreements and institutional approval processes. It has also necessitated development of a range of additional documentation to guide and support each stakeholder, including an operational handbook which identifies the responsibilities of each partner; a handbook to support the mentors; and legal agreements and contracts with both CGI and students.

It is hoped that the summary above of the successes of the Winchester/CGI collaboration provides insights into the manner and potential values of university/employer partnerships. The next section will consider the benefits to universities in moving beyond a single partner relationship to a much broader model of degree apprenticeship provision. It will then consider some new challenges that have emerged from moving to a broader model before presenting an overall summary of lessons learned.

Degree apprenticeships: advantages for the university

Reputation-building with employers across the region, and raising the
            Winchester Brand
In 2016, Winchester Business School will have ten years' experience in designing and delivering employer-led programmes. This experience of the distinctive challenges and demonstrable benefits meant that it was well-placed to contribute to the design and development of the national Degree Apprenticeship in Digital & Technology Solutions. Moreover, in pioneering the degree apprenticeship development, the Business School was able to gain both local and national recognition that in turn has raised the profile of both the Business School and the University.

New student numbers
The degree apprenticeship has created new student numbers at a time when it is a 'free-for-all' in the higher education marketplace with the recent removal of the cap on student numbers. Furthermore, this increase

in numbers is achieved without the associated accommodation demands that are faced by universities when they grow full-time student provision.

Comparatively low-cost delivery?

This is not as straightforward as it might seem. Part-time students who attend one day per week equivalent are regarded as comparatively low cost delivery. However, our experience shows that there are many hidden costs in establishing, building and maintaining effective working relationship. These costs are exacerbated when one considers the number of potential partnerships that may need to be developed to build a meaningful portfolio of degree apprenticeships.

Enhances graduate employability

The CGI programme produced students with excellent academic standards and no issues with 'graduate employability'. As employees as much as students, they are not only already in roles but have potential to move their career forward ahead of graduation. There is no reason to think otherwise but it will be interesting to see whether or not this pattern will be replicated in the wider degree apprenticeship schemes.

Funding award from the Skills Funding Agency

For universities, degree apprenticeships offer a new funding stream. However, the payment of fees for degree apprenticeships will not come from individual students. At present, the degree apprenticeship benefits from government funding that offers to double the investment made by sponsoring employers, up to a maximum of £18,000. For SMEs, there are a number of additional incentives to take on degree apprenticeships. The administration and cash flows of drawing monies from the Skills Funding Agency and individual employers will add to the administrative burden and costs faced by universities.

Government business/professions support - Tech Industry Gold degrees

Tech Industry Gold degrees are degrees accredited by the Tech Partnership's growing collaboration of employers working to create skills for the digital economy. The accreditation was designed to help both employers and young people choose degrees and degree apprenticeships that are recognised by industry for their quality and relevance to today's business environment.

The Tech Partnership led the process of developing national standards calling on the expertise of a range of universities and trailblazer employers. The University of Winchester contributed to that process through weekly, then fortnightly, conference calls. This enabled the University to learn and reflect upon the vision and development of degree apprenticeships across all faculties.

We recognise the expertise that the Tech Partnership (formerly e-skills) brought to this programme development and the support they offered our degree validation event. Gaining approval for this relatively unknown product, despite our ten years of learning and with the support of two leading IT companies, still presented a challenge. The Tech Partnership, with its credentials in trailblazing, was a welcome additional team member.

Degree apprenticeships: challenges for the university in developing these new programmes

Recruitment
One of the biggest challenges with regard to recruiting degree apprentice students has arisen due to the limited publicity around degree apprenticeships. We have found that schools and colleges, parents and many employers do not understand the concept of degree apprenticeships.

Cultural
Interestingly, in marketing all our programmes in the Business School, we recognise that for some in our audience there is a need to get over the stigma of the word 'apprenticeship' and the notion of equivalence to traditional honours degrees.

Funding award from the Skills Funding Agency
As identified earlier, this is a potential new income stream for universities. However, there is a requirement to make a case for the numbers and type of degree apprenticeships that a university plans to recruit. From those plans, universities have to bid for a share of a 'pot'. Not only that but they also have to complete SFA and HESA returns, as well as learning a raft of new regulations.

Achieving an holistic learning experience whether a large or small
      organisation

In launching a new degree apprenticeship, the University of Winchester
has also tailored the programme for Fujitsu and a number of SME
partners. Rather than having a cohort of students from one employer, we
have a mixed cohort of enrolled students onto the BSc (Hons) Digital &
Technology Solutions Degree Apprenticeship Programme. The challenge
is to identify and respond to the differing needs of each partner.

Both government and employers cite the essential need for degree level
apprentices to possess knowledge and subject-based skills with a high
economic value. Increasing the requirement for the degree apprentice
learning experience to be appropriate to their individual learning needs
extends the learning beyond the classroom and into the workplace. The
delivery and assessment of the programme must therefore incorporate a
design in which learning experienced both at work and within the
university environment is able to converge, in order to provide the
apprentice with the most beneficial learning experience. Crucial to this is
finding an effective mix of classroom and work-based learning where the
strengths of each distinct setting can be harnessed to advance
achievement of the learning outcomes.

Working from the Lucas and Spencer (2015) definition of an
apprenticeship (cited in Goodyer and Frater 2015) - 'a job with significant
in-built learning designed to prepare individuals for future employment,
employability and citizenship' - the design and delivery of the learning
programme must be centred on a student being employed in a real job role
within the business. The aim of a degree apprenticeship programme
should be to integrate academic learning at degree level and on-the-job
practical training. It should provide a holistic programme of education
and training to meet the skills needs of employers now and in the future.
Furthermore, as the above quote suggests, the programme should also
aim to produce well-rounded individuals who are able to make a positive
contribution to society.

Drawing together lessons learned

To recount every factor that had an impact on our journey leading to the
design and delivery of the University of Winchester degree apprenticeship

would be an onerous read. Instead it may be useful to those contemplating employer partnerships and/or developing degree apprenticeships of their own to identify a few key influences.

First, and critically important, is strong support from university senior management who want the partnership to succeed, and are willing to listen and invest. At the University of Winchester, both the Vice Chancellor and the Dean of the Business School were committed to moving the degree apprenticeship forward and instrumental in motivating staff to support this initiative.

Second is the importance of aligning the strategic vision of any collaborative partnership with potential offerings of a degree apprenticeship. Where a number of partners bring degree apprentices into the same classroom, this strategic vision needs to be compatible with the direction of each partner with mutual buy-in and commitment to delivering success.

The alignment of strategic vision has also to work for the university. The University of Winchester's strategic plan provides for an increase in the number of vocational programmes and to recruit from an increasing number of businesses locally, regionally and nationally. 'We will take bold steps to develop our curriculum, ensuring its continued relevance based on clear evidence from the market and the needs of employers.'

A clear strategic vision is essential for setting priorities and for coordinating action and resources. When combined with insight at the operational level, this has ensured that the University of Winchester's change agents got buy-in from senior management, heads of departments and the teaching teams, thus countering potential resistance to change.

CGI clearly demonstrated strategic vision and commitment to the partnership, initially through their innovative ideas to 'grow their own' graduates, and through establishing the sponsored degree programmes as their flagship undergraduate recruitment scheme.

Third is the need to develop a framework for a programme that will work for an employer, professional body and government accreditation agency.

Locating and working with a Trailblazer employer and/or professional body will provide access to specialist support.

Fourth is the need to recognise that there are clear benefits to be gained in the development of existing programmes in terms of curriculum and employability. There are also many opportunities for original research and knowledge exchange, all of which significantly enhance the proposition to universities.

Fifth is the vital need that the voice of the degree apprentice is heard. University of Winchester students already have a significant voice in our programme through regular satisfaction surveys and student representation on programme committees. Opportunities to assist academics and employers in trouble-shooting problems as they arise and to negotiate potential solutions should be afforded and recognised as a good learning experience.

Finally, open lines of communication should be maintained between universities and their partners, especially in relation to workplace mentors. Open lines of communication help to build relationships across levels of the university and collaborative partners. This not only raises awareness of degree apprenticeship programmes, but it creates synergy and commitment to shared goals, key to which will be supporting the degree apprentice.

Conclusion
While employers hold the key to creating the high-skills environment, it has to be recognised that collaborative partnerships with the university sector will play a vital role in developing the nation's skills. It is clear that for the future prosperity of the UK and to grow the higher-level skills workforce, employers and universities should work together with a shared sense of commitment and purpose. Employers and universities must therefore develop the mechanisms and expertise necessary to design and implement successful degree apprenticeship programmes.

Responsibility for achieving the UK's higher skills agenda must be shared between government, employers, universities and individual degree apprenticeship learners. Key to successful university and employer engagement is strategic commitment, shared values and a mutual desire to succeed. Established partnerships emerge where both the university and employer strive for high-quality, flexible provision and where each partner is prepared to invest much time and effort in tailoring provision to meet individual employer needs.

While much emphasis is given to relationship building between a university and employer in the 'traditional' sense of employer engagement, it should be recognised that it is today's degree apprentice who will be better able to bridge the gap between employment and education in the future. To break down prejudice and further grow the academic achievements and value to businesses and the economy, today's degree apprentices should be encouraged to contribute to the design of effective work based learning and degree apprenticeship models.

References

Covey, S. R. (1989) *The Seven Habits of Highly Effective People*, New York: Simon and Schuster.

Lambert, R. (2003) *Lambert Review of Business-University Collaboration.* London: HM Treasury.

Leitch, S. (2006) *Prosperity for all in the global economy*, Final report of Leitch Review of Future Skills.

Lucas, B and Spencer E. (2015) *Remaking Apprenticeships: Powerful Learning for Work and Life* cited in Goodyer, J. and Frater, G. (2015) *A report for the Tertiary Education Commission.*

Tudor, S. and Mendez, R. (2014) 'Lessons from Covey: win-win principles for university-employer engagement', *The Journal of the University Vocational Awards Council* 4.1: 223 – 27.

8.

TRAILBLAZER DESIGN AND DEVELOPMENT
'GRASPING THE NETTLE': A CASE STUDY

Dr Jacqueline Hall
Head of Assessment Services, Energy and Utilities Skills

———

Overview

Apprenticeships in England are experiencing a seismic shift from traditional supply-led provision to employer demand-led workforce skills development. In recent years, one of the major barriers to workforce competence has been the 'disconnect' between how UK industries operate and approaches to design or indeed implementation of skills policy.

Nowhere has this been more apparent than in the currency, interpretation and relevance of National Occupational Standards (NOS) and why, even to date, employers in the energy and utilities sector view NOS and Trailblazer development as connected rather than a separate silo activity.

This chapter will present the journey for those Trailblazer apprenticeships designed for the power, water, waste and gas industries under the Energy and Efficiency Industrial Partnership between 2013 and 2015. Nevertheless, whilst this has been a largely positive experience, this has also led to employer calls for pragmatism. While England continues to forge ahead with its new Trailblazer policy implementation, challenges around consistency, comparability, quality and funding remain for those transnational employers with apprenticeships across the other three UK nations. More fundamentally, the question about the shape and future of NOS also remains unresolved.

Alongside the trailblazers, the Energy & Efficiency Independent Assessment Service (EEIAS) was created following a successful Energy and Efficiency Industrial Partnership (EEIP) bid in 2013 to the Employer Ownership of Skills pilots in England. Employers had already welcomed

94

and embraced the government's reforms, particularly around trailblazer apprenticeships, and indeed there was shared and fundamental belief in their ambition to raise the quality of workplace learning and increase the opportunities for a shared talent pipeline across the sector.

The focus was to bring innovation and partnership to workforce development to ensure a sustainable, rigorous and robust approach to assessment in this safety-sensitive environment. To that end the EEIAS, through the EEIP, committed its expertise and its resources alongside employers to drive these new developments and act as a sustainable conduit for not only apprenticeships, but also the upskilling and reskilling of the whole EU sector workforce. Indeed today, despite on-going changes in Trailblazer guidance, and delays to the publication of standards and assessment plans, employers' commitment, enthusiasm and resolve remains undiminished:

> to be leading on such an important piece of work is really exciting. UU decided to take an active role in the development of these new standards, after we realised that our current frameworks were not as effective as they could be. By improving these standards we're able to raise the bar for future UU employees and make sure that our apprentices have all the skills they need to succeed.
>
> Technical Training Manager United Utilities

Trailblazer apprenticeship design - power gas, water and waste
Historically, policy documents often reveal an absence of consensus around the meaning of the word 'skills', and thus more often it is used as a synonym for different abilities, attributes, qualities and competencies. (Gilbert et al 2004). This is why a priority activity has been to support employers as they develop the new industry standards and assessment plans as part of the government's trailblazer apprenticeship reforms. More widely, the introduction of the Levy in April 2017 continues to further sharpen the focus and urgency of standard development through its impact on workforce development budgets. Wolf earlier stated 'employers are the only really reliable source of quality assurance in vocational areas' (2011:144). Sixty-seven employers have therefore been involved in developing ten Trailblazer standards and assessment plans within the energy and utilities sector between 2013 and 2015 – many involved in more

than one, demonstrating a huge commitment in terms of time and expertise.

A further eight Trailblazers are identified for potential development over the coming years, further highlighting the role of the EEIAS and employer partnership as a cornerstone for sustainability through its combination of cutting-edge technical and pedagogical expertise. This knowledge and experience has been harnessed in a way not essentially realised before – for employers and by employers. Moreover, critical to this proposition is the need for a highly-skilled and competent workforce – because arguably this is the greatest central asset of any organisation aiming to secure an economically sustainable future based on growth and productivity. Trailblazers therefore range in design from, for example, level 2 Smart Meter Installer and Gas Team Leader to level 7, Power Degree Apprenticeship.

A rationale for employers was simple - reform made not only good business sense to solve demographic skills deficits in the energy and utilities sector, but through an independent assessment service, it also enabled existing niche areas to be addressed, rather than subjugated to economy of scale factors. For example, employers expressed frustration that the skills to manage gas pressure within the network were ignored because of low numbers. This had resulted in stagnation around formal qualification development for roles intrinsic to the interconnection of upstream and downstream gas, and thus the supply to millions of householders.

The development was born in March 2015 around degree apprenticeships because employers identified that engineering graduate applicants lacked the depth of technical knowledge and experiential learning required to fulfil power job roles following graduation. Indeed, employers identified a critical need for high-level competent power engineering knowledge/competencies coupled with thought leadership and conceptual problem solving abilities. In a nutshell, there is a need for those to whom we will turn in the 2020s to solve both the 'known' and the as yet 'unknown' energy challenges.

These are but a few examples that demonstrate why employers are passionate about workforce competence and productivity. Moreover, it also reflects their desire to work both collaboratively and transnationally. For example, employers in Northern Ireland contributed to discussions on apprenticeship assessment, upskilling/reskilling and even graduate training, despite Trailblazers being an English-centric policy. Consistency and competency accord between employers and across industry therefore continue to be key drivers to raise the bar on skills, productivity and workforce development.

Employers equally view this innovation around trailblazers and EEIAS development as a once-in-a-generation opportunity. They take this responsibility in the same manner as the methods by which they must operate within their highly regulated environment. Job competence is critical – indeed it could be more appropriately termed as 'assessment with consequences'– those being risk to life of the employee, members of the public, litigation and damage to the brand. In practice, this means that all of the trailblazers are developed in conjunction with technical experts provided by the industry's employers and with the support of the appropriate Professional Engineering Institution (PEI) and Trade Unions.

The design journey
The design journey overview is provided in Figure 1, however, at a granular level this continues to evolve following each Trailblazer guidance document issue. At present, this means the submission of an expression of interest, a standard assessment plan, and a costing template that shows both the on-programme and end-point cost estimates. This facilitates the application of a funding cap.

It is still too early to evaluate the potential impact of the new 2017 levy, but it remains a source of concern amongst employer groups. Nonetheless, Trailblazer apprenticeships are not simply motivated by a desire to secure government funding, because often the 'real' cost and investment by employers is significantly higher. Instead, employers are taking the opportunity afforded by the introduction of apprenticeship standards to radically re-model the quality and approach to their apprenticeship programmes. This capitalises on existing industry best practice while utilising end-point assessment aligned to engineering technician registration requirements.

The government design journey may indeed end with assessment plan publication, but in this sector employers do not consider 'readiness' ends with a high level assessment plan, and thus have developed units/specifications in the public domain that can be used to support the apprenticeships. (www.eeias.co.uk). These will sustainably support existing workforce upskilling and reskilling needs alongside the apprenticeship and respond to 'business as usual' needs beyond government funding.

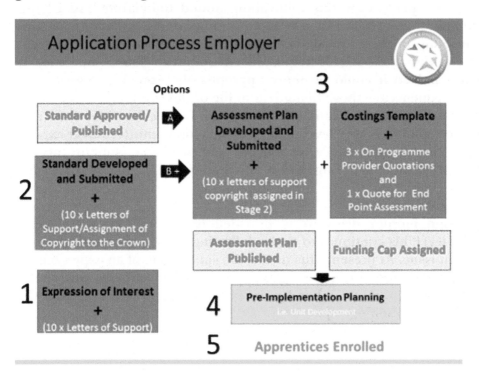

Figure 1. Development Process

The trailblazer design process
Labour Market Intelligence/Expression of Interest
Collaboration and consultation supported by workforce analysis was a key factor in identifying the need for standard development. Indeed, employers shaped their industry strategy by using reports which provided a high level view of workforce development needs, referring also to the human resource matrices within their own organisation. At this stage, ten

employers need to support an application to the Minister for Skills to develop a standard. Examples of the reports that drive the development of the Trailblazer apprenticeship include: Regulated Water and Waste Networks Workforce Plans PR14 EUSkills 2/11; and the National Skills Academy for Power Strategic Plan 2015-2018. An example of articulated need for Smart Meter Installers is provided below:

SMART

'Meeting the Government target for all homes and small businesses to have smart meters by 2020 is a significant challenge. Latest figures indicate that over 51 million meters will need to be changed, which equates to over 11.4 million meters at the height of the rollout. The industry is relatively immature with only 1,512 people currently working as smart meter installers. (Metering Network Yearbook, 2014) It is forecast that at its peak in 2018, the rollout will require over 10,400 installers (Energy & Utility Skills workforce requirements for smart meter installers 2015). The smart metering industry views the apprenticeship as the key mechanism to resource this requirement.'

The Standard

The standard details the job role activities, duration, pre-requisites, core and specific requirements including the skills, knowledge and behaviours required as a fully competent and productive worker, and identifies eligibility for future professional body registration. The standard is supported by a minimum of ten employers, including SME representation and professional body support, with the copyright assigned to the crown at the time of submission. The standard is more latterly subject to an external consultation period, following by a review by a government approvals panel. Following the monthly panel, the standard may be approved and published.

Employers have been integral to the interpretation of government policy directives, writing of draft documents and interpreting employers' collective thoughts and ideas, and distilling those into a two-page standard. The standard sets out the core and specific competence, skills, knowledge and behaviour requirements for a particular job role or occupation. The introduction of behaviours is controversial because of the potential for subjectivity which might undermine both reliability and validity of end-point assessment/final outcomes. This is arguably a key barrier to four-nation adoption. Other challenges include the language of industry, i.e. technician or craftsperson, versus pay scales.

Figure 2 provides an example of the Power Degree Apprenticeship to show how the design of trailblazers encompasses not only school leavers, but also those progressing within an apprenticeship career path. More crucially, all the Trailblazer developments do not sit in isolation but are connected to developing the eligibility for EngTech, IEng or CEng progression routes. This is achieved and enhanced by the relationship between employers, trade unions and the Professional Bodies/Engineering Council.

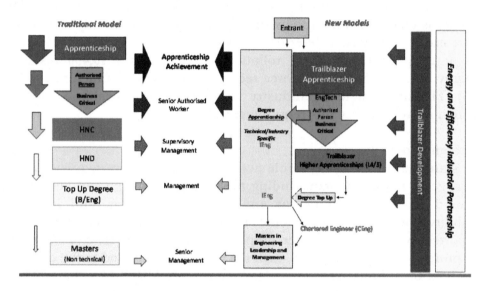

Figure 2. Power Degree Apprenticeship Routes

The Assessment Plan

From the outset the employers have undertaken this work with a view to ensuring that the outcomes, as well as meeting their requirements for comparability, credibility, high quality, and rigour, would also form part of their business as usual (BAU) activity. In pure business terms this has meant that processes and procedures have been designed to match the rhythms of the workplace and their patterns of delivery. Approaches that are tried, tested and trusted by industry are central to actively demonstrating the crucial role of both on-programme and end-point assessment to demonstrate workforce competence. The judgements around competence are business critical in terms of health and safety and productivity.

The development of the assessment plans has proved to be the most challenging due to evolving policy decisions about what constitutes end-point assessment. In the early stages, on-programme was considered a disconnected separate activity. However, more recent developments have enabled employers to construct a work log, detailing in some cases the progressive authorisations and assessment needed for the industry. Trade tests, technical interviews, and work logs can now constitute part of a portfolio which can be synoptically assessed within the end-point window. An example of the on-programme and end-point journey is provided in Figure 3.

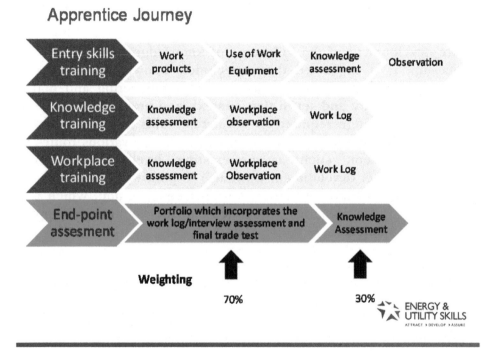

Figure 3. Example of Apprentice Journey

The Learning Journey
Core skills and knowledge (induction)
This induction phase establishes the common, underpinning competences, knowledge and understanding required to support the apprentice on-programme, The core skills and knowledge of the apprenticeship are delivered during the early stages of the programme to provide an initial foundation of industry-related skills and knowledge on which all subsequent modules of training will build. The end of module assessments will be central to facilitating progress through the programme and must be included as part of the apprentice's portfolio.

The aim of this phase should be to ensure that apprentices are trained to work safely at all times. It should provide insight on employer and co-worker expectations and how and where the apprentice can seek guidance and support.

Knowledge learning
To ensure development of the full range of knowledge required for this apprenticeship, it is suggested that a technical knowledge solution is developed by employers. Training may include engineering and maths relevant to the industry and set in that context, and provide the range of underpinning knowledge required to accelerate skills development and successfully tackle the end-point assessment. There are various routes to knowledge attainment, such as pre-existing level 3 engineering qualifications. Further development work also responds to the employers' desire to develop an Industry Standard Knowledge Solution, which will become the industry's recognised qualification, for example to support Water Process Technician.

Work-place learning – listening to employers – the work log
As apprentices progress through their training, they build up evidence on the full range of skills, knowledge and behaviours required by the standard and will be assessed on particular tasks or procedures or items of equipment, namely 'progressive authorisation assessments', for example, safe isolation of operational equipment for maintenance.

Industry practice requires authorisation assessments to be recorded in a work log. The work log evidences that the apprentice can apply skills, knowledge and behaviours required in a variety of tasks. Progress review documentation could also be included in accordance with company policy and procedures. A summative assessment of the work log therefore, in the eyes of employers, forms an important part of the end-point assessment portfolio assessment.

Employers lobbied hard for its inclusion in early 2014. Indeed, a fundamental aspect of the design has been the approach to both on-programme and end-point assessment in order to build upon employer's tried, tested and trusted working practices - by necessity robust and rigorous. Figure 4 therefore provides an example of the end-point assessment gates for apprentices for both Water Process and Utilities Engineering Technicians.

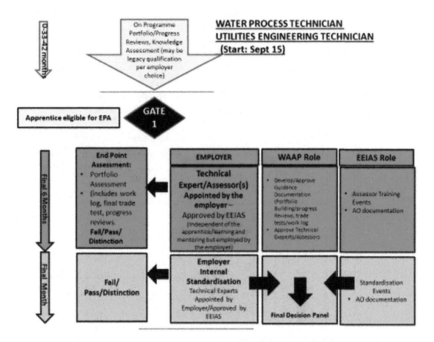

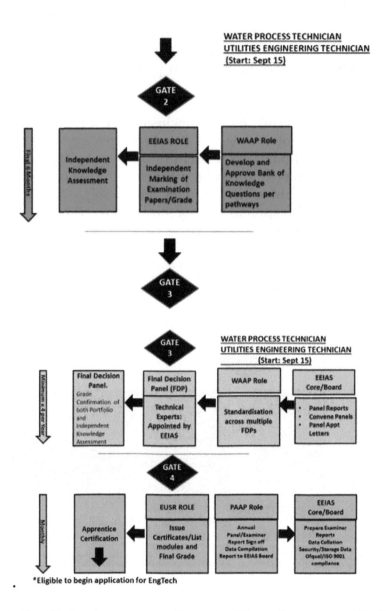

Figure 4. End Point Assessment Gates for Water Process and Utilities Engineering Technician

Questions around validity and independence have proved challenging constructs within an evolving policy context, although less so for employers themselves. Indeed within highly regulated industries, there are no short cuts, and assessment must, by necessity, be rigorous and robust, and indeed build on tried, tested and trusted working practices at the heart of the industry. However, to address the notion of independence (employers not marking their own homework), independent knowledge assessments (examinations) are a common feature in many assessment plans in this sector. Furthermore, a range of approaches to quality assurance have been designed to support end-point assessment. For example, final decision panels or external quality assurance by those technical experts at the cutting edge of technical practice in the sectors. An excerpt is provided below:

> Final Decision Panel
> Decision panels will consist of three people:
> - Technical expert from apprentice's employer
> - Technical expert independent of the apprentice and their employer i.e., not from their employer or training provider and
> - Another technical expert independent of the apprentice and their employer or from a relevant professional body.
>
> One of the independent panel members will act as chair of the panel. The decision panel will check all available evidence and discuss to enable the independent chair to make the final decision of whether to award a fail, pass or distinction. Therefore, someone independent of the apprentice and their employer will always determine the grade awarded. The assessment organisation will co-ordinate the final decision panels and observe and intervene where necessary to ensure they are operated in accordance with the guidance, ensuring comparable decisions consistently and comparably across panels and over time.

Employers in many plans opted for final decision panels that often also included members of professional institutions, not only to ensure comparability and validity of assessment, but also to adopt an innovative yet independent approach to cross-sector competency accord. This work can only be maintained and secured through this stewardship approach to assessment and standards. Assessment plans for the energy and utilities sector can be viewed at *Apprenticeship standards - GOV.UK.*

A second model is described below but also signals another flex in approach around sampling – a new feature that emerged but with later caveats for SMEs:

> Assessment Organisation Quality Assurance (Smart Meter Installer)
> An independent assessment organisation will moderate end-point assessments on a 'risk based' sampling basis. The final grade will not be confirmed until after moderation. There will be close scrutiny and audit to ensure the assessment organisation maintains confidence in the rigour and robustness of the employer's assessment decisions. Moderation will be based upon level of risk. Sampling will be dependent upon the level of experience of the employer's quality assurance technical expert, together with the consistency of the results from previous moderation outcomes.

More fundamentally, it arguably serves to illustrate the risks for government from individual Trailblazer assessment plans developed at different stages between 2013 and 2015. Assessment plans will now need to address the need for consistency in the external quality assurance between different end-point assessment organisations, to assure not just validity of the apprenticeship, but also to avoid a dilution of the technical and competence outcomes between end-point assessment organisations. This will likely require retrospective adoption for early Trailblazers.

Grading – listening to employers' needs
Grading was a new feature of the Trailblazer. However, employers within a safety-sensitive sector were concerned that there should be no dilution of competence, for example in managing gas supplies, and that end trade tests remain binary - fail/pass. However, attitudes and behaviours were seen as a great asset to employers and indeed to productivity, and whilst arguably more subjective, could be differentiated alongside knowledge tests/examinations. Employers were supported in the design of grading criteria and points scoring systems upon which to build a solid foundation to award a pass or distinction for example. An excerpt is provided below:

> Grading will be standardised to ensure consistency across the sector. The apprenticeship will be graded fail, pass and distinction. The final grade will be determined by collective performance in the

end-point assessment's two assessment tools. The weighting of the apprenticeship is 70% on the portfolio, which incorporates the work log and final trade test, and 30% to the independent knowledge assessment. A points system will determine if the apprentice has achieved a pass or distinction and is described below:

Pass – 5 Points (3.5 Points Portfolio + 1.5 Points Knowledge Assessment)
Distinction – 10 Points (7.0 Points portfolio + 3 Points Knowledge Assessment)

Portfolio %	Points	Grade	Knowledge Assessment %	Points	Grade
<69	0.0	Fail	<69	0.0	Fail
70 – 84	3.5	Pass	70 - 89	1.5	Pass
85-100	7.0	Distinction	90 - 100	3.0	Distinction

To achieve a 'Pass' the apprentice will be demonstrating competence across the standard. To achieve 'Distinction' the apprentice will be demonstrating performance over and above the standard.

Figure 5. Gas Network Craftsperson

Cost of end-point assessment
Employers were integral to the completion of the costing template because of the high level of investment already in existence within their workforce, but also in the supply of the three quotations needed via their own academies or third party providers. The costing template introduced in wave 2 Trailblazers provided some challenges to completion because of the difficulty in comparing old and new world apprenticeships. In particular, technical certificates of similar programmes were sometimes not fit for purpose and were a poor base for a funding calculator. As a result, pre-Trailblazer it was not uncommon for additional training post - apprenticeship in the water sector due to a failure of existing qualifications to ensure safe solo working post-certification.

Serendipitously, the development of additional levels of detail derived from the 'detail' within occupational pathways, which are now referred to as Rules of Unit Combination (EEIAS February 2016) served to provide a strong foundation upon which to compile a costing to secure a funding cap. This could only be achieved because of the involvement of employers' subject matter experts who assisted the design of the Rules and the writing of the specifications. This enabled an understanding of the volume of learning hours needed to deliver the training to ensure the Trailblazer was fit for industry purpose. It also supported any funding appeal or query to the Skills Funding Agency through the provision of a sound rationale.

Journey's End?
Following government panel approval and ministerial sign off, the Trailblazer is deemed 'ready for delivery'. Employers would argue otherwise and the development journey continues today, learning from 'real-time experience'.

Employers deployed their own internal subject and technical experts to support the development of high quality units/modules of learning in pre-implementation planning, beyond what was required by the Trailblazer approval process. This level of unprecedented commitment and engagement provided a footprint not just for apprenticeships but also all their future upskilling and reskilling needs. In summary, employers have demonstrated their commitment to this important area by the fact that around 500 units or specifications have been designed to date to support seven of the most advanced Trailblazers in this sector. These may be viewed on the EEIAS website (www.eeias.co.uk) as new units for new Trailblazers continue to be developed, signed off and uploaded. The crucial aspect of this work is that it uses a transparent approach to avoid the ambiguities of interpretation so often commented on within other traditional apprenticeship frameworks. Again it illustrates the mantra 'for employers and by employers', yet within a robust pedagogical construct.

Conclusion
There are significant challenges ahead, yet Trailblazer apprenticeships offer a unique opportunity to build the future strength and resilience of workforce development to bridge the gap between industry and policy. There remains a need for standard rationalisation, as standard proliferation and overlap increasingly emerges, presenting new challenges for collaboration and the unintended consequences of decisions made by government at the embryonic stage of policy implementation.

Employers, however, under the apprenticeship Trailblazer banner, have actively demonstrated their commitment to go 'the extra mile' in terms of commitment, creativity, effort and expertise, adding tangible value to the emerging high level occupational standards. Nevertheless, these developments are not considered in isolation, because the business reality for transnational employers mean skills and competences must not only be portable, but widely recognised. This remains an unfulfilled aspect of the reform to date and one that must be addressed through an acknowledgement of the meta-competences required for specific job roles, irrespective of nation perspectives about the inclusion of, for example, behaviours within synoptic assessment judgements. Indeed in the energy and utilities sector, employers have moved government rhetoric to a tangible reality with actual apprenticeship enrolments from as early as September 2014 and obtained its first achievers.

Nevertheless, it is perhaps appropriate, as employers 'grasp the nettle', that the last word rightfully belongs to a new Trailblazer apprentice:

> Being part of these new apprenticeships is a great opportunity to develop my knowledge and skills. Apprenticeships are a great, hands-on way to learn and I'm really glad I chose this programme. One day I hope to be in a management position and I'm confident that this apprenticeship is going to arm me with the skills I need to succeed.
>
> Emily, Water Process Technician Apprentice, United Utilities

References
Brooks, B. (2014) *Apprenticeships – Managing the talent pipeline. A paradigm shift in developing workforce skills.* Tribal Publication (accessed 7 November 2015).

Gilbert, R. Balatti, J. Turner, P. Whitehouse H. (2004). *The generic skills debate in research higher degrees.* Higher Education Research and Development 23.3: 375-388.

HM Government. (2013) *The Future of Apprenticeships in England.* (www.gov.uk).

HM Government. (2015). *The Future of Apprenticeships in England: Guidance for Trailblazers from standards to starts.* (www.gov.uk).

Richard, D. (2012) *The Richard Review of Apprenticeships.* (www.gov.uk).

Wolf, A. (2011). *Review of Vocational Education – The Wolf Report.* (www.gov.uk).

9.
GROWING THEIR OWN AT E.ON:
A CASE STUDY ABOUT TRAINEESHIPS AND APPRENTICESHIPS

David Way CBE

The introduction of traineeships was an important and bold move by Matthew Hancock, Skills Minister 2012-2014. He was responding directly to the consistent message from employers that many young people were not quite ready for work and needed something extra to prepare them.

This chapter looks at the introduction of traineeships through the eyes of one large employer, E.ON. They are one of the UK's leading power and gas companies and employ 17,000 people in the UK.

The context was that E.ON is in a sector that suffers from skills shortages and usually has problems recruiting young people, and especially those with good school achievements. They take recruitment and training extremely seriously but had problems in introducing an apprenticeship programme that worked for them. When traineeships were launched, they saw this as the opportunity to try again.

This chapter begins with a report of my visit to E.ON's offices in Nottingham.

Meeting the trainees at one of E.ON's offices, it was difficult to believe that these young people had been raw recruits six weeks before. Most had been unemployed and worried about their futures.

It was much easier to see why E.ON's innovative programme, aimed at 16 to 24-year-olds who are about to leave school or college, or currently not in education, employment or training (NEET), was attracting interest from other energy companies. Here was a group of fifteen young people keen to build their careers in their chosen sector. While it was a competitive process, with each of the trainees in pursuit of a full-time paid apprenticeship, they were relishing the opportunity to gain fresh skills and enhance their job prospects.

Tammy Bristow, Industry Qualifications & Standards Manager at E.ON and a previous apprentice, was given the task of introducing a new apprenticeship programme by E.ON UK HR Director, Dave Newborough. The previous attempt had not 'landed' very well in E.ON, with a lack of job-ready applicants. With excellent customer service at the heart of the business, individual managers faced a tremendous amount of training before they could put the new apprentices in front line contact roles.

E.ON saw the opportunity to design traineeships and apprenticeships that were tailored to their needs and provided a talent pipeline into the business. Designing both programmes together was one of the keys to success and to 'growing our own' according to Apprentice Coordinator, Darren Cook.

Results from the seven-week traineeship have been highly encouraging. Only one of the initial cohort of thirty had not got an apprenticeship with E.ON or another company in a customer service role. The retention rate of the apprentices who went through the traineeship was very high. Rachel Barnaby was one of those trainees now in the business and described it as 'a privilege'.

Alison Patrick-Smith, who was the line manager for ten of the apprentices, is delighted with their skills, knowledge and positive attitude. Because of the breadth of their training, Alison says she has learnt from the apprentices who are encouraged to network widely across the business.

The new programme has had great support from the top with E.ON's Chief Executive, Tony Cocker, regularly seeking progress reports and senior managers seizing the opportunity to be mentors to the apprentices. This meant a lot to the young people.

Rachel said that being mentored by a senior manager with twenty years' experience in the company made a big difference and she felt more confident about looking up and forward to a career at E.ON.

The programme itself is demanding and covers financial awareness, team working, data protection, diversity, CV writing and more. The trainees also have to undertake a project and present their findings to a panel of managers. About 60 per cent of the trainee's time is spent in the classroom and 40 per cent in the business.

Most of the trainees were NEETs prior to joining. Some had already tried other career paths but they had not been what they wanted. The trainees recognised that they have grown considerably through their training so far, both at work and outside, and have become much more confident and responsible.

Some were very honest about their limitations when they joined, and found much of the training challenging. Now they were clearly enjoying the experience, especially of working in teams. They looked forward to their presentations on subjects such as the importance of good communication at work. For some, this would be the first presentation of their lives.

While they all hoped to progress onto the apprenticeship with E.ON, they appreciated that they had learnt so much that this could only benefit them in whatever they did in the future. A number talked about university aspirations if things went well with the company. All were extremely positive about the energy sector.

This was very gratifying for everyone running the programmes but Tammy was very clear that getting jobs for all those who completed the seven-week traineeship was the objective that really mattered.

There were a number of learning points from E.ON's experience so far. Tammy highlighted the importance of securing commitment to the ambition of the programme across the whole organisation. This had translated into active and passionate support from everyone in the business.

E.ON is not only tackling local employment problems such as high turnover and competition for young talent, it is also giving much needed career opportunities to young people with potential that was at risk of remaining undiscovered.

The second tip from Tammy was to be prepared for an important counselling role as well as training. Many of these young people do not have the same life or work experience as many other colleagues within the business. More needs to be explained, and advice given that often ranges beyond the workplace. However, the personal growth and productivity that is seen from those young people as a result of this extra time, attention and patience is hugely gratifying.

E.ON benefits greatly from the EU Skills Talent Bank that has supported the placement of the minority of trainees that E.ON does not recruit itself onto its apprenticeships, as well as from the Employer Ownership pilot that provided public funding until March 2016.

Having tested out traineeships and apprenticeships and found they work so well, E.ON has now expanded the programme to include other operational areas within the business that are facing skills gaps.

By October 2015, ninety traineeships had been successfully completed; sixty had been offered apprenticeships by E.ON and more than twenty had gained employment elsewhere.

Other parts of the business have been lining up to participate. The strong support from above and across the organisation, and the talented young people who are now moving into the business as energy apprentices, means that the challenge of expanding the scale of the apprenticeship programme is being successfully achieved.

Apprentices are moving on in the business with permanent contracts. Many of the apprentices are outperforming other employees in the Key Performance Indicators and NPS scores used by E.ON. Retention rates are at 85 per cent.

In summary, E.ON has achieved a very successful apprenticeship programme as a result of introducing traineeships that provide the flow of employment-ready individuals who can be successfully integrated into the business with its customer-facing workforce.

Without this preparatory stage, E.ON may well still be struggling to find an apprenticeship model that works for them. Instead, they have found a solution that is enabling them to grow apprenticeships in the business and recruit more young people who otherwise would otherwise boost the number of NEETs in the area.

While this is just one example of how traineeships are working for an employer, it was highly encouraging to find that they offered a good solution to a very real problem. Not only has the employer benefited by being able to redesign the whole recruitment and training process but those young people who are most in need of extra help are benefiting directly from government investment and E.ON's help.

It is especially interesting to me in thinking about the three million apprenticeships target that without traineeships this company may well not have offered apprenticeships at all. So in considering the future growth, the effectiveness of feeder arrangements such as traineeships needs to be part of any 'end to end' consideration.

The chapter was prepared with the help of Tammy Bristow at E.ON.

PART TWO

PROMOTING APPRENTICESHIPS

INTRODUCTION TO PART TWO

David Way CBE

I have selected Promoting Apprenticeships as the second theme for this book about apprenticeships because so many people are bemused that something that is seemingly so rewarding for employers and young people is taken up by relatively few people. Return on investment statistics for employers are impressive and satisfaction ratings for apprentices are very high. Public endorsement of apprenticeships is very positive.

Indeed, people would often say to me on hearing that I was the head of the National Apprenticeship Service that I must have the easiest job in the world.

This section of the book looks at the appeal of apprenticeships to businesses, young people and parents. Apprenticeships evidently do not sell themselves. Why is this and what do we do to ensure that positive attitudes translate into increased uptake of apprenticeships?

The first chapter looks at why apprenticeships matter to the UK economy as well as to one of its largest businesses, ensuring skill shortages are avoided that otherwise would threaten the UK's digital economy. The scale of skilled recruits that will be needed in this important sector is startling. It is therefore vital to win 'hearts and minds' to ensure employers invest in apprenticeships especially in those technical and practical skills that we will need for the future.

Sir Mike Rake is not only one of the most influential business leaders in the UK but the company he currently chairs, BT, is one of the biggest employers of apprentices. In his time as Chair of the UK Commission for Employment and Skills, he confronted many of these issues and is excellently placed to advise on approaches that will help make a real difference. His expectations are not just that apprenticeships will help ensure BT's continuing competitiveness but that they will keep BT at the forefront of the industry internationally.

When I ran NAS, I gained a great deal from talking to marketing professionals about how to run successful campaigns. The next chapter taps into their expertise. I have asked Stu Youngs, who was heavily involved in a very successful apprenticeship campaign, to reflect on this experience and advise on how he would approach the promotional challenges today.

Stu considers what needs to happen if we are to change the nature of campaigns from bursts of marketing activity to approaches that would seek to positively affect attitudes and behaviours permanently. How do we make apprenticeships irresistible to employers, young people and their parents? How do we ensure that people don't automatically follow what he describes as the 'royal route' but make the best choices for them. There are some strong and deep-rooted messages that have been heard in favour of higher education. How can we make space for messages about apprenticeships to be heard? How can apprenticeships be 'normalised' as a good choice and not seen as an alternative that feels risky?.

This theme of needing to get positive messages out to those who need to hear them is picked up in the next two chapters.

When Gordon Brown was Chancellor of the Exchequer, he invited an impressive group of business chief executives to come together to advise the government on apprenticeships and importantly to be Apprenticeship Ambassadors. The rationale for this being that a message explaining why take on an apprentice has more force if it comes from someone who has real experience of employing apprentices in their own business.

David Meller is Chair of the Apprenticeship Ambassadors Network and he explains their important role and work, including spreading the ambassadorial approach into regions and locally. This work is proving vital in taking messages out to more employers and in influencing the many local partnerships, such as Local Enterprise Partnerships, that can be effective allies in achieving the three million apprenticeships ambition.

When I first attended conferences about apprenticeships a decade ago, it was common to hear people say apprenticeships were a great idea for other people's children, just not for your own. They had somehow lost the high professional status they once enjoyed (and found in most other European countries) to those who went to university.

Inspiring the Future is an impressive organisation that has done great work initially in getting employers into schools to talk about apprenticeships. This is, of course, much needed and is usually much appreciated by the schools that welcome support on apprenticeships, about which they often know very little. I recall visiting an all-girls school in Birmingham and not one of the 15-year-old pupils knew anything about apprenticeships. A female apprentice from Land Rover then told them about her experiences, her car and her salary. Everyone sat up and took note. Inspiring the Future has facilitated thousands of such visits and have gone on to broaden their work into primary schools and inspiring more young women to aim for more ambitious careers as they prepare for work.

This chapter looks at whether these attitudes are changing and successful approaches being taken to promote apprenticeships to schools. Anthony Mann is a respected writer on such matters and is the Director of Policy and Research at Education and Employers. He is well placed to comment on whether attitudes towards apprenticeships are changing and whether parents are becoming advocates for the vocational route.

Apart from national organisations like Inspiring the Future, many individual employers have seized the initiative and gone out into local communities to find candidates for their apprenticeship schemes who will better reflect the experiences of their customers. The most ambitious project that I have encountered is run by Barclays Bank. The results have exceeded all expectations. The Barclays Apprenticeship programme is recognised as one of the UK's leading programmes, having won the BITC Inspiring Young Talent Award in 2015, as well as the prestigious Target Jobs Best Apprenticeship Programme of the year 2015.

Mike Thompson is the Head of Employability Programmes for Barclays and led the development and introduction of their ambitious new programme to attract more young people from a wide range of backgrounds as Barclays Apprentices. It has been extremely successful

121

and attracted the interest of many other UK employers. Mike has now been appointed to the government's Apprenticeship Delivery Board.

This chapter explains the journey that Barclays has been on since launching its first ever apprenticeship programme in April 2012; the drivers behind the programme; how it has developed and evolved; and how it has had impact both on the organisation and on the individuals who have gone through the programme.

This section then takes a wider look at how UK apprenticeships are presented to ourselves. This stems from my belief that, while we should not be blind to improvements that are needed, we should do more to promote the many positives if we hope to attract more young people to become apprentices. Our pride must surely be conveyed to those young people and employers who we are seeking to attract.

When I was with the National Apprenticeship Service, there was great interest from other countries in what we were doing and the excellent progress being made then to grow the number of apprenticeships. They were interested in our approach, in higher apprenticeships, the central vacancy filling arrangements, the marketing, and the existence of a dedicated apprenticeship service, for example. Employer and apprentice surveys gave impressive results and the National Audit Office was supportive because of the excellent returns on public investment. Even the Public Accounts Committee was complimentary!

Chris Jones is a respected and well-travelled writer and commentator, most recently as head of the UK's leading awarding body. City & Guilds have been restoring pride in the UK apprenticeship system as sponsors of many high-profile skills events and the Top 100 Apprenticeship Employers list. He sets out why we should do more to appreciate and celebrate the strengths of apprenticeships in this country and the importance of learning from one another internationally. This reflects the fact that businesses and individuals are more mobile than ever before.

Finally, this section examines the promotion of apprenticeships on the ground. This is mostly undertaken by colleges and training providers. The final chapter in this section looks at the approach taken by one successful training company that promotes services and relationships well beyond a narrow training delivery arrangement. Why does it take this approach and what benefits does this bring? Are employers interested in longer-term partnership arrangements? What are the key requirements that employers want training providers to satisfy? Roger Peace is the Chief Executive of the UK's biggest employment and training provider. They provide employment and training services to more than a thousand businesses in the UK. Roger shares his insights into what encourages employers to work with them and to train more apprentices.

10.
WINNING HEARTS AND MINDS

Sir Michael Rake
Chair, BT Group

———————

Introduction
The UK's future as a technology leader hinges on young people getting
the skills, support and training they need to create successful careers in
science, engineering and IT. As a major player in the UK tech sector, BT
wants to see the UK's digital economy thrive, but there is already a skills
shortage that will affect future productivity and competitiveness. The
nation's tech economy could grow by £12billion in the next ten years, and
the sector will require nearly three hundred thousand recruits at higher
skills levels by 2020, but nearly 20 per cent of vacancies are already difficult
to fill due to skills shortages.

Ensuring that the UK has a well-educated and highly-skilled workforce
relative to the rest of the world is vital to ensuring that it remains well
positioned in an increasingly competitive global environment. Given the
pace of skills development and education in other parts of the world, the
UK will need to invest significantly in education and workforce skill
development to retain its current position. And young people don't just
need academic qualifications; we need to be able to ensure that they have
the practical skills that mean they can apply the knowledge in business
and the right positive attitude to, and enthusiasm for, work.

From BT's point of view, skilling future workers in cyber security, digital
media, data science and specialised diagnostic skills will enable us to
remain not only competitive, but at the forefront of customer and business
global requirements. But our economy won't generate just more specialist
tech jobs; almost every job in the future will have some elements of tech.

Our potential as a country can't be fulfilled unless young people are 'tech literate'. Every child needs to be able to access and use everyday technology, be confident with the concepts of how it works, and embrace its impact in shaping society. With all sectors increasingly become technology driven, virtually all will need to have a basic competence across a broad set of common digital skills (use of communications devices and software, internet navigation, social media, etc.) if they are to be fully effective in their chosen vocation. Young people are familiar with using technology and are avid consumers, but they often lack the ability to create – and don't always understand the full range of career prospects associated with tech skills.

This is a pressing problem with many dimensions – including issues around connecting classroom learning with tangible opportunities in the world of work and parents' misconceptions of tech careers and a tech industry that sometimes struggles to display its character beyond its products.

The BT story
Apprenticeships play a key role in ensuring that BT maintains and develops a highly-skilled workforce, and are undoubtedly an effective part of our talent strategy. More importantly, for young people they're a great way to transform raw enthusiasm into valuable skills that will serve them well wherever their careers take them.

Apprenticeships have been a core part of BT's talent strategy for over fifty years. In the last five years we have recruited over 2,300 young people and expect to have around 2,000 apprentices at any one time, working in fields such as technology research, engineering, IT and TV, helping to create and build the next generation of communications technologies for the UK, as well as in business administration, customer service, financial services and HR. Our approach is broad; we offer both advanced and higher apprenticeships and from September 2015 we now have a full degree honours programme in digital media. The mix of academic learning and practical application brings real benefit to all parties.

We invest a lot in recruiting new people to our business, and hiring apprentices is extremely cost-effective with over 90 per cent of BT apprentices still with us after five years, a much higher percentage than with other externally recruited employees at similar grades. And our apprenticeships are highly sought after; we receive more than twenty valid applications for each job, giving us a huge talent pool to choose from. We attract ambitious young people who are keen to earn and learn. They enjoy the scale of BT and the variety of career paths it can offer. By coming into contact with senior leaders who began their career as an apprentice, they quickly understand that hard work and the right attitude can take them to the top.

Issues
The business case for us and many other large companies is irrefutable. But there is a need to actively focus on how to bring the benefits of apprentices to SMEs so that they can inject this type of young talent into their businesses and grow the skills and attitudes that they need. Rightly or wrongly, many are currently reliant on immigration. SMEs face different challenges from the likes of BT. By their very nature they are small, focused, organisations that could benefit from an apprentice to grow but struggle to find the time to engage with the concept of hiring and training up an apprentice. They are likely to view taking on a young person with the potential to be a real asset in time, but who may lack basic literacy skills, as a luxury they can ill afford.

Many young people come out of school without any exposure to what work means and therefore don't know how to present themselves in a way that demonstrates what they are good at, and in a way that is going to impress an employer. A number of companies, BT included, offer young people traineeships that ensure they are ready for work. Our traineeships are targeted at young people not in education, employment or training and provide them with essential skills and a qualification so that they are better equipped to secure a job – and this includes making sure that they have basic literacy and numeracy. This can then be a stepping-stone to an apprenticeship or direct employment with on-the-job training.

A bigger issue is with too many young people coming out of university not only with debts, but also without some of the necessary skills to get them appropriate jobs. An alarming fact is that, while the assumption is that graduates will end up with higher paid jobs, 50 per cent of graduates end up in non-graduate roles, a rise from 37 per cent in 2007 according to UK Commission of Employment and Skills (UKCES), and thousands are out of work six months after graduation. Too often the degree courses are like a silo, ignoring the wider skill-set that businesses need. Apprentices learn both on and off the job, giving them a wider experience than a traditional full-time education route.

However, parents, teachers and school pupils themselves still do not yet see apprenticeships as a viable alternative to university. We need to give our young people better advice about the options available to them at the critical points in their school life when they are thinking about what they want to do and what their interests are. Parents who have gone down the higher education route tend to think this is the only credible path to take, and teachers are too often focused on university applications as a way to demonstrate the worth of their organisation. Myths are perpetuated: apprenticeships are for those who don't have the grades to go to university, a mark of failure – but apprentices can get higher qualifications; professional occupations are only achievable through university – when chartered accountancy status can now be reached more quickly via an apprenticeship or you also could become a qualified solicitor via an apprenticeship.

Apprenticeships through a different lens
Large and small companies in partnership. SMEs make up over ninety nine percent of the UK's 4.5 million businesses. In 2014, there were 1.8 million more registered businesses than in 2000, an increase of around 51per cent. For SMEs with limited resources, finding the right staff can be especially challenging. More are recognising that apprenticeships offer a cost-effective route to securing highly motivated and ambitious staff who can be trained on the job, and that fully trained apprentices can be more effective than new graduates who sometimes lack the skills for a specific job role.

Whatever the size of the business, finding the correct skills when employing new members or staff or upskilling existing employees is integral for any company to grow and prosper. That is why, when I was chair of UKCES, I developed talentmap - a simple framework for employers to engage with the education, employment and skills systems across the UK. Perhaps a similar initiative is needed to support businesses to offer apprenticeships.

Logic now says that in order to fill the skills shortage we have in the UK and with the government introducing the new large employer apprenticeship levy, we need a fresh look at how we approach skills development. Small companies could be crying out for potentially talented young people but do not themselves have the resources or flexibility to commit to offering a full apprenticeship. Could larger companies work with them, or a number of small companies together, so that between them they could provide an apprenticeship?

Blurring of schools and work. There is a need to expose kids to the world of work very much earlier in the education system – before they rule out career options. Young people need to have opportunities to see exciting workplaces and talk to apprentices and other employees about their career paths and experiences. We need to look at how we can integrate apprenticeships with the sixth-form more effectively.

In September, at BT we ran a crowdsourcing event to explore the challenges we face in our ambition to help build a culture of tech literacy for the nation. Leading thinkers and practitioners from technology, business, education, government, parenting and youth groups came together to share insights and come up with solutions to problems identified from talking to kids, parents and teachers. Asked how we build a skills pipeline for the digital economy, experts came up with the big idea of creating a learning experience for 16-year-olds in a model similar to the Duke of Edinburgh Awards scheme, providing a programme of tech skills development, industry placements and extra-curricular activity. A scheme along these lines would open up the eyes of young people to alternative career pathways and show that, while university is an option, it isn't the only one that will lead to a successful and fulfilling career. This could be particularly important in encouraging more girls to think about apprenticeships in tech and engineering.

Conclusion
The government aim to reach three million apprenticeships for this parliament is both bold and ambitious and to be welcomed. To achieve the goal, hearts and minds need to be won over the business benefits – which are significant and demonstrable – as well as in showing parents and teachers, who influence the choices of our young people, that apprenticeships are a route to a successful career and are to be aspired to

as much as a university place. A graduate will come into a company with the academic qualification but may not be work-ready with the skills a business needs, and an apprentice will be learning to be work ready alongside the gaining of qualifications and skills, but ultimately both will get to the same place - where they are competent and capable employees and an asset to the company. That is certainly my experience from the talent that comes through the door at BT.

These are exciting times for BT with further advances in high speed broadband, mobile, sport and digital television. We want young people to be inspired and we'll continue to put our new recruits at the heart of this, investing in their future, and that of BT and the nation.

11.
SUCCESSFUL APPRENTICESHIP CAMPAIGNS

Stuart Youngs
Executive Creative Director, Purpose Ltd

———————

When I was with the National Apprenticeship Service, we ran a number of campaigns to try to raise awareness of apprenticeships and to boost the number of opportunities. The evaluation after each campaign showed that we were steadily making progress with the message that apprenticeships were back and that this would be a great time to become an apprentice.

I worked on these campaigns with a number of top agencies and brilliantly creative people. It made me appreciate the special marketing and communications skills and techniques that are needed if attitudes towards apprenticeships are going to continue to shift favourably and if there is to be positive endorsement of apprenticeships based on a deeper understanding of what they truly offer.

I therefore returned to Stu Youngs, who had been at the heart of one of the best apprenticeship campaigns and sees the challenges and opportunities in respect of apprenticeships through a different lens from that of skills policy experts.

DW: What first goes through your mind when you think about promoting apprenticeships?

SY: A job unfinished. We flew out of the traps, but are still stumbling at the first hurdle of a very long race.

The world has changed significantly from when we began to now. While the traditional methods of communication initially showed signs of impact, now is the time to consider new methodologies – with greater scientific rigour. To achieve this, we need a much deeper understanding of the systemic, cultural and attitudinal challenges. We can then design interventions that respond directly to these on a micro level.

This challenge is much more complex than any level of broadcast communication can change. This type of communication alone will not change attitudes and behaviour. Whilst it contributes to the solution, a more rigorous programme of behaviour change is required.

It is a hard sell to SMEs in particular. They're difficult to reach, engage and persuade. They need to see immediate value and how it can work within their business without becoming a burden.

For learners, it's about finding the right approach, and sensitively saying that apprenticeships are an excellent route to success. To date, we have only tip-toed around the subject; debt, studying time and opportunity. We need to learn more from other countries like Germany and Switzerland and build the kudos of apprenticeships, so that they are perceived to be of equal value to university and not the poor cousin.

DW: How did you try to make a difference when you orchestrated the previous apprenticeship campaign?

SY: We tried a lot of different approaches: showcasing apprentices and employers, celebrating their successes to inspire others; organising the apprenticeship brands to make the entry and exit points to all audiences clear; enabling flexibility to talk to specific audiences, without causing confusion or overlapping with other messages; peer to peer communication – better use of case studies, telling real stories in a way that was engaging, informative and inspiring; instilling greater pride and raising the esteem (of those already engaged); and highlighting the impressive returns on investment (ROI) that employers can expect.

Whilst all of this was successful, it has not been sustained and so perceptions ebb back to those that are more typical. The only way to compete with the dominant beliefs and attitudes is to take an 'always on' approach, which requires constant investment.

DW: Was it easier or harder than you expected and why? What seemed to work?

SY: It was much harder than expected to reach the audience. We had not appreciated the underlying systemic problems and how difficult (even impossible in some cases) it was to affect the various parts of the puzzle. Careers advice in schools for example.

For those already in apprenticeships, we instilled greater pride and advocacy. As a result, their messages were amplified, their engagement increased and subsequently they influenced a lot of people.

It was much easier than expected to get the right creative work agreed by BIS and the National Apprenticeship Service (NAS.) The appetite, bravery and willingness to support from the internal apprenticeship team made the process much easier than it could have been.

At the time, it felt as though we were making progress – numbers were on the up and perceptions were shifting. We put a lot of pressure on the BIS/Agency communication teams and they now recognise that further progress will only be seen if there are creative interventions across the whole system of decision-making.

Looking back, what we should have done was look at a deeper programme of behavioural change. That is definitely the main challenge for the future.

DW: Everyone seems to love apprenticeships but take up remains stubbornly low. How do you promote something so that this positive attitude towards apprenticeships turns into action?

SY: You have to understand what and where the barriers are. Where are the points of conversion? What are the alternatives?

To confront the challenge head on will not solve the problem. It's endemic. There are many points, especially for young people, where the typical 'royal route' is seen as the norm and optimal journey – and yet little appears to be done to change careers advice, to show the breadth of options that are available, including apprenticeships.

Getting people to like apprenticeships should not be the objective. That's too soft. They need to work for the head and the heart. To achieve this requires greater interventions much earlier on in the decision making process. Apprenticeships need to be normalised – not feel like an alternative. Alternatives are perceived to have much greater risk.

DW: Are there particular challenges in converting SMEs to apprenticeships?

SY: Although we have worked hard to position the National Apprenticeship Service (NAS) as the 'Go To' for apprenticeships, unless you are a large employer the relationship only lasts a phone call. Then you are handed over to a provider. Providers are private businesses that make their money from volume, so it becomes difficult for them to invest considerable time in creating bespoke solutions for minimal return.

It just isn't as easy as it needs to be, especially for SMEs. It takes too much time, commitment and investment – it needs to be much easier for employers to start their programme.

Personally, I'm a great fan of apprenticeships, but the frameworks are difficult to apply effectively to our type of business. Consequently, we would have to invest even more time to make them relevant. So, the investment of time wouldn't match the return for us. I'm sure this is typical for many small employers.

DW: How do we appeal to employers? Can we promote both social responsibility and business productivity at the same time?

SY: You can absolutely promote both messages. They are both relevant and important to employers today. But they shouldn't necessarily be communicated at the same time - consider the right person, at the right time, in the right place, in the right way.

Employers generally get the concept and are aware of the possible benefits, but need a nudge. What this nudge is depends very much on their personal situation. Therefore we need to take an adaptable approach to drive people across the line. More work and research needs to be done on proving the ROIs. They don't currently seem to relate to SMEs as much as to large employers.

DW: Are there obvious barriers for employers such as complexity and constant change that make promotion harder?

SY: Complexity is most definitely a barrier. While everybody recognises there will be some complexities, it currently feels as if employers have to fit in with the system, as opposed to employers being the consumer.

It's easy to make something complicated, very hard to make it simple yet effective. But that has to be the challenge if it is to cut through. Constant organisational change, products changing off the back of new reports and new initiatives being launched make constant, coherent communication very difficult. Whilst it's important to keep refining the product, it cannot be at the expense of consumer understanding.

DW: Similarly, what, from your experience, can get the attention of employers?

SY: Speak to them in the language that resonates.

This means messages must be commercially focused. They need to consider employers' opportunities for long-term growth, as well as their short-term needs. They must outline the hard and soft impact. Messages need to be personal to those receiving them. They should only highlight organisations to which they can relate and respect. There is no point showcasing a multinational to an SME. Finally, speak to them in the appropriate context.

DW: Whose attitudes and actions are you trying to change? Employers, young people, parents, advisers? Is any group more important?

SY: They all affect the outcome, so they all need coherent and complementary strategies to solve the problem.

All the audiences are as important as each other, as without one of them the whole eco-system fails. Because they are all intrinsically linked, you need to create interventions concurrently. We are trying to change attitudes and actions – communication alone won't change these so we need to look at a different methodology, creating the right interventions at the right time to change behaviour.

DW: Presumably you need a segmented approach or is there a way of collectively addressing this problem of a lack of take up?

SY: There may well be a blanket approach to help address aspects of the challenges, but this would have to be supported by much more tactical work.

The message and proposition is completely different for different audiences. So too is how you reach them. Interventions must be designed to address the barriers for conversion, which will be very different for each audience.

DW: Most people's views of apprenticeships are outdated i.e. male, manual occupations. It is presumably a problem if people are positive about apprenticeships but their sense of them is not up to date?

SY: Absolutely. While I think there is better understanding of the diversity of apprenticeships on offer, there is a long way to go before their breadth and depth is fully understood.

The millennials who are now becoming parents are much better educated about the apprenticeships offer and have an up to date view. This may help to shift perceptions in the long-term, but only if apprenticeships remain front of mind and they continue to see them as a credible and effective way to employment. With activity dropping off significantly, we run the run the risk of regressing back to where we were before 2011 because the dominant pathways still have a strong voice.

DW: How do we achieve a change of attitudes that is massive and irreversible rather than incremental?

SY: Unfortunately, I'm not sure there is a 'silver bullet'.

Communication can help, but unfortunately even a multimillion-pound advertising campaign will not solve the problem. What we need is cultural, systemic and behavioural change. To achieve a more profound shift, we need to influence more pieces of the puzzle e.g., careers advice. Ultimately, it's near impossible to create something irreversible.

You have to stay relevant and front of mind. That requires constant investment, activity and promotion.

DW: Any sense of a reasonable timescale or scale of change programme that would be needed?

SY: There is not a quick win. This is a long game.

Communication alone will not create the change and cultural shift that is required. It needs to work in tandem with other interventions.

12.
GOING ONE STEP FURTHER:
THE VITAL ROLE OF APPRENTICESHIP AMBASSADORS

David Meller
Chair of the Apprenticeship Ambassadors Network

Summary
This chapter looks at the important and evolving role of employers as ambassadors for apprenticeships in recent years. It illustrates the power of employers at regional and national levels working together and giving confidence and advice to others new to apprenticeships, especially SMEs. Ambassadors help deliver apprenticeships, challenge myths, and present the reality that apprenticeships are key to business growth and productivity. Ambassadors continue to have a vital role to play in helping achieve the government's ambition for three million more apprenticeships by 2020.

Introduction
Over the past twelve years, we have seen a number of significant changes within apprenticeships policy, leadership, and delivery – and I am sure that many of the contributors to this book could argue over which of these have been the most and least successful!

However there has been a key constant throughout this time, and that is the necessity for employers to be engaged with, and supportive of apprenticeships.

At first glance this may seem like a given. Of course we need employers on board with apprenticeships – as apprentices need to be employed in real jobs – so where would we be without engaged employers who believe in this system? Whilst that is of course true, and employers are absolutely critical to the delivery of apprenticeships as a whole, I am talking about the need for a number of those employers to go one step further – and become ambassadors in the truest sense of the word.

The role, remit and format for apprenticeship ambassadors has, over time, evolved – just as the apprenticeships landscape and priorities have. Today the Apprenticeship Ambassador Network is made up of some of the largest employers in the country, who actively engage with other businesses about the benefits of apprenticeships and building a skilled workforce. For this reason the AAN will be absolutely critical in supporting the delivery of the three million apprenticeship starts during this parliament.

History

In 2003, Gordon Brown, then Chancellor of the Exchequer, launched a new 'National Modern Apprenticeship Taskforce' to provide strategic support on apprenticeships policy and delivery.

Mr Brown argued that the launch of the taskforce heralded 'a step change in (the government's) campaign to bring together employers with government, trade unions, the voluntary sector and other partners, to improve the skills of our workforce.'[1]

Formalising the relationship with employers in such a way also highlighted the importance of apprenticeships within the wider skills agenda, at a time when upskilling the country's workforce was seen as 'essential for the wider health of (the) economy'. At that time, skills were very much seen as Britain's Achilles heel, and had been allowed to flounder for too long. Eight million people had below level 2 qualifications, including 20 per cent of 18- to 24-year-olds. With this in mind, the government stated their ambition to create 'nothing less than a revolution in standards across the education and skills sector' – and the Apprenticeships Taskforce was viewed as a key part of that. The Chancellor accepted that, just a few years before, apprenticeships had been 'dying' and as such the taskforce was part of a significant effort to improve on this front and ensure that apprenticeships once again became a serious pillar of the skills agenda and future success.

---

[1] HM Treasury. (2010) Modern Apprenticeships Taskforce Launch webarchive.nationalarchives.gov.uk.

The taskforce was chaired by Sir Roy Gardner and it delivered its final report in 2005. This concluded that employers believed the delivery system was too complex, and that there was a need for greater collaboration and partnership between key players. The report went on to note that there were areas of common purpose in relation to the structure of frameworks, and the further linking of employers' training arrangements to the requirements of apprenticeship frameworks.

A ministerial-led steering group had been set up to address these issues which, in my mind, essentially indicated the need for more employer ownership of the apprenticeship programme even at that early stage – something we absolutely saw happen during the last parliament.

Crucially, the recommendation was also made to create an Apprenticeship Ambassador Network – which was launched the subsequent year, and also chaired by Sir Roy Gardner.

The AAN when it was first established in 2006 was a very different beast from the AAN of 2015. The original remit was to promote apprenticeships and also advise the government on issues which prevent or discourage employers from getting involved. In this sense they played a significant role in terms of influencing policy. They commissioned research, including an important and much-needed piece on the return of investment in training apprentices, so that they could continue to extol the benefits of hiring apprentices to employers suffering from the economic downturn.

The benefit of employer ambassadors quickly became accepted as a workable and beneficial way of expanding apprenticeship numbers, and supporting meaningful engagement with employers.

AAN Today

The AAN changed significantly in 2013/14. I took over as Chairman, and we strengthened our reach into different areas of the UK by bringing together Local Apprenticeship Ambassador Networks (LAANs).

The LAANs have shifted the ambassador landscape significantly. They make an incredibly important contribution, are constantly active, full of fresh ideas and often in the best position to drive local demand. The

LAANs have also enabled us to capture the SME market far better than we were previously – and can use their powerful local influence to really shape thinking on apprenticeships within their areas.

We know that traditionally apprentices tend to stay and work relatively close to home, and as such the apprenticeship market is still very much shaped and led by local industries and the demand of employers in that area. The LAANs enable us, as employers, to support that growth in the best possible way – by ensuring that there is leadership from local employers helping to shape the local apprenticeship agenda.

As localism and devolution have become an increasing feature of political life, the LAANs have forged essential relationships with Local Enterprise Partnerships, local authorities, chambers of commerce and others to ensure that as these bodies become more integrated in the apprenticeship world – that they do so with an understanding of what employers want and need from quality apprenticeship schemes.

Many regional ambassadors are not only forging strong local industry partnerships and supporting other businesses to set up or grow their schemes, many are also working with local schools to attract more students into the apprenticeship world – something which is becoming increasingly important. Ambassadors are incredibly well placed to do this. They speak with passion about their own schemes and wider apprenticeship opportunities – and really believe in the benefits that apprenticeships present to students.

Indeed many ambassadors speak on this with experience having started as apprentices themselves. George Ritchie MBE, for example, who chairs the North East Ambassador Network started as an apprentice. Stories like George's, and so many others, highlight that apprenticeships really can take you anywhere – they offer an opportunity which, when young people are inspired to take it up, really can change their lives. It is this reality which all of the Ambassadors recognise and totally believe in that leads us all to do what we do.

The changes in 2013/14 not only allowed us to bring together stronger regional networks but also to refresh the national network by bringing in some new members and introducing a fresh focus on delivery as opposed to policy. With the shift towards employer ownership of apprenticeships during the last parliament, including through the introduction of Trailblazers, I felt very strongly that employers must take up this mantle – and actively support the delivery of more high quality apprenticeships.

Employers are in the driving seat more than ever before on apprenticeships, and as such we need to play our part to spread the word – and encourage more businesses to get involved and start delivering. We also have a duty and responsibility to do this as employers. Many of our industries have reported skills shortages now – or projected skills shortages in the future. We need to retain our international competitiveness and protect the future of our businesses and industry by ensuring that we put the work in now to train skilled individuals for the future. Current ambassadors therefore play a very active role in engaging other businesses, and helping them to set up or grow their programmes with support from colleagues at the Skills Funding Agency.

Through business networks and existing relationships, ambassadors are often well placed to open the door to an employer who is new to apprenticeships, and show them the significant benefits of establishing a strong apprenticeship programme.

Many employers still hold back from hiring apprentices due to the myths that exist around them, the belief that apprentices aren't right for their industry for example, or that they would have to deliver all the training themselves – all myths which are easily bust by a conversation with a strong apprenticeship employer who knows just how beneficial the schemes can be, and how easy they can be to deliver. Hearing this from a fellow employer can often secure the level of engagement that we need, and it does work.

Many ambassadors go much further than engaging individual businesses. Hayley Tatum of ASDA, who is an absolute exemplar of an Ambassador, has set up and run events to target multiple employers at any one time, including supply chain events to engage SMEs in the apprenticeship world and bring them on board. Other ambassadors have led from the front to showcase new developments in the apprenticeship world, giving confidence to other employers to also come on board – for example the delivery of 5,000 digital traineeships by the BBC.

We have a great deal more to do, but also a great deal to celebrate – and we do celebrate and recognise excellence in addition to working to increase demand. As well as the Apprenticeship Awards, a recent highlight was the centenary event in 2014 which bought together employers who have employed apprentices for 100 years or more – with speeches from HSBC and Terry Morgan of Crossrail, himself an ex-apprentice. The reason apprentices have been a staple of British industry for so long is because they make sense – employer-led, on-the-job training, is a clear pathway into a future career. One hundred years ago, today and, I have no doubt, in the future too – high quality apprenticeships have proven their worth. Apprenticeships are here to stay, and both now and in the future the employer voice will continue to be a critical part of delivering that in the best possible way.

Over the next five years, as we head toward the three million target, the apprenticeship landscape will shift once again. The introduction of the levy from 2017, growth of higher and degree apprenticeships, changes to public sector procurement and recruitment, the new Careers and Enterprise Company working to improve careers advice, the introduction of new Standards, and indeed a continued shift towards employer ownership – there is change on the horizon.

I believe this is welcome change that can benefit us as employers, and the wider British economy. It also means that it is more important than ever to have a strong employer voice in the apprenticeship world, heralding the benefits that these changes can bring us – and encouraging take up; employers going one step further and acting as Ambassadors in the truest sense of the word. As such, the Ambassador Network has a key part to play both now and in the future.

13.
PROMOTING APPRENTICESHIPS
TO YOUNG PEOPLE AND SCHOOLS
Dr Anthony Mann
Director of Policy and Research, Education and Employers

One of the most significant trends of apprenticeship provision in England in recent years has been the increasing age of the average apprentice. Over the past decade, while total numbers of apprentices have grown rapidly, participation by young people under the age of 19 has struggled.

This chapter considers the flow of young people into apprenticeships. It offers an oversight of literature on young people's (often negative) perceptions of the value of becoming an apprentice and draws on studies of how young people make sense of the labour market and make decisions relevant to it whilst in education. Such literature highlights the significance of teenage exposure to authentic experiences of the workplace in shaping career thinking.

The chapter also describes a programme which connected volunteers with first-hand experience of apprenticeships to a quarter of a million young people. The study showcases ways in which schools responded positively to the opportunity to enable pupils to learn more about apprenticeships. This shows how the attitudes of young people towards apprenticeships changed following interactions with volunteers, with a significant proportion going on to make active moves towards becoming an apprentice themselves.

Apprenticeships and the modern school to work transition
The association between apprenticeship and youth is long standing and charged with meaning. Historically, the period of the apprenticeship has served as a convenient shorthand for describing the journey of a young person, even the rite of passage, from adolescence to adulthood. The

novice who is growing to maturity, learning at the shoulder of the experienced worker, and practising skills under a guiding eye.

In political discourses, the association between apprenticeship and the youth transition into employment is familiar: '16,000 apprenticeship jobs introduced to tackle youth unemployment' read a typical headline in the run up to the May 2015 British general election (Hall 2015). And yet, in 2014/15, 43 per cent of new starters on apprenticeships in England were aged 25 years or older. More than that:

> The number of starters aged under 19 as a proportion of all starters almost halved between 2009/10 and 2014/15 (going from 42% to 25% of all starts). More generally, those aged under 25 went from representing 82% of all starts in 2009/10 to 57% of them in 2014/15. (Delebarre 2015).

The current apprenticeship can no longer be seen as synonymous with the school-to-work transition. The numbers of young people aged 16 to 17 entering English apprenticeships has flatlined at around 100,000 a year since the 1990s; growth has been among employers taking on older, often relatively experienced workers. The 2014 Apprenticeship Evaluation Learner Survey commissioned by the Department for Business Innovation and Skills polled nearly 6,000 apprentices and found that 64 per cent had worked for their current employer for at least a year before beginning their apprenticeship. Of the remaining 36 per cent under the age of 24, barely half (51 per cent) had come straight from school of college.

Whereas at the beginning of this century, the majority of apprentices were aged under 19 at the start of their programme, by the middle of the 2010s, fewer than one in five starters can be said to be following in the footsteps of the classic school-to-work transition apprentice in entering their training programme straight out of education (BIS 2014, 22-25).

The declining proportion of apprenticeships being followed by young English people has rightly been the subject of attention and concern for both researchers and policy makers. Government acted, for example, in 2013 to reform funding, trialling loans for older apprentices in order to increase the attractiveness of apprenticeships for young workers. Rather than make the programme of training less attractive to older workers, policymakers have devoted more energy to increasing demand from young people for apprenticeships through media advertising and programmes of activity. The latter serves as the focus of this paper.

What young people think about apprenticeships
Recent government marketing has aimed strongly at trying to persuade young people to become more interested in apprenticeships. A 2014 Department for Business Innovation and Skills press notice issued to launch a new advertising campaign 'calling on young people to 'Get In. Go Far' by choosing an apprenticeship' quoted Skills Minister Nick Boles:

> 'As another group of young people achieve their GCSE and A Level results, there has never been a better time to consider an apprenticeship. Through an apprenticeship young people can achieve a degree and work at some of the biggest companies in the country.
> The new campaign features some great success stories which show exactly how far an apprenticeship can take you. I would recommend any young person that isn't sure what to do next, to look at some of the new and exciting apprenticeship opportunities available to them' (BIS 2014a).

The aim of the government campaign was explicitly to change the way that young people were thinking about apprenticeships, urging them to give such training programmes serious consideration as a pathway out of education and into work. While such a message underestimates the significance of competition from older workers in accessing apprenticeships, it certainly speaks to survey data and academic studies undertaken over the preceding decade suggesting lack of youth demand to be an issue requiring attention.

Apprenticeships: limiting, risky and probably second best, but intriguing
A series of recent studies has revealed a consistent picture of how teenagers commonly view the idea of becoming an apprentice (See Mann and Caplin 2013; Beck et al 2006; Beck et al 2006a; YouGov 2010). First and foremost, understanding is hazy of what it is to follow an apprenticeship. If pressed, there is a widespread instinct that apprentices are stereotypical male, white, engaged in manual labour and that the programme of training on which they are embarked is second best to academic study.

For very many young people consequently, there is a widespread instinct that apprenticeships are probably not right for them, that 'people like us' don't do them.

A common conception among young people is that apprenticeships may limit their future career choices and opportunities for academic progression. One of the great attractions of university study is that career decisions can be delayed while knowledge and skills, widely assumed to increase employability, are developed. Asked what would make apprenticeships more attractive, most young people say that they would be more likely to apply if they thought they would not be tied down to a particular job in the future or if the apprenticeship provided an ultimate route to university.

The perception that apprenticeships can limit future employment opportunities can make them appear a risky prospect to many young people. This is reinforced by a widespread viewpoint that apprenticeships are what you do – as a teenager – if you do not get the grades to progress academically: the second best option. Moreover, polling shows many young people to be unsure about how positively employers really see apprenticeships. A specific risk looms for young women. With apprenticeships seen as largely a male preserve, and participation by framework heavily structured by gender, a key concern among potential female apprentices is that non-traditional workplaces will be unfriendly towards them.

And yet there is a high, and growing, interest in the idea of the apprenticeship. Large majorities of teenagers surveyed like the idea of jobs which have structured training, where they can learn as they earn and want to know more. In an era where students are expected to pay the full costs of university tuition, the burgeoning attraction of going straight into the workforce from school or college, bypassing higher education, is understandable.

Schools and apprenticeships: improving young people's access to authentic information

Since the publication of *Triumphs and Tears* by Hodkinson and colleagues in 1996, a growing academic consensus (ably summarised in Tomlinson 1993) has been reached on how young people come to make career choices. It shows that pupils' decisions reflect their own social circumstances (gender, ethnicity, social class, immigration status) and the growing complexity of the labour market. Scholars have drawn notably on the concept of 'habitus', developed by French sociologist Pierre Bourdieu to understand the ways in which the attitudes and assumptions of young people both emerge out of their social situations and serve to shape their future trajectories. In such a way, certain jobs or careers can be seen as 'unthinkable' or instinctively undesirable, as has been observed in the case of young people's views of apprenticeships.

It was such a viewpoint which influenced thinking behind New Labour's Aim Higher programme – intervening to encourage academically able young people to think afresh about higher education, challenging assumptions over who university is for or not for. Aim Higher operated through the organizational structure of the school. It enabled state actors to reach over the heads of parents and families to connect directly with young people, so bypassing the originators of socially located assumptions perceived to be limiting and misplaced.

Parallels with work-related learning approaches are clear. Scholars have highlighted the significance of labour market experiences perceived to be highly authentic and accessed through programmes of school engagement with employers (Mann and Dawkins 2014). Following a detailed study of young people on work experience placements, for example, sociologists Carlo Raffo and Michelle Reeves observed:

> What we have evidenced is that, based on the process of developing social capital through trustworthy reciprocal social relations within individualized networks, young people are provided with an opportunity to gain information, observe, ape and then confirm decisions and actions with significant others and peers. Thus, everyday implicit, informal and individual practical knowledge and understanding is created through interaction, dialogue, action and reflection on action within individualized and situated social contexts. (Raffo and Reeves 2000)

In such an experience, while young people may have limited opportunity to significantly develop skills relevant to the workplace, what they do have the chance to get in abundance is access to authentic, trustworthy information about the workplace and its cultures, to be assessed from an individual's personal perspective.

This informational element of social capital (a resource of ultimate economic value stemming from the character of social networks) has been isolated by a number of recent studies and been seen to influence later employment outcomes. Mann and Percy (2014), for example, have evidenced a statistically significant relationship between school-mediated episodes of teenage employer engagement and the earnings of young adults, finding evidence of wage premiums of up to 18 per cent. They argue, following US sociologist Mark Granovetter, that interventions which are often of short duration, episodic, and unintegrated into programmes of learning, are of high value to young people if they offer young people, whose understanding of the labour market is commonly highly limited, access to new and useful information to guide more informed decision making through educational journeys.

In seeking to address the negative attitudes and assumptions of young people about apprenticeships, the research literature suggests the need for increasing the level of authentic exposure young people have to the apprentice route. Work by Louise Archer (2014), Becky Francis (2014), Carlo Raffo (2000, 2006), Steven Jones (2015) and others has highlighted the difficulty that British teenagers have in making sense of the world of work, and in securing access to first-hand information for themselves about its opportunities and challenges. They highlight the importance of real-world encounters in broadening aspirations, challenging assumptions, and giving confidence in thinking about future economic identities.

In a world where both universities and training providers are incentivized to recruit the maximum number of candidates onto programmes of study or training, young people are presented with a swirling marketplace of information of uncertain relevance and trustworthiness (Archer et al. 2010). Such confusion is witnessed in a 2013 publication which compared the career aspirations of 11,000 teenagers against the labour market projections of the UK Commission for Employment and Skills. A statistical analysis of teenage ambitions, from age 13 to 18, found that they had nothing in common with the UK's best quality skills demand with routinely one-third of young people expressing interest in ten narrow occupations (Mann et al 2013).

In the case of Aim Higher, the all-graduate teaching workforce was exceptionally well suited to support government initiatives. Teachers were well placed to share personal experiences in advising and encouraging pupils to explore options. No comparable widespread resource exists within schools, of course, in terms of apprenticeships. In an unpublished 2012 survey undertaken of 518 secondary school teachers by Education and Employers with the Times Educational Supplement (TES), participants were asked how they felt about advising young people on what an Apprenticeship was and how someone might go about getting on to one: 52 per cent were 'not at all' confident of providing such advice and only 15 per cent 'very confident'.

Earlier work commissioned by the Edge Foundation highlighted the fact that teachers routinely underestimate the extent to which parents, young people and employers value apprenticeships as a realistic alternative to A-levels (YouGov 2010). Such polling reveals a teaching workforce with limited understanding of such training pathways but very importantly with a clear desire for young people to have access to relevant knowledge.

In the same Education and Employers/TES survey, 80 per cent of teachers responding felt it 'very important' for pupils to hear directly from employers/employees about jobs and the best routes into them. The polling was consistent with other studies, which have shown the great majority of teaching staff to be very open to the idea of greater employer involvement in school life (YouGov 2010).

Employer engagement to enhance apprentice awareness: a worked example
In 2012, London-based education charity Education and Employers began work with the National Apprenticeship Service to significantly increase demand for apprenticeships from young people. In developing a programme of activity, three key assumptions (drawing on research insights discussed above) shaped thinking:

> Schools are willing to help: while teachers collectively lack sufficient information themselves to offer young people informed

advice about apprenticeships, they will strongly support pupil access to relevant information about apprenticeships

People matter: enabling interactions between young people and workplace volunteers able to provide authentic insight about apprenticeships will enable young people to gain access to new and useful career-related information

Action follows interaction: a meaningful proportion of young people can be expected to undertake some career-related action related to apprenticeships following their exposure to new and useful information.

The delivery model: Inspiring the Future
Launched in 2012, Inspiring the Future was developed by Education and Employers in partnership with many national organisations representing employers and the workforce, schools and teachers, to revolutionise the ways in which educational bodies connect with local employers.

Whereas historically connections had been enabled by brokers liaising between the two sides, Inspiring the Future harnessed online technology to enable employee volunteers to signal their willingness to work with teaching staff. Schools, in turn, would be presented with lists of potential volunteers who had already indicated their willingness to be approached, providing brief biographical details about their areas of experience. Volunteers could be approached at the time of teachers' choosing. With costs and barriers to entry minimised, Inspiring the Future quickly became a national programme with schools and volunteers registered across England.

With this infrastructure in place, it provided a potential mechanism to connect, through a national campaign, those people willing and able in schools and places of employment to work together to improve young people's understanding of apprenticeships.

The apprenticeship campaign
The campaign ran from April 2012 to December 2014, over which period the number of volunteers who self-declared themselves willing and well placed to speak in local schools to young people about apprenticeships

grew from 564 to 3,084.[2] With the volunteers, on average, selecting two different local authority areas in which they were willing to connect with schools, an effective new community of 6,000 volunteers became available to state schools and colleges. Volunteers were primarily recruited through supportive intermediary organisations (large and small employers, professional bodies, trade unions) which agreed to raise awareness of the volunteering opportunity.

Schools are willing to help
Over the period of the campaign, some 1,062 state schools and colleges (representing more than one quarter of all secondary educational institutions in England) responded to the invitation to invite an apprenticeship volunteer to speak with pupils. Volunteers willing and able to speak about apprenticeships proved to be the single most popular category of volunteers available to schools through Inspiring the Future over the campaign period: 91 per cent were approached by at least one local school within three months of registration.

Support from the national organisations representing schools and college leaders was strong with the two bodies representing 90 per cent of head teachers – the Association of School and College Leaders (ASCL) and the National Association of Head Teachers (NAHT) – becoming formal partners in the campaign, actively encouraging members to engage with volunteers through Inspiring the Future. Between April 2012 and December 2014, the number of unique state schools and colleges registered to use Inspiring the Future grew from 1,535 to 4,416 with comfortable majorities of English secondary schools and further education colleges signed up.

People matter
The campaign was designed to make it easy for large numbers of teenagers to interact with volunteers with first-hand experiences of apprenticeships through events taking place in their schools or at a community location. In all, 244,172 young people engaged with volunteers, of whom 180,988 were aged between 16 and 18 at the time of their event.

---

[1] All numbers presented were independently verified by the ISOS Partnership on behalf of the National Apprenticeships Service/Skills Funding Agency.

Testing whether teenage participants were accessing informational social capital through events, 2,169 pupils (52 per cent female and 48 per cent male) completed the survey:

> 85% agreed that they better understood apprenticeships following the event.

> 71% agreed that they planned to find out more about apprenticeships.

> 65% agreed that they had learnt something 'new and useful' during the event

> 62% agreed that they could now imagine doing an apprenticeship themselves.

Teachers also provided generic feedback on what they felt young people gained from their interactions with volunteers. In 2014, 150 teacher users responded to a survey request that was exploring impact. Teachers argued that they had observed improvements across a wider range of outcome areas related to attitudes and informational assumptions, notably:

Following activities with Inspiring the Future volunteers, teachers observed the following improvements in students:		
Aspect	'A lot of improvement'	'A little improvement'
..Aspiration	57%	37%
..Understanding of career pathways	50%	37%
..Understanding of the world of work	47%	48%
..Understanding of the value of education and qualifications	47%	40%
..Motivation	42%	49%

Table 1. Teacher perspectives of what young people gain from interactions with Inspiring the Future volunteers (Education and Employers: Annual Survey. N = 150.)

Action follows interaction

Unusually, the opportunity emerged to monitor whether young people attending events had gone on to take active steps towards securing an Apprenticeship. Midway through the campaign, data on 4,796 young people aged 16-18 attending events was submitted to the National Apprenticeship Service/Skills Funding Agency.

Through a process of 'fuzzy matching', officials calculated that up to six months following an Inspiring the Future Apprenticeship event, 11 per cent of participants had registered on the national Apprenticeship Vacancies system and that, of these, 64 per cent (or 7 per cent of the original 4,796) had gone on to apply for an apprenticeship vacancy.

Over the period of the campaign, the numbers of young people under the age of 19 beginning apprenticeships rose from a low of 115,000 in 2012 to 126,000 in 2014, reversing a long term trend of decline (Delebarre 2015). With a high proportion of young people signalling that something new and meaningful had happened to them as a result of their interactions, the causal character of the observed correlation proved persuasive, driving further state funding for activity in this area over the 2015-16 financial year.

Conclusions

This chapter argues that when it comes to young people thinking about their career options and whether they might give serious consideration to pursuing an apprenticeship, people matter. Many young people bring negative assumptions about apprenticeships into their schooling, shaping their career-related decision-making.

As the apprenticeship product itself becomes more diverse, by occupational frameworks and levels, interest can inevitably be expected to grow. However, this can be expected to be slowly because few young people possess easy access to sources of trusted advice needed to give the knowledge and confidence to pursue one.

Research literature suggests that access to authentic, trustworthy information about the reality, or rather realities, of apprenticeships, can challenge assumptions, making young people think again.

This worked example shows that schools can be seen as vehicles happily harnessing the willingness of volunteers to share their knowledge honestly with young people, a majority of whom left with positive experiences and having gained some 'new and useful' information.

Such initiatives help young people to see through the decision-making fog. In achieving three million more apprenticeships, people matter.

References
Archer, L. (2014) 'Conceptualising Aspiration' in Mann, A., Stanley, J. and Archer, L. eds. *Understanding Employer Engagement in Education: Theories and Evidence.* London: Routledge.
Archer, L., Hollingworth, S. and Mendick, H. (2010) *Urban Youth and Schooling.* Maidenhead: Open University Press.
Beck, V., Fuller, A. & Unwin, L. (2006) '*Increasing risk in the 'scary' world of work? Male and female resistance to crossing gender lines in apprenticeships in England and Wales',* Journal of Education and Work, 19 (3): 271-289.
Beck, V., Fuller, A. & Unwin, L. (2006a) '*Safety in stereotypes? The impact of gender and 'race' on young people's perceptions of their post-compulsory education and labour market opportunities'* British Educational Research Journal, 32 (5): 667-686.
BIS. (2014) *Apprenticeship Evaluation: Learners.* London: Department for Business Innovation and Skills Research Paper Number 205..
BIS. (2014a) '*Get In and Go Far with new apprenticeships'* London: Department for Business Innovation and Skills Press Release: 20 August.
Delebarre, J. (2015) *Apprenticeship Statistics: England (1996-2015).* London: House of Commons Briefing Paper 06113.
Hodkinson, P.M., Sparkes, A. C. & Hodkinson, H. (1996) *Triumphs and Tears: Young People, Markets and the Transition from School to Work.* London: David Fulton Publishers Ltd.
Hall, M. (2015) '*16,000 apprenticeship jobs introduced to tackle youth unemployment'* Daily Express: 9 April .
Jones, S. & Mann, A. (2015) '*The 'Employer Engagement Cycle' in Secondary Education: analysing the testimonies of young British adults'* Journal of Education and Work.
Mann, A. & Caplin, S. (2013) *Closing the Gap: How employers can change the way young people see Apprenticeships.* London: PwC.

Mann, A. & Dawkins, J. (2014) *Employer engagement in education: Literature review.* Reading: CfBT.

Mann, A., Massey, D., Glover, P., Kashefpakdel, E. & Dawkins, J. (2013) *Nothing in Common: The Career Aspirations of Young Britons Mapped against Projected Labour Market Demand 2010-2020.* London: Education and Employers and UKCES..

Mann. A & Percy, C. (2014) '*Employer engagement in British secondary education: wage earning outcomes experienced by young adults*' Journal of Education and Work 27:5, 496-523.

Norris, E. & Francis, B. (2014) 'The impact of financial and cultural capital on FE students' education and employment progression' in Mann, A., Stanley, J., and Archer, L., eds. *Understanding Employer Engagement in Education: Theories and Evidence.* London: Routledge.

Ofsted. (2012) *Apprenticeships for young people.*

Raffo, C. (2006) '*Disadvantaged Young People Accessing the New Urban Economies of the Post-Industrial City*' Journal of Education Policy 21:1: 75 –94.

Raffo, C., and Reeves, M. (2000) '*Youth Transitions and Social Exclusion: Developments in Social Capital Theory.*' Journal of Youth Studies 3:2, 147 –166..

Tomlinson, M. (2013) *Education, Work and Identity, London:* Bloomsbury Press.

YouGov. (2010) *EDGE Annual Programme of Stakeholder Research: Business in Schools.* London: Edge Foundation.

14.
SHARE AND SHARE ALIKE:
IMPROVING APPRENTICESHIPS ACROSS THE WORLD

Chris Jones
Group Chief Executive, City & Guilds

————————

Summary
This chapter sets out why we should do more to appreciate and celebrate the strengths of apprenticeships in this country, and the importance of learning from one another internationally when businesses and individuals are more mobile than ever before, and the challenges for young people especially to secure employment are expected to become more acute.

It reports the great interest and positive response to our domestic apprenticeship arrangements when I undertake overseas visits. We need to be more proud of our own apprenticeships while remaining keen to learn from other countries.

Finally, it raises questions about how we cope with the growing demand for transnational standards in skills and encourages greater international collaboration on apprenticeships.

Introduction

'It was a matter of general discussion in the engineering profession that the Admiralty found it very difficult indeed to get skilled artificers and artisans.'

Sir Hudson Kearley, Liberal MP for Devonport, said this during an 1897 parliamentary debate[3] on the strength of the navy. Nearly 120 years later,

————————————

[1] Strength of the Navy, HC Deb, 05 March 1897.

the specifics have changed, but we are still discussing skills shortages and skills gaps in the UK and how we can best overcome them.

Now, I am not one to be quiet on the topic of how we could do a better job in the UK of creating quality apprenticeship places in important growth industries, but many of the ways we can achieve this have been covered by others throughout this book. So, instead of beating that particular drum, I'm going to explore why there is reason to celebrate how far apprenticeships have come in recent years and also how important it is that we share our successes and learn from others. Perhaps you can chalk it up to British self-deprecation and cynicism, but we spend little time celebrating our apprenticeship system, particularly in comparison to Germany and Switzerland – countries that unabashedly champion their apprenticeship models abroad.

Since we were established in 1878, the City & Guilds Group has helped people get the skills they need to contribute to economic prosperity and growth. We help them to get into a job, progress on the job, and move into the next job. Through the years, we have witnessed many, many changes that have affected the employment market - from war, to industrialisation, to women increasingly participating in the workforce. Today, we operate in more than 100 countries around the world, which means I spend a considerable amount of my time travelling and speaking to employers, learners and government officials about vocational education and, of course, apprenticeships.

On my travels I've heard loud and clear that the UK is certainly not the only country wrestling with how to develop the right skills for the future. Nor are we the only country looking very carefully at our apprenticeship system. This is because the trends that are creating these skills challenges are global. Skills gaps are being exacerbated by globalisation, the changing nature of work, the evolving needs of industry and, of course, an ageing developed world coupled with booming youth populations in the developing world.

This discussion about how to develop the right skills is being had, in one form or another, all across the world and rightly so, because these are big and important questions. What will the future labour force look like? What skills will be needed in the coming decades? Where will growth come

157

from? What should we be doing now to prepare for the future? This is why I believe sharing knowledge on what's working and what's not, for both apprenticeships and vocational education and training more broadly, has never been more crucial.

Apprenticeships in the UK – out of the cold

It is sometimes easy to forget that ten years ago, no one was talking about apprenticeships in this country. These days, rarely a day goes by without a story about apprenticeships in a national paper. This 'apprenticeship renaissance' has been driven by businesses looking for work-ready employees, and by politicians from every party picking up apprenticeships as a smart policy choice that incidentally plays well in the polls. The change started when the Labour Government began to reinvigorate the brand in the early 2000s – but still Tony Blair's call for 50 per cent of 18 year olds to go to university lingered, and reinforced the perception that apprenticeships were 'second best'.

Facing high youth unemployment and economic decline during the global financial crisis, the Coalition Government accelerated the focus on apprenticeships through a concerted campaign to promote them to young people, parents, businesses and teachers across the country. During the 2015 General Election, every party had a pledge for apprenticeships – something that I called the 'apprenticeship arms race' - as politicians tried to one-up each other with promises about apprenticeships. Most recently of course, David Cameron has pledged to deliver three million apprenticeships by 2020.

I think it is fair to say that apprenticeships have finally come in from the cold, to be understood as a valid and vital educational choice in the UK. This is encouraging because apprenticeships make good economic sense. Our research with the Cebr[4] found that a 10 per cent increase in vocational education enrolment by upper secondary school pupils would lead to a 1.5 percentage point drop in the youth unemployment rate. The Cebr predicted that a 1 per cent increase in the UK's vocational skills base would uplift GDP by £163 billion in a decade's time – and of course, while apprenticeships are not the only way to develop these skills, they certainly

---

[2] City & Guilds. (6 July 2015) 'Chancellor must support vocational education to boost productivity', City & Guilds.com.

play a vital role and provide a strong contribution to economic growth. In fact, our Cebr research found the annual productivity gains from training an apprentice are £10,280 per year per apprentice.

How the UK apprenticeship system stacks up
In summer 2015, a few of my colleagues and I went to Washington D.C. to talk about apprenticeships. We spoke to policymakers on Capitol Hill, think tanks, charitable foundations and academics about the aspects of the British apprenticeship approach that could be adopted across the pond. Time and again, we heard about how impressed they were by British apprenticeships. And it's not just the Americans; I've had similar conversations in India, South Africa, Malaysia and the United Arab Emirates.

One of the key strengths of our apprenticeship system is the vast number of pathways we have – with apprenticeship frameworks for everything from robotics to wind-turbines to social media. The Americans were surprised to hear that companies like Microsoft and PwC offer apprenticeships as a key part of their talent strategies. The availability of apprenticeships beyond the perceived core vocational subjects is something to be proud of and provides a distinct advantage of our apprenticeship system over many others – particularly the well-promoted 'dual systems' in Germany and Switzerland.

Comparatively, our apprenticeship system also offers more flexible options for career progression than many others. In my opinion, the Dutch model, which enables young people to opt for interchangeable pathways that don't restrict what they do at eighteen, is still ahead of us here. But we have been moving towards more flexibility and transferability between academic and vocational pathways in recent years. This includes expanding access to technical options via university technical colleges and career colleges – and the City & Guilds TechBac – which allow progression into apprenticeships and from apprenticeships into university or employment.

Gender parity is another key strength of British apprenticeships. According to the latest statistical first release from the Department for Business, Innovation and Skills, 49.5 per cent of English apprenticeship starts in the 2014/15 academic year were female.[5] This is much higher than the 34 per cent in Australia, 17 per cent in Canada, 7 per cent in the US and just 2 per cent in Ireland. England also ranks highly in terms of apprenticeships for people with disabilities: the proportion of apprentices with a physical or learning disability is 8 per cent in England; this compares to only 2.2 per cent in Germany and 1.5 per cent in Australia.[6]

The increasing level of employer engagement in our apprenticeship system is also encouraging. The development of the Trailblazer programme is bringing together large and small employers from key industries to design the standards for apprenticeships. We are making strides towards a system that works as it should. In fact, the US has developed an employer engagement programme for apprenticeships called LEADERS, based on the Trailblazer model. It is still early days, of course, but it is encouraging to see innovation that results in a more employer-led apprenticeship system.

All that being said, there are also areas where we can learn and improve our apprenticeship system, based on what is working in other countries. The UK vocational education system has been plagued by extensive change and instability, with sixty-one secretaries of state and more than thirteen related major acts of parliament in just three decades[7]. This has created a situation where colleges, training providers and employers have to constantly adapt and evolve based on ever-changing policy. There have been plenty of good ideas, but they have not been given time to be systematically tested and improved before being scrapped and politicians moved on to the next big policy announcement. This constant change has undoubtedly affected the quality and take-up of apprenticeships in the UK.

---

[3] Statistical First Release, Skills Funding Agency, 18 November 2015.

[4] Towards a model apprenticeship framework: a comparative analysis of national apprenticeship systems, International Labour Organisation, 30 January 2014.

[5] Sense & Instability: three decades of change in skills and employment policy, The City & Guilds Group, 13 October 2014.

Demos recently reported that there are just eleven apprentices for every 1,000 employees in England, compared with thirty-nine in Australia, forty in Germany and forty-three in Switzerland. Yet the same report found that 54 per cent of 16- to 24-year-olds in England say they would take an apprenticeship if one were available.[8] The demand is there, but the supply of quality apprenticeships needs to be improved. I don't think it is a huge leap to assume that because Germany and Switzerland benefit from a more steady apprenticeship system, it is easier for employers to engage with it.

Nevertheless, my point is that there is no perfect apprenticeship system. As a country with a strong commitment to apprenticeships today, and a long legacy of apprenticeships dating back hundreds of years, there is no question that we have insights and learning to share with the rest of the world. It's important that we appreciate what is working and what is successful in our current approach. And as we continue to invest in and develop our apprenticeship system, we need to learn from what is working elsewhere too.

Building skills in a mobile, global labour market
'A generation ago a British Prime Minister had to worry about the global arms race. Today a British Prime Minister has to worry about the global skills race' – Gordon Brown, 2008[9]
You may be wondering why, if our apprenticeship system is working rather well compared to other countries, we would want to share our secrets? If we were Machiavellian about it, couldn't our apprenticeship system become a competitive advantage? The answer is maybe, in the short term - but it would not be in our interests in the longer term.

This is simply because the labour market is becoming increasingly global. The International Labour Organisation estimates there are 232 million international migrants today.[10] Since 1990, this number has increased by 65 per cent (53 million) in the Global North, and by 34 per cent (24 million) in

---

[6] 'Up to the job', *Demos*, 26 February 2014.
[7] 'Brown pushes apprenticeships in global "skills race"', *Politics.co.uk*, 28 January 2008.
[8] 'Labour migration', www.ilo.org 17 December 2015.

the Global South.[11] As the global economy becomes integrated, people are becoming increasingly mobile. And findings from the Boston Consulting Group (BCG)[12] make it clear that this is a trend people are on-board with; almost two thirds of job seekers around the world told BCG they would be willing to move abroad for work.

The reality of migration, which is not covered as often, is that while people are migrating to Britain, the British diaspora numbers some five million people, located all across the world. This makes it the largest diaspora of all rich countries, and is roughly the same size as Scotland's population.[13] Better apprenticeships across the world, both entry-level and higher level, will not only improve the quality of technical skills for people who migrate to the UK but also for those who leave, work and learn abroad, and who eventually come back.

In a more interconnected, interdependent world, the challenge ahead is to develop and provide training that acknowledges the global nature of the future labour market and offers recognisable transnational standards. In fact, at the City & Guilds Group, we are seeing an emerging movement around transnational skills standards. It's early days yet, but I do think there is a growing recognition (based on business and economic expediency in many ways) of why qualifications need to be portable and skills standards need to be global. Apprenticeships are no exception to this.

Many countries, such as South Korea, India, Singapore, China and the member nations of the EU, are actively seeking to formalise and validate their systems of learning and awarding occupational credentials, especially industry-recognised certificates. The goal is to ensure the certificates their citizens earn are 'portable and stackable' (i.e. widely recognised), accepted nationally or internationally, and offer clear and consistent pathways from student or apprentice to trained worker and,

---

[9] Global Migration Trends: an overview, International Organisation For Migration, 18 December 2014.

[10] Decoding Global Talent, Boston Consulting Group, 3 October 2014.

[11] 'And don't come back', The Economist, 7 August 2014.

eventually, technical mastery of specific skills. If this does happen, it will benefit British workers, both in the UK and those based abroad.

The make-up of the workforce is also changing, with people working for longer and more women with children staying in or re-joining it. According to UKCES research[14] by 2030, we'll have four generations in the workforce at any one time. In light of anticipated demographic changes such as longer life expectancy and an ageing population, pooling our expertise becomes especially key. Alongside huge levels of cross-border migration, the global population is surging; according to the World Economic Forum, 600 million jobs need to be created over the next decade alone. By 2018, about 215 million people worldwide are expected to be unemployed.

Crucially, population growth is not balanced, presenting a vast employment challenge. A quarter of those aged between 15 and 24 live in developing countries, while developed populations are ageing; globally, the number of over 60s is expected to more than double to over two billion in 2050.

Compare the UK and India. The latest ONS projections suggest the UK population will reach 70 million by mid-2027, but also that by mid-2039, more than one in twelve of us will be aged 80 or over[15]. Meanwhile India is predicted to have the largest workforce in the world by 2025,[16] amounting to a potential surplus of around 47 million skilled workers. Simply put, if you think we have a mismatch of skilled labour to available jobs now, then, as the Americans say, 'you ain't seen nothing yet'.

More people necessitate more jobs, and it is incumbent upon us to lay the groundwork and make sure that those jobs are in the right places, the right sectors, and that the next generation is trained and ready to fill them. Even now, we know that a big reason unemployment falls so heavily on the

---

[12] The Future of Work, UK Commission for Employment and Skills, 3 March 2014.

[13] National population projections, 2014-based Statistical Bulletin, ONS, 29 October 2015.

[14] 'Reaping India's promised demographic dividend', EY, January 2014.

younger generation - around 40 per cent of those out of work around the world are under the age of 25 - is the lack of relevant, high-quality skills amongst young people. How much worse will that be in twenty years' time if we fail to act?

In our interdependent global economy, no country will be able to escape the demographic challenges of the future. But no country needs to start from scratch either. Instead of attempting to navigate the skills questions of tomorrow as individual actors, governments and policymakers need to work together. Failing to assess and, where appropriate, replicate the solutions applied elsewhere is as senseless as us failing to consider the lessons from our various experiences.

Apprenticeships are not a silver bullet, but a sensible place to start
It is important to clarify that I do not believe that sharing learning on vocational education should start and end with apprenticeship systems. I also do not believe that improving access to and the quality of apprenticeship provision alone will somehow magically produce perfect solutions to the huge challenges created by the evolving labour market. However, I do believe apprenticeships are a great place to start. This is partially because of the political and public profile they are enjoying in so many countries, but also because, by virtue of being work-based, they provide a fantastic vehicle for developing the most up-to-date and relevant technical skills that businesses need.

Over the next few years, we need to focus on identifying how we can best share knowledge, and what aspects of this country's skills system – from apprenticeships to technical colleges – are transferable.

This is no mean feat. But there are ways. Technology is changing the way we work, but it can also be harnessed to share knowledge across the world, breaking down geographic and cultural barriers.

Multinational businesses are well positioned to play a key role too, through their global footprint and their supply chains. Organisations with positive records of supporting training in the UK can inspire their colleagues abroad by communicating the value of good quality skills education to a business. Indeed we are already seeing this start to happen with some of our employer customers.

Meanwhile, as they seek to achieve the three million apprenticeship starts target, the government has a responsibility to look beyond our borders at what the UK can learn from other countries. By the same token, they must take pride in celebrating the inherent strengths of the UK system as a way to showcase and export excellence and ensure expertise is shared.

And other actors in the education and skills world, the City & Guilds Group included, must look to forge links around the world. Already, we are proactively working to establish transnational apprenticeship standards in key sectors and we invite employer partners to discuss this with us.

Skills shortages won't respect national borders now or in the future. It's time we all got better at sharing best practice. It's time to raise our voices about the lessons we can offer to other countries – and open our ears to them.

15.

TAKING APPRENTICESHIPS TO THOSE WHO WILL BENEFIT MOST

Mike Thompson
Director Early Careers, Barclays

———————

This chapter shares the journey that Barclays has been on since launching its first ever apprenticeship programme in April 2012. It will look at the drivers behind the programme, how it has developed and evolved and how it has impacted both the organisation and the individuals who have gone through the programme. It will also look at the internal and external factors that have impacted our programme and in particular give a perspective on the reforms of the apprenticeship system from an employer's perspective.

In November 2015, Barclays recruited it 2,600[th] apprentice and is now one of the largest employers of apprentices in the UK. Nearly 200 apprentices recently attended a graduation event at the House of Commons alongside proud line managers and parents. A truly special day, and very symbolic of how far the programme has come since beginning in 2012 with just twelve apprentices.

The common bond between all apprentices graduating was the fact that prior to joining Barclays they were all classified as being NEET (Not in Education, Employment or Training), had struggled to find employment, and had often been written off by society as failures. This is what makes our programme unique and those who come through it truly special to our organisation. Many have overcome adversity and are now thriving and successful because they are Barclays Apprentices. This is a legacy of which, as a company, we are truly proud.

So how did the apprenticeship programme at Barclays begin?
At the beginning of 2012, all the major UK banks received a letter from the Mayor of London, Boris Johnson, challenging them on what the banking sector could do to help tackle rising youth unemployment and help grow apprenticeships.

At the time, youth unemployment had topped over one million in the UK, and the term NEET had been coined. The Work Foundation had published its *Missing Millions* report outlining the long term risks to the economy and young people if rising unemployment levels were not tackled. So the external context and environment when the letter was received was clear and pretty gloomy.

At the time of receipt of the letter, Barclays had only 300 employees under the age of 21 in the UK and, in parts of our business, large sections of the workforce were approaching retirement age. In an organisation facing pressure to modernise and adapt to the new digital age, the lack of young people with digital skills was a concern.

The letter was timely and made our response a very easy one. Something had to be done, not just to tackle a broad societal issue, but also to tackle a challenging demographic.

The very clear direction that flowed from the follow up was that our response had to be both meaningful in scale but also meaningful in impact, to tackle both the issue of youth unemployment, and our own HR challenges. In other words, we had to look at how we could reach those young people who needed the greatest help and how we could do this in a way that led to long-term opportunity not simply short-term 'fixes'. We also had to do this at a scale that would quickly re-balance the shape of our workforce. We set a target of 1,000 apprentices by the end of 2013 and then set about working out how to climb what was a proverbial HR Mount Everest.

Having never set up nor run an apprenticeship programme before, not least one that targets long-term unemployed young people, we set about learning fast how the system works and how it should be geared up to support our target audience. Our early insights and experiences demonstrated some of the strengths but also some inherent weaknesses in the system.

The fundamental strength we found was that the significant government support and funding that lies behind apprenticeships opened so many doors to support, guidance and advice and allowed us to get our programme up and running quickly and easily.

But what was also very quickly revealed was a system that was not driven by employers nor one that provided a great deal of flexibility or choice to employers. The frameworks presented to us were outdated and complex; the funding rules as to who could benefit from apprenticeships were complex to understand and administer. Thankfully, recent reforms have made great progress in addressing these issues. I will touch on more on this later.

One early challenge we encountered was the provision to support the pre-employment training our young recruits required to make them job ready and successful at interview. It was patchy and limited with very few providers in the marketplace having the scale or capability to prepare young people for work wherever in the UK we wanted to employ them.

This lack of quality scale national provision remains a weakness in the system for employers such as Barclays where our training requirement stretches across the whole of the UK. The system is very well geared up for local provision of local skills via FE colleges but this isn't replicated nationally. With the new government target to achieve a goal of three million apprenticeships, it must be of concern as to whether there is enough quality delivery capacity in the current system.

The early focus of our scheme was on bringing long-term unemployed young people into our high volume entry-level roles in branch banking and telephone banking. When we researched the target audience, we very quickly identified the need to completely change our approach and criteria to hiring.

Our traditional requirements of hiring against previous work experience and academic performance would simply not work for a group where academic attainment was often limited or non–existent, and where few young people had any meaningful work experience.

So we tore up the rulebook and started again. We removed all academic criteria and need for previous experience, and focused on attitude, aptitude, and motivation to work for us and develop. We moved from a world of online applications and screening to a more tailored 'hand held' supportive recruitment model.

This model was designed to prepare the young person for success at interview and in their first job, whilst recognising that they were far from the 'finished article'. Over time, this up-front investment in developing employability skills has grown to include both classroom-based training and work experience and now forms part of what is the biggest traineeship programme in the UK.

Being able to work with young people intensively in advance of interview allows us to create an early bond and identify those that are a very good fit for Barclays. It also enables us to support those who aren't right for us and help them into jobs with clients, suppliers and other companies. It has led to significant improvements in retention levels in the first twelve months where now we lose less than 10 per cent of new hires versus over 25 per cent previously.

Recruiting 1,000 young people over a short time frame brings with it both challenges and opportunities. Much of our focus in the early days, for instance, was acclimatising our management team to the new type of recruit they would be working with, and creating a supportive, nurturing environment. This proved less challenging than we might have imagined as we discovered that many of our middle and senior managers had joined Barclays through the YTS (Youth Training Scheme) programme and were therefore strong advocates of a vocational pathway.

Perhaps the biggest challenge, beyond the perception/environmental issues that inevitably arise when making a change of this magnitude, was managing the expectations of the apprentices themselves. Inevitably, once their apprenticeship is successfully completed and a full time job secured, the next question we receive is 'what next?' The ambition to progress both in career terms as well as academically is huge amongst our apprentices and we have had to work hard to build pathways to meet their needs. We have gone from two high volume entry-level programmes to nearly twelve progression/higher apprenticeship pathways.

Our goals in creating these new programmes have been simple and consistent. The first goal has been to give any apprentice the opportunity to obtain as high a level of academic attainment as possible but with a particular view to offering those who have not previously succeeded academically a 'second chance'.

The second goal is to use this development to foster progression amongst apprentices in terms of salary and career. Whilst still early days, evidence suggests that we are succeeding against both goals, with a number of apprentices already moving onto our degree programme and progressing to management positions or more senior roles having joined Barclays with limited or no qualifications.

A striking example of this is Constance Nafuna, who joined Barclays having been job seeking for over a year. Constance had come through our immigration system (and detention) from Africa with no qualifications or work experience and struggled to get an interview even for basic entry-level roles. Constance was desperate for any type of work and when, through the job centre, it was suggested she could apply for Barclays she laughed at the adviser. If she couldn't get work cleaning dishes or waiting tables why would a bank offer her a job?

Constance came through our pre-employment programme and stood out so much at interview that her recruiting manager stated that she was the best candidate he had interviewed in his time at Barclays. Constance went on to work at one of our London branches, where not only did she learn to master English but also the local languages of the predominantly Indian and Bangladeshi community, making her the 'go to' community banker for these customers. Having won numerous awards for service and performance, Constance completed her level 2 apprenticeship including passing her maths and English tests before moving onto a level 3 apprenticeship in Business Banking.

Constance excelled in this role, supporting many local small and medium sized businesses and quickly becoming a role model banker in her area.

This led on rapidly to Constance applying for our degree level Higher Apprenticeship in Relationship Management for which she was one of only six successful candidates from a pool of 4,000 and she recently moved into our Investment Bank dealing with the trading needs of some of our largest global clients. A powerful illustration of the value of apprenticeships.

Our belief that there was huge untapped talent amongst the NEET population is borne out by this and many similar success stories. It re-emphasised our perspective that academic achievement is a poor predictor of talent and that we were right to scrap academic criteria from our recruitment practices and focus far more on the attitude, aptitude and motivation of the individual.

Constance's progression would not have been possible, however, without the wide-ranging reforms of the government's Trailblazer programme. These have allowed the financial services sector to create a range of new 'standards' that allow progression to higher levels of skill and knowledge. The first of these was the Relationship Management Higher Apprenticeship that Constance is now studying. These qualifications have been developed together with our professional bodies, and have played a substantial part in raising the bar of our apprenticeships by aligning them to industry recognised qualifications.

Despite the Trailblazer process at times being slow, fundamentally it has instigated a profound and far-reaching set of reforms that will allow employers to take ownership of the skills development of their workforces. This can only be a positive thing.

Perhaps the greatest benefit of the reforms was the removal of the age cap on funding which meant we were able to look at offering apprenticeships to any age group and led to the launch of our 'Bolder Apprenticeship' programme targeting older, long-term unemployed recruits. This programme targets job seekers who are over the age of 24 and have been over twelve months unemployed.

The growth in long-term unemployment amongst the older population is a big concern for Barclays and is something we as an organisation want to do something about. It has an impact on our customers and colleagues in every community in which we operate.

The opportunity to address the issue via an apprenticeship programme, and be the first in the UK to do so, was appealing to us, not least because we could again be the first to tap into a talent pool as we had done with young people. Whilst this programme is relatively new, having launched in Summer 2015, it is already proving very popular within the business and allowing us to deliver talent that our business values and will nurture.

Over the past three and a half years, we have learnt many valuable lessons. The most important of these is to challenge traditional thinking both within an organisation and in wider society. Over the past three decades in the UK, recruitment practice has evolved to the point where it has become very difficult for many talented people to access the workplace. This could be because of their age, disability, and lack of experience or qualifications. Having an apprenticeship programme provides the opportunity to create pathways for all and, in so doing, help build more diverse organisations and tackle societal issues.

To achieve these goals, however, the second lesson we have learned is that it is important to have a strong working knowledge of the system and understand how to maximise it for the benefit of your business. The rules have been simplified and new standards created which mean that apprenticeships are more flexible than they were and the system more employer-focused. This flexibility allows you to deliver a development journey that is tailored for your business, and qualifications that are relevant and add value. Having strong and knowledgeable delivery partners who can help you navigate the landscape is vital.

Finally, it is important to recognise that apprenticeships are a long-term investment in talent and that often it can take years for someone to develop to their full potential. In modern business, results are often demanded quickly, and it is easy to take short-term recruitment and development decisions. By taking a longer-term view, the benefits are greater loyalty from the individual and often higher business performance and engagement.

16.
THE POWER OF PARTNERSHIP WORKING:
A CASE STUDY

Roger Peace
CEO learndirect

This chapter looks at the approach that we in learndirect take towards ensuring more employers take on apprentices for the long-term and ensure a high quality training experience. It shares some of our experiences of what attracts employers to work with an experienced and respected training provider.

We've all seen the headlines about skills shortages, the ageing workforce and employers unable to fill vacancies. It seems obvious that for the UK to improve its productivity, it needs to invest in training and development.

We're supportive of the government's drive to get more apprentices in this parliament than it has before, and believe apprenticeships are critical to improving the UK's productivity and competiveness on a world stage. As the largest apprenticeship provider in the UK, we can be at the forefront of this drive to make apprenticeships front and central in improving the nation's skills.

However, we need to make sure it doesn't just become a numbers game – we all have a responsibility to focus on quality, developing and delivering programmes which are needed and that will improve the productivity of the UK.

We're very proud of the contribution learndirect makes each and every day to identifying, inspiring and developing young talent. Pivotal to this is how we approach our relationships with our employer customers.

The way we work with them is about true partnership – common goals and shared ambitions sit at the heart of our employer relationships. We invest time in getting to know how a business ticks, its culture and values, and what it feels like to be an employee. Only then can we develop apprenticeship programmes that add value, and dovetail with organisation-wide talent management strategies.

Long-term relationships reap rewards – if we get under the skin of an organisation, if we understand its rhythm and pace, as well as its challenges and opportunities – then together we can evolve our delivery. A perfect example of this is some work in the financial services sector where the programme we deliver has grown from eleven to twenty-two frameworks, and expanded to include higher apprenticeships.

Our partnership approach cuts across all aspects of our work with major customers – and I've outlined some learning which can be applied to developing and delivering future programmes.

1. Programme design
learndirect is a key player in the review and development of vocational education – and is involved in a number of groups and consulted on changes affecting the sector. For example, we contributed to the Richard Review and the approaches currently adopted within the Trailblazer approach. Dereth Wood, our Director of Learning, Policy and Strategy was also appointed by BIS to be a Commissioner on the review of Adult and Vocational Teaching and Learning (CAVTL, 2013).
This insight and understanding underpins our work with employers.

Considerations:
*   In an ever-changing policy and funding landscape, it is important you keep on top of the latest policy and funding requirements, such as Trailblazer standards, so you can interpret when designing and developing your programme with your employer. As your delivery partner, your employer can support you with this. For example, members of the learndirect team have been asked to provide advice and consultancy to a number of groups on the development of the standards and assessment models within trailblazers. We work with

our employer partners to get them ready for the introduction of new standards – developing programmes of learning which take them along the journey from SASE Framework to Trailblazer standard.

- It may seem obvious but you need to invest in your relationships with each and every employer delivery partner. This is important, particularly at the outset of a relationship when you're developing a new programme, often looking at what else is going on in the business, organisational objectives and working practices. Within learndirect, in the first weeks of beginning a new contract, we invest in a familiarisation programme – and this continuous exploration and collaboration continues throughout the life of a contract.

Example: a hybrid Trailblazer programme for a major retailer

Due to the delay in the release of the Retail Trailblazer, learndirect developed a hybrid apprenticeship for stores in England. The twelve month hybrid programme uses the trailblazer methodology, allowing staff to familiarise themselves with the new method of apprenticeship delivery. The new methodology comprises participants' learning and building competencies to agreed milestones for the first ten months of the programme and then completing assessments in the final two months of the programme, in line with trailblazer expectations.

As we develop a hybrid delivery model, encompassing the legislative requirements of the SASE framework, we will, over time as agreed with the customer, migrate away from SASE units and integrate the new trailblazer units to ensure a smooth path towards new standards.

2. Delivery

Size and scale matters to many employers who are looking for the ability to deliver apprenticeships in a consistent manner the length and breadth of the UK, in the communities in which many of our employer partners operate. Our methodology has enabled us to effectively deliver apprenticeships to more than 20,000 learners each year. Our approach to delivery includes the application of blended learning methods which offer continued support, guidance and re-assessment, meaning that learner success is maximised.

Considerations:

- Assessment of existing knowledge, skills and prior achievement against those required to meet the needs of their job role
- Provision for structured learning (both on-the-job and off-the-job) to enable the development of the relevant skills and knowledge, providing employees facing difficulties with the support, time and flexibility they need to overcome these difficulties in their qualification
- Assessment and accreditation of existing and new abilities to the competencies of the qualification and therefore the job role

Example: The Co-operative Group
learndirect delivers an apprenticeship programme on behalf of The Co-operative Group, one of the world's largest consumer co-operatives which is owned by more than eight million members. To date more than 2,000 apprenticeships have been delivered across England, Scotland, Wales and Northern Ireland.

We initially helped to develop and implement a completely new apprenticeship framework, jointly approaching a range of sector skills councils before finding the best fit with Skillsmart Retail as part of their specialist retail remit. The development of the qualification was based on The Co-operative's internal training best practice processes and procedures which we mapped to existing national occupational standards.

Currently, a range of apprenticeship opportunities is offered through learndirect nationally across the Group's diverse range of businesses including banking and legal in sectors such as retail, management, business administration, customer service, funeral operations and IT.

3. Quality
Quality assurance and continuous improvement are central priorities for learndirect and really matter to employers. Employers frequently use externally validated quality marks as markers of quality. This is particularly true with Ofsted, and the importance of our Grade 2 rating with a Grade 1 for Leadership and Management. In particular, Ofsted noted 'inspirational leadership at all levels and exceptionally good team working' and said 'learners clearly enjoy their learning and the learndirect

experience, showing significant improvements in self-confidence and the ability to learn'.

At learndirect, we work with our customers to develop an overarching quality improvement cycle. This underpins a programme of observation of teaching, learning and formative assessment.

To enable a high quality programme, you should consider the following:
- Overall quality strategy
- Sign-up procedure
- Initial assessment procedure
- Assessment/review procedure
- Observation procedure
- Quality assurance process, e.g. internal verification procedure
- Supporting in the production of self-assessment report

4. Apprentice-focus
Although apprenticeship programmes need to be tied back to organisational priorities and objectives, you should keep your apprentices in mind as you develop and implement your programme. We know if we approach delivery from the learner's perspective, providing them with a quality experience, they're more likely to persist with their learning and therefore achieve success. Underpinning this should be robust quality processes, which are regularly reviewed and evaluated, and take learner and customer satisfaction and feedback into account.

Some considerations:
- Your apprentices may be young people with limited or no experience of a working environment. Think about what support you need to provide to them or to their line managers
- Think about how you can make the most of the latest – and familiar – technologies in your programme
- Involve your apprentices in the development and evaluation of your programme. Undertake regular satisfaction surveys or involve in them in focus groups
- Acknowledge and celebrate success

Example: Lois McClure, The Co-operative Group

When Lois McClure left school, she had her heart set on going to university. But during her second year at college, she realised it wasn't the path for her and began to look into doing an apprenticeship. After applying for a learndirect apprenticeship in business administration at The Co-operative, she started with the company in 2012 and has since progressed from a level 2 to a level 3 apprenticeship. Lois secured a permanent job in the team and won Intermediate Apprentice of the Year at the National Apprenticeship Awards in 2014:

> In April 2013, I was appointed to the first ever Co-operative Young Members' Board which is a board of 15 young people across the UK brought together to help the Co-operative gain younger members and to develop products and services for young people. I was also elected as Vice Chair.

5. Talent strategy beyond apprenticeships

Employers value services beyond training and are very pleased if they can get a range of high quality services from one supplier. While apprenticeships are at the heart of what we do at learndirect, they're not the only thing we offer. If you put apprenticeships into the mix with our range of employment related services, English & maths delivery, traineeships and more, employers benefit from solutions which can be sized and shaped according to needs – whether that be location, workforce characteristics or seasonal demands.

Consider your apprenticeship programme as a pivotal part of your overall talent management strategy. You may want to explore:

- Traineeships as a route to get young people into your business; helping them prepare for an apprenticeship or employment
- Career progression for your employees – your apprenticeship programme can be a feeder for higher, degree-level qualifications, including apprenticeships
- Pre-employment training as part of your community or corporate social responsibility commitments
- Work experience or internships for school and college students

Conclusions

We have found that attracting employers to a single product has limited value for them and for us. It takes time and resources to tackle the often deep-rooted issues with which an employer is grappling. We therefore seek to impress employers with our deep interest in their workforce development issues and our readiness to work with them for the long term.

We are the biggest training provider. This means that we have the scale and expertise to meet most needs but we also recognise the importance of knowing where we need additional skills and expertise. We are therefore very happy to work in partnership with others in the best interests of the employer.

We believe that remaining close to policy development is critical to ensuring we are able to support the journey of our employers so that this aligns well with the direction of government policy and funding. Sharing best practice, through mechanisms such as the learndirect Business Exchange, also adds value to employers.

We are committed to helping employers understand the new levy arrangements and to ensure they are able to maximise the benefits of these new arrangements for themselves and for their future apprentices.

Reference
CAVTL. (2013) *It's about work...Excellent adult vocational teaching and learning*.

PART THREE

APPRENTICESHIPS AND BUSINESS

INTRODUCTION TO PART THREE

David Way CBE

If we are to achieve three million more apprenticeships then the number of employers employing apprentices has to increase sharply. This part of the book looks at what could bring about that change.

It starts with an explanation of the appeal of apprenticeships to business, some of the barriers to further expansion and how these might be removed.

I was fortunate enough to chair a series of employer events in 2015 hosted by local chambers of commerce. Many of the issues that the employers raised, such as demystification of the skills system are picked up in this part of the book and in the following part concerning navigating the apprenticeship and skills systems.

Chambers of commerce are strong supporters of skills and are highly influential on UK businesses with more than 5,000,000 employees. John Longworth was the Director General of the British Chambers of Commerce from 2011 to 2016. John shares his insights from his close engagement with employers across the UK. He identifies the importance of clear routes for young people to see and follow into apprenticeships from school and the greater help needed by SMEs as critical barriers to remove, while quality is the key to sustained growth.

It was my great pleasure to work closely with Jason Holt on his report on how government could make apprenticeships more attractive to small businesses. He brought his small business owner perspective to the task and produced a report that was admirably practical and a very good blueprint to follow. He came up with a range of fresh ideas such as a Trip

Advisor-style facility for employers looking to find the right training provider. Since leaving the NAS, I have attended many presentations to employers as an Apprenticeship Ambassador for learndirect. I have seen the truth and relevance of Jason's conclusions and recommendations. Employers, especially SMEs need much more help if they are to feel confident in taking on apprentices.

Jason is now the government's Ambassador for encouraging SMEs to take on apprentices. He is the Chief Executive of the Holts Group and Founder of the Holt Academy. Jason was awarded the CBE in the 2015 New Year Honours list for his services to apprenticeships.

In his chapter, Jason revisits the findings in his report to underline those recommendations that he feels will make the most difference. These include the need to provide more tailored evidence on the financial returns from apprenticeships to which SMEs can relate, backed up by messages in language they understand as small businesses. Jason is optimistic about current reforms and achieving current ambitions but makes clear what more needs to happen to take SMEs on this journey.

Industrial Partnerships are the latest organisational creation to take responsibility for skills development in a sector. Eight Industrial Partnerships were created as part of the Employer Ownership of Skills pilot overseen by the UK Commission for Employment and Skills. Initial reviews have concluded that the Partnerships are starting to deliver ambitious programmes, but they are now expected to be self-funding.

This chapter looks at the added value and potential of Industrial Partnerships from the perspective of the critically important energy and utilities sector. Its skills will be essential if the government's multi-billion pound plans for the National Infrastructure Plan are to be delivered.

Steve Holliday is experienced not only in business but also in working with government. He has been the CEO of National Grid as well as the Chairman of the Energy and Efficiency Industrial Partnership and vice-chair of Business in the Community and the Careers and Enterprise Company. He sets out why Industrial Partnerships need and deserve the unequivocal backing of government and sets out what is being achieved to forge and deliver a long-term skills strategy to, at last, overcome long-standing issues.

The next chapter benefits from contributions from three strong advocates of skills and quality training - Frances O'Grady, the General Secretary of the TUC, Tom Wilson, the Director of Unionlearn until 2015 and Matthew Creagh, TUC Apprenticeships Policy Officer. Trades Unions have been very supportive of training in the UK. I first saw their value in supporting efforts to tackle literacy and numeracy. I have subsequently had many productive discussions about their support for apprenticeships

This chapter looks at the important question of how to raise the status of apprentices. It underlines the importance of apprenticeships that are sufficiently broad to open up career opportunities rather than being restricted to a single job. Alongside this is the importance of young people receiving better information about career prospects, not least so that young women can see the implications of their over-representation in poorly paid sectors. Progression is a further issue explored here with thoughts about opportunities for everyone who has the ability to be able to be upskilled to at least level 3. Finally the chapter explores various international approaches to ensuring that apprentices are suitably rewarded as they move through and beyond their apprenticeship.

Apprenticeships in the UK will get a major boost if they contribute significant levels of skills to the major infrastructure projects planned that will invest over £400bn over the next decade or so. Initiatives such as Crossrail in London have shown what can be achieved. However the potential benefits to apprenticeships from current and future projects such as HS2 are substantial.

Neil Robertson is excellently placed to set out how the UK and apprentices can benefit. Having been involved in drawing up the National Infrastructure Plan, Neil is now the CEO for the National Skills Academy for Rail Engineering. Neil examines supply chain initiatives and voluntary approaches to ensuring apprenticeships through procurement, such as The 5% Club.

The next chapter concerns arguably the most important question for businesses. How do I get the best out of the apprenticeship? How can

185

employers ensure that their apprenticeship programmes are productive and benefit the business and the individual?

Andy Smyth shares his experiences about how apprentices can be supported to be more productive at work, drawing particularly on his experiences with TUI, but also as chair of the Industry Skills Board for City & Guilds. Andy underlines the importance of doing the basics right and not being distracted from developing what the business needs. He underlines what made a difference for TUI, including using mentors and buddies.

The concept of 'employer ownership' has been the single most influential driver of apprenticeship policy and practice in recent years. Employers are increasingly 'in the driving seat' in all policy documents, especially when it comes to setting the standards required by employers from those completing an apprenticeship and directing public funding that sits alongside their own company investment.

The UK Commission for Employment and Skills has been in the vanguard of employer ownership, launching a series of important pilots in 2012. Simon Perryman was the Executive Director at UKCES with responsibility for the Employer Ownership pilots until his departure in 2015. Simon recounts the history of attempts to engage employers in the development of skills, a challenge beyond many of our international competitors who look enviously at UK arrangements. He underlines the importance of linking industrial strategy to the delivery of skills and shares important lessons learnt from the Employer Ownership pilots.

There is a growing momentum behind the notion of a clearing-house arrangement for apprenticeships that compares with that currently run centrally for those applying to university. Adopting such an approach could help raise the status of apprenticeships and conjures up a wonderful image of talented young people mixing top universities with the best employers apprenticeships on their preference forms: BT, Oxford, Rolls-Royce, Bristol, PwC, Edinburgh.

A more modest but nevertheless important approach has been explored and implemented across a number of employers who typically were over-subscribed with interest from individuals for their apprenticeship opportunities but did not want to lose the skills and interest of the applicants to the sector or to their supply chain.

One of these employers is Siemens, and Martin Hottass provides a chapter in which he shares their experiences of trying to make a clearing-house system work for them in the engineering sector in great need of replacement talent for an ageing workforce. Martin has been one of the main advocates and drivers of the clearing-house approach, preferring to redirect rather than reject talented and ambitious young people. Having introduced this approach within power companies and wider, the energy and utilities sector increasingly competes as a sector to recruit entry level talent rather than simply competing against each other. The results have benefited the sector, businesses and their supply chains.

A further chapter follows EDF's experiences of attracting and recruiting technician apprentices and how they have invested heavily in new apprenticeships that give recruits wide-ranging experiences that have produced outstanding results.

The final chapter in this section of the book looks beyond the UK to the much-admired Swiss system. Dominik Furgler is the current Swiss Ambassador in London and he continues the Embassy's collaboration with the UK to understand, appreciate and learn from our common interests in apprenticeships and supporting business.

This chapter summarises the Swiss system for apprenticeships. It describes the individual and collective responsibility taken by employers to absorb the number of young people leaving school into apprenticeships. This comes from recognising the vital importance of apprenticeships both for economic prosperity and for social stability. The Ambassador explains some of the different approaches taken in Switzerland compared with the UK and the opportunities to collaborate on areas of special interest, such as their belief that every qualification should have a follow-up route.

17.
THE APPEAL OF APPRENTICESHIPS TO BUSINESS

John Longworth
Former Director General, British Chambers of Commerce

Introduction
Apprenticeships have long been recognised across the world as a way of developing employee skills and increasing productivity. But while apprenticeships are appealing to business, there are still barriers to expanding the apprenticeship programme.

More and more businesses in the UK are taking on apprentices. This is despite the downward trend in investment of UK employers in training over the last two decades. Businesses are overwhelmingly positive about apprenticeships and regard them as an effective way of meeting training needs. They also recognise that work-based training can help boost skills and productivity, as well as the career prospects of young people.

The government is right to keep the spotlight on apprenticeships, and the increasing focus on vocational training is the correct approach. In the UK we have a history of placing a higher value on the academic pathway from school, to A-levels, to university. However, one size does not fit all, so it is vital that we develop high-quality vocational pathways that get the same respect other educational choices receive. Apprenticeships can play a big part in meeting important ambitions to boost skills and drive-up productivity.

Despite recent increases in apprenticeship numbers, we still lag behind our international competitors when it comes to the number offered. This is because there are still fundamental barriers that are holding back the successful development and growth of the apprenticeship programme. Business, government and the education sector all need to work together to address these, and it is important that all parties work in a policy environment that promotes quality over merely aiming to increase numbers.

Why apprenticeships are appealing to businesses

Apprenticeship numbers have steadily increased over the last decade, with half a million starts in 2014/15. There are 250,000 workplaces across the UK using apprenticeships to attract new talent, upskill existing staff and tackle skills shortages.

The BCC's research shows that businesses recognise apprenticeships as an effective way of training people for roles in the workplace. A relatively high proportion of our members offer apprenticeships and our research shows that many small employers, who don't currently employ apprentices, are considering doing so in the future.

Apprenticeships are appealing to business for three main reasons. Firstly, apprenticeships reduce training and recruitment costs for business. For many firms, training apprentices can be more cost effective than hiring high-skilled staff. Also, businesses are increasingly aware that potential employees don't always want to go to university, and firms don't want to miss out on recruiting this talent. This means that hiring apprentices can lead to lower overall training and recruitment costs for companies.

Secondly, apprenticeships can help develop a skilled and motivated workforce. Properly designed schemes, built around the needs of the business, provide workers will the skills needed by the firm. Apprentices who provide fresh ideas and benefit from training often prove to be motivated and loyal to the company, which can lead to significant reductions in staff turnover costs for firms. Businesses also report that having apprenticeships in the company and developing a culture of support, learning and training has wider benefits for staff across the firm.

Thirdly, setting up apprenticeships can have wider corporate social responsibility benefits for companies. It sends a strong signal to potential employees that the business is serious about training and developing staff. There is also now a proliferation of companies that use their apprenticeship programmes to build public trust and develop their profile as responsible corporate citizens. This is especially relevant given the context of persistently high youth unemployment.

Barriers to growing the apprenticeship programme

Despite these benefits, apprenticeships in the UK still have comparatively low market penetration when compared to other countries. We need a two-pronged approach to addressing this issue: grow the number of under 25-year-olds going into apprenticeships; and support more small and medium sized businesses to take part in the apprenticeship programme.

Much of the growth of apprenticeships in recent years has been driven by over 25-year-olds undertaking training. This group now represents around four in ten apprenticeship starts. This is no bad thing in itself – the apprenticeship programme should be accessible to people of all ages who are training for a new role. However, we can't escape the fact that apprenticeship numbers for under-25s have only seen a modest increase.

It is sometimes assumed that employers are overwhelmed by apprenticeship applicants. Actually this often isn't the case. We hear of employers offering well-paid good quality apprenticeships and still struggling to recruit. Part of the problem is that schools often don't appropriately signpost their pupils to apprenticeships. There is a blockage when it comes to supporting routes into apprenticeships and high quality vocational education for young people.

The school funding system means they largely encourage pupils to stay on and take A-levels – there is little incentive to promote apprenticeships. As schools are primarily judged on their academic results, this often means they overlook alternative training and vocational pathways. We even hear of cases where schools choose not to send their pupils to careers events where they know that apprenticeships will be promoted, out of fear of this competing with their sixth-form. This needs to be tackled head on. To begin with, we need to start judging and assessing schools on the outcomes of their pupils, including progression into apprenticeships. This will help incentivise them to promote these alternative pathways.

For a business to thrive, it needs access to the right skills, in the right place, at the right time. Firms need educational establishments that prepare young people for employment, showing them the wide range of career options available, and that work with businesses to address skill shortages and equip the next generation to start the enterprises of tomorrow. To achieve this we need to develop much stronger pathways into apprenticeships, where the incentives for schools, colleges, training providers and employers are aligned.

Government policy also needs to focus more on supporting SMEs to engage with the apprenticeship programme. Recent policy has been squarely focused on large employers, many of whom are already heavily involved in training. Our research shows that it is actually the smaller businesses that are most likely to be thinking of taking on apprentices for the first time.

But apprenticeship funding policy needs to better reflect the greater costs for smaller businesses to offer apprenticeships. These firms, with limited human resource functions, have much less capacity and resources to run apprenticeship programmes, yet they receive hardly any additional support. Some help is available on a very small and restricted scale through the apprenticeship grant for employers, but additional financial support needs to be much simpler and easier to access for all SMEs.

Chambers of commerce and apprenticeships
At the BCC, we are doing our bit to tackle some of these barriers. Chambers are well placed to bridge the gap between education and business. Local chambers sit at the heart of their business communities and have strong links with their education and training sectors. Currently chambers broker or deliver over 4,000 apprenticeships per annum and have over 1,500 schools in membership.

Chambers of commerce across the UK are vocal champions of apprenticeships. They deliver support events for hundreds of small and medium-sized businesses that want to find out how to take on their first apprentice. Chambers are also delivering 250 'Your Future' careers events across the country, reaching 70,000 pupils who get the chance to interact with local businesses and find out about different career pathways, including apprenticeships. But our activities and those of others need to be underpinned by effective government policy.

Government policy
The Conservative government, which assumed office in May 2015, has set itself a target to deliver three million apprenticeships in the next five years. There is nothing wrong with setting ambitious goals, but we should be

cautious of focusing too much on numerical targets. Having consulted with our business membership, we think the biggest risk with this arbitrary target is that it focuses on quantity, potentially at the expense of quality. The danger is that it could risk undermining the efforts that have gone into increasing the quality and profile of apprenticeships.

The focus on hitting this arbitrary figure could lead to a conveyor belt model, where apprenticeships are churned out and quality suffers. We have seen it time and again that when government sets rigid self-imposed goals, quality of delivery can suffer as it is trumped by the imperative to reach the target. For example in 2010/11, the credibility of the apprenticeships suffered as Train to Gain – a vocational training programme primarily for over-25-year-olds without level 2 qualifications - was essentially subsumed into the apprenticeship programme. This had dire effects on the apprenticeship brand.

Businesses are also concerned about the government's proposal for funding this apprenticeship increase through an apprenticeship levy, set to take effect in 2017. This is a relatively blunt instrument, and effectively a payroll tax on large firms to help the government reach its apprenticeship target.

What will the effects of the apprenticeship levy be? Certainly, the fact that businesses have to pay this levy will mean they will redouble their efforts to engage with the programme. However, one of the dangers is this could create a perverse incentive for businesses to shoehorn and re-package perfectly valid existing training into apprenticeship frameworks to recuperate their levy funding investment. Successful delivery will also largely depend on the electronic voucher model which is developed to run the levy system. Government should keep a watchful eye on the implementation of planned reforms to make sure unintended consequences don't undermine the delivery of the levy and the quality of apprenticeships.

Conclusion
In conclusion, there has been some good progress in increasing the number of apprenticeships and raising the profile of this training pathway. But there are still some fundamental barriers that we need to address to achieve a step change in the UK apprenticeship programme. We need to

develop much clearer pathways from schools through to apprenticeships to increase the number of younger people taking part. We also need to provide greater support for small and medium sized businesses to engage with the apprenticeship programme. While we all want to see an increase in the number of apprenticeships, this should not come at the expense of quality.

Ultimately, the best way to encourage firms to take on apprentices is to increase their quality and relevance to business. If this is achieved, the demand from both employers and potential apprentices will naturally follow. Only by prioritising quality over quantity can we forge a credible alternative to the typical academic pathway, which businesses and future apprentices can fully buy into.

18.
THE EXTRA HELP THAT SMEs NEED
Jason Holt CBE
Chair Holts Group and Apprenticeship Ambassador for SMEs

Jason Holt (JH) responds to questions from David Way (DW)

DW: Your report to government on the special circumstances and needs of small businesses considering apprenticeships was very well received at the time. Looking back now, what were the most important messages?

JH: I would highlight three main messages. The first was about the perception of apprenticeships. Too few SMEs are really aware of the benefits of apprenticeships. Their knowledge and appreciation of why and how apprentices can make a real difference is patchy.

There is a general misunderstanding of the meaning of apprenticeships: what they are; who they are for; and how they work, including the employer's own responsibilities and how training providers or colleges will support them.

I also found that there is little appreciation of what apprenticeships could do for a small business relative to large businesses like Rolls Royce. We need more messaging and compelling evidence targeted for SMEs.

My second message is that it is incredibly difficult for SMEs to navigate the apprenticeship system, let alone feel they are empowered to control and achieve the outcomes they need.

The types of questions they have are fundamental. As a business interested in employing an apprentice, where do I go? How do I get the best from a training provider and what do they really offer? Navigation (on and off-line) needs to be easy, clear and coherent.

Thirdly, SMEs were not confident that there were sufficient really good young people out there looking for apprenticeship opportunities. Smaller businesses simply aren't able to carry young people who are not reasonably work ready and they worry about what might happen if they try.

DW: Have those messages changed in the time since the report was published? Has it become any easier for SMEs to engage with apprenticeships?

JH: Taking each theme in turn:

1. Perceptions:
Yes. The greatest improvement has been around awareness and improving the image of apprenticeships. I've noticed significant traction of apprenticeships with fast growing SMEs. They are increasingly seeing engaging apprentices as an effective, and in some cases necessary, means to secure the best talent.

For instance, Optimity, a fast speed broadband business, employing approximately sixty people, now employs 20 per cent of their workforce as apprentices whilst growing in excess of 20 per cent per annum for over two years. This fast growth Gazelle company is run by their MD, Anthony Impey, who correlates the two factors as follows:

> Hire apprentices to grow, and grow your company to hire apprentices.

In general terms, the feeling I get from attending a number of business events in my role as Government Apprenticeship Ambassador for SMEs is that the proportion of businesses being aware of the true nature of apprenticeships has grown significantly since 2011.

Purely anecdotally, in 2011, it would be rare to see more than a couple of hands being raised when I asked a room of business owners if they employed an apprentice. Now, the numbers are far greater. Perhaps as many as 50 per cent of owners of fast-growing businesses raise their hand.

Why has there been a perceptions shift? I think the government's commitment to apprenticeships has been strong and consistent through the last parliament and as illustrated by the PM in the May 2015 election. One of their main headline pledges was three million apprentices by 2020. This ambition has resonated with people and is making a difference.

We have also seen better and more active engagement by key intermediaries. There are better message holders than government, especially organisations and people who are trusted advisors. We have seen the four big employer bodies, the Chartered Institute of Personnel and Development (CIPD), Federation of Small Businesses (FSB), BCC and the CBI, all actively working to raise awareness of the benefits of apprenticeships with their members. Accountants and professional bodies too are great advocates of the benefits of employing and developing young people through apprenticeships.

Employers are not just listening, they are talking to each other to learn what works and get tips and advice such as under the Apprenticemaker peer-to-peer scheme.

In the last three years, I've also seen improved joint working across government departments. The manifestation of that being more young people are aware of the opportunities provided by an apprenticeship and understand where to find them. However of course there's still progress to be made.

2 Navigation and empowerment
When I published my review in 2011, I said that for some small businesses navigating the system was like sailing from Britain to Brazil without a compass.

At this juncture, just prior to the implementation of the reforms, and during this transition phase, I think there is little material change as the benefits of the new system are yet to be felt. Navigation of the system remains patchy and, in too many cases, difficult for many smaller businesses to be converted to apprentice employers.

Having said this I remain optimistic for a number of reasons:

1. The government's new digital service has the potential to make a paradigm shift to navigating the system by:
   a. moving away from the .gov portal (something employers – particularly smaller ones – do not feel at ease with due to its governmental feel);
   b. being a stand-alone inspiring site – taking employers and would-be apprentices through a simple journey;
   c. keeping it easy and simple with the wiring hidden;
   d. being as easy to 'buy' from as Amazon, and transparent as Tripadvisor. Ambitious I know, but that is the benchmark employers will measure this service against. Get it right, and we have a chance to engage the 85 per cent of businesses not currently engaged. Get it wrong and we risk losing many of the 15 per cent currently involved.

2. I am pleased to learn that many of the recommendations in the Holt Review, for instance the training provider search tool, and real time provider rating mechanism, are being incorporated into the new digital apprenticeship service.

3. While I still hear the usual moans of poor service from providers about: giving a 'take it or leave it approach'; inability to tailor provision around the needs of the business; lack of responsiveness; poor quality of candidates, etc., I think progress has been made. It is now rare to meet principals of providers who don't understand that to survive the reforms they must refocus the college/provider to be far more customer-centric.

4. Many more businesses now have a relationship with a specific training provider or intermediary who help them with brokerage and crucially, have gone to the trouble to understand their customers' needs.

5. Following on from the above, businesses seem more motivated to preserve and make it work with providers. They're giving them more leeway to get it right. I think in part because apprenticeships are gaining currency, they know it's not easy, and so are more forgiving.

6. The impact of apprenticeship ambassadors like me. If I am known to them, employers will persevere, if nothing else to avoid my nagging! I am known amongst my peers as "Mr Apprentice who doesn't let go until you hire". What choice do they have?

3 Supply of young people

A key factor in why businesses don't engage is their lack of confidence in the candidates' ability to be work-ready and productive in the work place. I regret to say I think progress is limited here despite the passage of time. For instance, in an Apprenticeship Stakeholder Board meeting in December 2015, the common complaint from the large employers, CBI and FSB, was the lack of applicants for apprentice vacancies. For two principal reasons, progress here has been limited:

1. the raising of the age of participation to 18 has resulted in schools hanging onto their young people for longer; and
2. routes into apprenticeships - there remains a culture within many schools where apprenticeships are considered second-rate compared with going to university. While this problem is being addressed, largely through the new Careers and Enterprise Company, it will take many years to win the hearts and minds of the 400,000 secondary school teachers in England and Wales.

DW: Why do you champion apprenticeships so strongly?

JH: Quite simply, it's in my DNA!

I have felt the benefits of being an apprentice (lawyer), employing them in my businesses and, since my Government review, helping form policy (through the Holt Implementation Board, Apprenticeship Ambassador for SMEs and Chair of the Apprenticeship Stakeholder Board.) Quite frankly, I can't believe my luck in being able to give back in this way. Equally, what an extraordinary time to be involved in a force for good of such importance.

On a personal front, I am fortunate enough to come from a family business background whose heartbeat is the passing on of skills from one person to the next and from one generation to another. Embedding learning through doing is what we've done from the beginning. It is in our DNA. It is how we have prospered over seventy years.

In relation to my government role over the past few years, I am lucky enough to have stumbled into the apprenticeship world through the review I was invited to write. One thing leads to another and I am now in the thick of the team that is developing apprenticeship policy.

I can't think of a better way to devote my time. I believe we're in the process of creating a world-class apprenticeship system. I can imagine that in ten to twenty years' time, people will see apprenticeships as commonplace, desirable and equally valid as university education. What an honour to have been part of the team to shape it.

DW: What has given you most cause to celebrate as an employer of apprentices as well as an ambassador?

JH: I love so many aspects of employing apprentices. It's the sense of discovery. Discovering a hidden talent of somebody who up to that point had the uncertainty of what to do post-school. To whom you gave a break in life and whose potential you've just unleashed. You've helped someone find themselves; their sense of purpose; and given them wings to be free from whatever contained them prior.

I have hundreds of examples. For instance, imagine the satisfaction I had when I heard that Thomas, our new apprentice software developer at Holition, was working on a project for one of our clients, Hugo Boss, within six weeks of starting – and undertaking technical work at a standard equivalent of a PhD. Rare undiscovered talent.

Or the time that we took on Caitrin who desperately wanted to be a jewellery designer and who a year later is one of the outstanding talents at Holts Jewellers.

DW: Why did you decide to set up the Holts Academy?

JH: Like many small businesses – especially family-owned ones – I had a seminal moment when I realised that I needed to diversify rather than try to change the jewellery business. So I created a number of independently run small and medium-sized businesses as part of the Holts Group.

My pride and joy is Holts Academy, the not-for-profit training enterprise I set up to reinvigorate the jewellery industry, and in which so far we have seen over 500 jewellery, business or technology apprentices go through its doors.

At the early stages, we set up a training arm to try to secure talent for Holts. We could see the shortage of new blood coming into the industry and it was obvious that without action, our ageing workforce would retire and with them the demise of our gemstone cutting and jewellery manufacturing business. So initially the training was the only way we could find new team members – as our skills are not available in the market. The only option is to home-grow our own talent.

We cherry-picked the best learners but then other firms asked if we could supply them with those we didn't employ. This scaled into what is now the de facto FE provider for the jewellery sector in the UK.

More recently, on the tech side, at Holition – a digital creative agency specializing in augmented reality - we've experienced similar difficulties in recruiting talent. This frustration has led to a new venture – Tech Up Nation - in partnership with a tech employer – Optimity - using the model of Holts Academy, to address a shortfall of apprentices in technology.

London has 24 per cent youth unemployment, yet 85 per cent of employers say they can't grow due to skills shortage. In Tech City, a tech hub of mainly SME tech companies, there are expected to be 300,000 vacancies by 2020. Tech Up Nation's ambition is to resolve this paradox.

DW: What is your reaction to the government's ambition for three million more apprenticeships?

JH: I applaud government's commitment to apprenticeships and this bold ambitious number of three million makes for a powerful and positive statement. It gives the sector and stakeholders something to aim for and is a stake in the ground.

To me the optimism and positivity the numbers provide give more weight than the validity of the number.

In my view, the stability and sustainability of the system – to get it right first time – to ensure routes to apprenticeships are improved – and fundamentally that the system receives cross-party support – are the more critical issues to agree on.

If the question is do I think we can achieve such a number? Yes I do.

Accounting for the fact that we're talking all ages and the possibility of large numbers in the public sector that could materially escalate numbers – I think we'll get there.

DW: What more do you hope government (and others) will do to make it easier for businesses and especially SMEs to take on more apprentices?

JH: There is much more that can be done to keep up the momentum. I would highlight the following:

1. Keep shining the light on SMEs. It's too easy to listen to big business. Don't fall into the trap of just listening to big business for the sake of numbers.
2. This is a particular threat given levy roll out and that the SME voice may be lost while large business lobby government about the levy's impact.
3. Make the case on return of investment:
   a. The case not been made yet. Government needs to do more to demonstrate ROI
   b. They need to show evidence which shows if you invest in an apprentice in your sector then business growth will happen
   c. The Swiss do it well!
4. Segment the SME Market:
   a. First of all no business identifies solely itself as an SME. Sector specialism, growth, location – not simply SME
   b. Also a word of caution in using the SME term. This is not an effective way to describe the vast range of organisations that employ 1-249 people
   c. The term SME combines the kind of fast-growing gazelle companies (or scale-ups) that create most net new jobs with lifestyle family firms
   d. Start-ups future is necessarily opaque
   e. Focus on those who want to grow their business.
5. Leverage Apprentice Champions
   a. There are many people like me who are evangelical about apprenticeships.

     b. Are we leveraging those people and networks like the AAN, chambers, as best we can? Government need to seek them out; employers listen to them more than to officials

6. Look to blend delivery with bite-size opportunities to involve the employer at every stage:
    a. the more involved, the more excited employers become
    b. Strangely the more you ask, the more they want to give!
    c. Visibility of employers in delivery – they know best. It is truly amazing how much they will contribute – for FREE

7. Ensure flexibility from providers. Inevitably providers will try harder where there are large volumes. So the challenge is to have flexibility such that if an employer had to pay for it they would do so because they see value.

8. Young people who aren't work-ready or don't have the necessary attitudes and behaviours: this is a major issue. Small businesses can't afford to carry them. They expect more than we can give. So preparing young people for the world of work is key.

9. Less meddling and constant change – this seems to be unique to the UK. It is essential that there are many fewer changes. My visit (with you David) to Switzerland really bought home the power of a stable system that has introduced few major changes in the last thirty years, let alone the last five.

10. Language: it is really important to choose our words wisely.
    a. If giving financial support, we need to be precise in our language. Language is essential. For example, it's about calling it a training grant or training subsidy rather than a wage subsidy. That reframes the conversation with the employer. It puts a stake in the ground with the employer receiving the benefit of a person who's in training.
    b. Employer of small business. Communications need to be more than good prose – they need to be catchy!
    c. Employers are always short of time and short of resource and usually not open to anything which doesn't reflect on the bottom line.
    d. There needs to be clarity on what the deal is, with a clear and consistent message and without any nasty surprises in the small print.
    e. Avoid education-speak - employers don't understand and don't value messages that use the jargon that is so easy for

everyone to drop into, such as Frameworks/Standards/Levels. Need to reframe in commercial context.

DW: What message would you give to SME employers thinking of taking on an apprentice for the first time?

JH: The key messages for me are:
1. The supply of good people is even more important for an SME. So if you want to grow your business how will you source the right skills? Do you want raw talent from a different talent pool? If so, taking on a graduate isn't the only option.

2. Look at the great young people seeking apprenticeship opportunities. I have the firm view that if employers see great young people, they'll buy. Employers can do more to highlight how great it is to work for a small business. It provides greater variety and personal responsibility.

3. Focus on business growth not on Corporate Social Responsibility (CSR). Yes, it's great CSR - great for the community. It's the right thing to do. However, this is secondary to the value apprentices bring to business. I'm yet to find a buyer who doesn't want to buy again. Good for business. Focus on that message.

DW: Any final thoughts?

JH: It's certainly an interesting time for apprenticeships given the pace of change. There are major policy changes of immense importance coming in including the levy, digital service and Institute for Apprenticeships. It will be more important than ever to listen to SMEs and ensure that we take them with us in achieving the new ambition for apprenticeships.

19.
THE FUTURE OF INDUSTRIAL PARTNERSHIPS

Steve Holliday
Chair, Energy and Efficiency Industrial Partnership

———

Introduction

I think it's important to start by giving a context in which the energy and utility businesses operate to explain the reasoning for an Industrial Partnership. National Grid, with its peers, faces some of its greatest challenges to date. Delivering clean energy is not only critical for our long-term future, but also for those societies in which we operate. To achieve our goals, we must be trusted and recognised as a business that understands this challenge and does something about it. We, as a collective of like-minded organisations, must take each and every opportunity to explain how we're building a responsible and sustainable future. We must inspire the creative and bright minds of today to design the infrastructure of the future, allowing us to preserve those aspects of life that people value.

The energy and utilities sector is our country's invisible army. The highly-skilled men and women working tirelessly to deliver vital energy services are invisible the majority of the time to the majority of people. And this is how we want it to be – our invisibility is proof of our success. It means that our energy assets and infrastructure are operating like clockwork, delivering services on demand; day in, day out. It's expected!

But this invisible army is also our single biggest point of failure, and the reason we have great expectations. Without our people, this sector will become visible – for all the wrong reasons – because we are unable to keep pace with our growing demand for energy. Without a workforce equipped with the skills to manage a changing landscape, we risk being unable to keep the lights on, the gas flowing, and our water and waste systems

204

functioning in the future. We need future recruits with layers of skills and experience, ready to question, create and deliver solutions

Across all energy & utility businesses, we need to recruit around 200,000 people over the next ten years. The power sector alone accounts for a quarter of this, meaning between 45,000 and 55,000 new employees will be needed, and by 2028 we will have had to replace more than 83 per cent of our workforce. This is a huge challenge!

However, the sector does not face this challenge alone! Between 2010 and 2022, Engineering UK anticipates that the engineering sector as a whole will need to recruit 2.56m people - that's 47 per cent of the engineering workforce (5.4m) of which 1.82m will need engineering skills from level 3 upwards - that's crafts, technicians and engineers at degree level. Much is being done across the engineering profession to address this challenge and it is almost exclusively built around collaboration – all stakeholders teaming up and working together.

Why did we come together?
At the heart of the matter was, and is, a desire to ensure we have a sound workforce plan that looks to the future - a long-term strategic collaboration and not a three-year project. It is now over two years since we formally launched the Energy and Efficiency Industrial Partnership (EEIP) that brought a wide group of employers together.

When the opportunity to come together as a sector was proposed by the UK Commission for Employment and Skills (UKCES) back in 2012, we collectively embraced the opportunity. At the time I was confident that the strength and scale of leadership by committed CEOs would at last build consensus on how to meet the huge skills challenges for new, innovative power and energy solutions, and at the same time help to create a more secure and sustainable system for the future.

There were eight industrial partnerships awarded funding under the previous Coalition government's Employer Ownership pilot in the automotive, aerospace, tunnelling and underground construction, energy, digital, nuclear, science industries and the creative industries. The Energy & Efficiency Industrial Partnership being the largest, collectively joined forces in 2014 to co-invest with government (70 per cent by employers) to deliver a £115 million change programme to transform the skills system for the sector, tackling some of the longstanding, fundamental issues:

- A lack of clarity over career paths and poor careers advice for young people;
- Poorly outlined industry skills standards;
- Inefficient and restrictive recruitment; and
- A bureaucratic, complicated and inflexible skills system.

We committed to provide a joined-up and simplified approach on careers, maximise the talent pool for the sector, and improve the take up of apprenticeships and traineeships with employers and individuals. One of our main priorities was to improve our performance on working with younger candidates by taking a structured approach to work experience and recruitment of our apprentices. We also pledged to target those people furthest from the labour market, including training over 6,000 learners at risk of exclusion from the job market, to generate real benefits for our economy and society.

It was clear that a more flexible approach to training and assessment would drive better value for money and open up a wider talent pool by retraining the current workforce, as well as providing high quality apprenticeships and traineeships. It's why I committed National Grid to be the lead employer.

The support from government at the time created a more conducive environment to collectively leverage our individual investments and we have begun to deliver and to see the benefits of our efforts. In the first full year of operation the sector embraced a new way of supplying careers information, delivered reformed Trailblazer Apprenticeships, trained and employed young people who would not normally meet our entry level employment requirements, and delivered funding more efficiently directly to employers to further bolster employer-led investment in skills.

Even more impressive was the creation of an Independent Assessment Service with well-respected high quality assessments and performance management practices as an integral process to our apprenticeships. This approach put us on track to double apprentices in the sector by 2017.

Without doubt, the delivery of high quality apprenticeships is the primary objective here, but there are five underpinning areas of activity and achievements to stimulate supply that stand out and would just not happen without collaboration on the scale of Industrial Partnerships.

1. Ensuring effective careers advice

Careers Lab created by National Grid is a new partnership between schools and relevant local business to bring the world of work alive to young people. The programme commenced in 2013 in the Midlands in response to the growing gap between the skills young people have and the needs of industry. Essentially, Business Ambassadors from industry deliver a set of four inspiring lessons alongside class teachers, sharing their career journey and real-life experiences. Supported by United Utilities, Thames Water, EON, Scottish Power, FCC, Viridor, Southern Water, Northumbrian Water, UK Power Networks, British Gas and SSE, the objective is to inspire young people to think about the choices they will have to make and the paths they might take to get the best chance of a job when leaving education or training. Now coordinated by Business in the Community, it is a sequenced programme starting at age 11 through to 16 that is now being scaled-up to around 600 schools and will be signposted in the DfE's new Careers and Enterprise Company tool kit.

2. Supporting younger candidates and those furthest from the labour market

As a partnership we committed to support young people into employment and we trialled a number of approaches to ensure our future activity delivered the best results both for the individual and for business. Led by United Utilities, employers in the north west have worked together providing 'traineeship' opportunities for young people who were not in education, training or employment to fast-track them into full-time work.

The programme enabled over 93 per cent of participants to gain a job or apprenticeship in the sector. This is the very first collaborative scheme of this type, led by employers, and is setting the pace among utility providers in getting the young unemployed back to work. This pilot demonstrated how our businesses could access the skilled people they need whilst offering opportunities to individuals furthest from the job market.

In a similar vein, E.ON created a traineeship programme in customer service, aimed at 16- to 24-year-olds and SSE have fast-tracked twenty young people from a wide geographical area stretching from Thurso and Wick to Inverness, who are considering a career in engineering. The ten-week intensive course includes students working alongside SSE engineers. Successful candidates are given the opportunity of an apprenticeship or further training.

3. Developing employer led Trailblazer Apprenticeships
The power sector took a leading role in shaping new Trailblazer Apprenticeships, the first of their kind. They are entirely employer owned, and result in a vastly improved experience for students and employers alike.

In its first year, the level 3 Power Craftsperson Apprenticeship enrolled over 100 apprentices with employers including National Grid, SSE and UKPN, who collaborated to develop these bold, ground-breaking apprenticeships.

4. Engaging the supply chain
To truly succeed, we need parity across the sector's entire supply chain, not just the principal asset owners. The Energy & Efficiency Industrial Partnership established a contractors' forum, which identified the opportunity of procurement policies as a lever for addressing skills issues. Currently a contractor seeks to remain competitive by a silo approach to its respective workforce, often resulting in under-investment in skills due to short-term, contract-by-contract mentality.

Procurement policies provide a means of delivering change at scale: introducing the same requirements for all suppliers, adopting a consistent approach to managing relationships, and measuring performance fairly and transparently.

Led by Thames Water, National Grid, Amey and SSE, we have created an industry accord where employers will sign up to promote investment in skills through positive procurement action and subsequent supplier relationship management.

The objective of this accord is to increase investment in skills sufficiently across the supply chain to ensure that the supply of competent workers meets projected future demand.

5. The Independent Assessment Service

This unique quality assurance process gives employers in our partnership absolute confidence that the skills being trained and developed are rigorously and consistently tested by peers who really understand employment and skills within the sector. It also means that anyone passing the IAS assessment has the passport to working right across the sector, confident that their apprenticeship or qualification is welcomed and respected everywhere by all employers - small, medium and large.

The added value of Industrial Partnerships

Whilst there are many benefits common across all IPs, the EEIP is differentiated by its breadth of engagement, with over sixty-seven employers across the energy and utility sector; and by its ability to use the industry's current robust quality assurance standards and regulations to ensure the quality of all training and apprenticeships through the Independent Assessment Service. Importantly, the relationship with the supply chain and the skills procurement accord are integral to the IP's value proposition - we are undoubtedly stronger from working together.

The Industrial Partnership brings together government ministers and business chief executives to act as a powerful force on tackling the skills needs that hamper our sector productivity. This opportunity for direct engagement means that ministers are listening to business and not having everything filtered through department officials. This is especially important at times when we face major challenges or need to respond quickly.

The timing for Industrial Partnerships to thrive could not be better, particularly in light of the critical role energy has in the National Infrastructure Plan (NIP) of which around 59 per cent is energy. It has never been more important to work closely with government and to raise the bar on the stock and supply of skills to our sector. Industry Partnerships are referenced in the NIP as being central to delivery. So it's important to build on the hard-won gains of IPs. Otherwise there is a real risk to the ambition for UK infrastructure that will result in internal, cross-sector competition for skills, increasing, not decreasing, skills shortages.

While it is early days, the vital ingredients are there for us to be able to secure quicker and smoother routes into employment, enabling us to attract more talented young people into our businesses and to start developing their skills. This is important since the process of skilling people for the type of jobs that are both technical and safety critical to the industry takes many years, and requires them to progressively build experience and competence over time.

While I am pleased with the new skills infrastructure that we have built so far, much has still to be done to address systems that are too complex, and to reduce duplication of effort. We are determined to have more streamlined and tailored arrangements that everyone understands and can navigate easily.

The future of Industrial Partnerships
My simple message here, to government, policymakers and business leaders is that we need Industrial Partnerships now more than ever!

We have put in place hard-won alliances that have the confidence and backing of employers. They are increasingly engaged with the strategic picture that requires us to link skills and employment across the wider energy and utilities landscape, and position us to deliver our part of the National Infrastructure Plan.

In spite of recent unhelpful skills reform around funding and the apprenticeship levy, the EEIP is intent on sticking together – it's a game changer. We are determined to build on this and I hope the government will continue to support this active demonstration of employer leadership of skills, for which they have been calling for many years. The government's role should concentrate on creating the conditions for economic success and to foster collaboration – joining up the dots and helping to galvanise our efforts towards higher productivity and economic growth.

We need government to be unequivocal in their support for Industrial Partnerships and all that they entail. Never before have our sector's employers felt so engaged in influencing and delivering a workforce strategy, but it requires regular reinforcement by the messages and actions of government at a time when the abrupt termination of Employer Ownership pilots has sent confusing and negative messages.

We need to avoid any possibility of business leaders feeling that government is no longer taking seriously the Industrial Partnership arrangements to which they have committed so much time, effort and creativity. For many, this is the first time they have come together collaboratively and we must not lose the momentum and traction that this change in approach has produced.

Call to action

So as we look ahead, it is vital that we look beyond the government's 2015 comprehensive spending review cuts, to 2020 and beyond to ensure our collaboration endures and delivers. There are just three things I suggest we should all keep in mind:

Firstly, teaming up is not a 'nice to have' it is a 'must have'. Only by working together can we inspire and deliver the workforce of the future.

Secondly, Industrial Partnerships are not just delivery vehicles - they need to be put on a stronger strategic footing within government (they are within business) to ensure they align with the longer term UK Industrial Strategy and Infrastructure Plans.

And finally, we must continue to involve the entire supply chain to ensure that we increase participation and increase cooperation on standards, competencies and training.

Not only is this the right thing for responsible employers to do, it is responsible government too! Industrial Partnerships must be part of the fabric of our nation's skills and workforce development if we are to secure sustainable commercial advantage and drive the prosperity of our industries and the UK in general. We all need to play our part!

20.
RAISING THE STATUS OF APPPRENTICESHIPS

Frances O'Grady
General Secretary, TUC

Tom Wilson
Formerly Director of Unionlearn

Matthew Creagh
Apprenticeships Policy Officer, TUC

———————

1. Introduction
This chapter outlines the Trade Union approach to apprenticeships,[17] what unions see as the key strategic issues, and how they should be tackled.

Youth unemployment and promoting skills have always been important issues for trade union members and their families. Young people are particularly vulnerable to exploitation, and trade union representatives have a strong track record in encouraging employers to create good apprenticeships, and in supporting apprentices to make the transition into secure employment.

The TUC is committed to ensuring that young people have access to high quality learning and skills opportunities that lead to secure, sustainable, fairly paid employment.

Unionlearn, the learning and skills arm of the TUC which was set up in 2006, has played an important role, supporting union representatives to promote and negotiate high quality learning opportunities for young

---

[17] The TUC standpoint is used to describe the general union view, unless there are differences within unions on particular issues, in which case this is made clear.

people. Where unions are recognised, they are able to negotiate learning agreements that solidify a commitment from both the employer and the union to support young people in the workplace.

While this chapter is primarily about apprentices, it is worth including some reference to traineeships, which are becoming an accepted route onto an apprenticeship. Unionlearn has published charters of best practice on both traineeships[18] and apprenticeships[19], as well as guidance on the components of high quality work experience[20].

2. Equality and diversity

It is worth beginning with diversity, which should receive far more attention in discussions on apprenticeships. Segregation by gender and race is still widespread, arguably more so than in any other area of employment. Ethnic minorities are under-represented in some apprenticeship sectors, including engineering and construction, and over-represented in others such as leisure, travel and tourism, and public services.

Jeremy Crook, chair of the BIS Apprenticeships Advisory Group (on equality) has highlighted that the proportion of ethnic minority people who apply for an apprenticeship is far higher (to a much greater degree than for white applicants) than the proportion who start one[21]. In 2011/12, around 25 per cent of applications made via the central Apprenticeship Vacancies system were from ethnic minority people but only 10 per cent of the starts in that year were by ethnic minorities.

Only 1 per cent and 3 per cent of apprenticeship starts were by women (aged under 19) in construction and engineering respectively[22]. Although these figures are slowly improving, the pace of change is glacial. Whilst the overall ratio of male/female apprentices is broadly balanced, many young women are working in sectors synonymous with low pay (such as retail, social care or catering) and do not have as much opportunity to

---

[18] Unionlearn. Charter for Traineeships: www.unionlearn.org.uk.

[19] Unionlearn. Charter Apprenticeships: www.unionlearn.org.uk.

[20] Unionlearn. Work Experience: www.unionlearn.org.uk.

[21] BTEB. Time Employers: www.bteg.co.uk.

[22] Alison Fuller and Lorna Unwin. Apprenticeships and Traineeships for 16 to 19 Year-olds: data.parliament.uk.

progress through apprenticeship frameworks which overwhelmingly stop at level 2. Careers guidance should be reformed to ensure it can play the vital role of ensuring that young women are made aware of better quality apprenticeships and encouraged to undertake apprenticeships in better paid sectors such as engineering.

Reforms are needed in the collection and monitoring of data on apprentices. For example there is no specific data recorded on the number of disabled apprentices (only those 'with a learning difficulty/disability' are recorded). Again, there is no data relating to other protected characteristics such as a young person's sexual orientation. These young workers may face higher levels of prejudice and harassment than heterosexual colleagues. But before under-represented young workers can be offered the support they may need, it is necessary to understand the scale of under-representation in apprenticeships amongst groups with protected characteristics.

There are also a number of practical reforms that could be considered, such as improving careers guidance, amending recruitment practices, and providing further childcare support for parents who are undertaking apprenticeships. For example, apprentices are not eligible to receive free childcare funding under the Care 2 Learn scheme, which other FE students can benefit from. This obviously has a disproportionate effect on young women and single parents, and may prevent these young people from embarking on the apprenticeship route. Similarly, the lack of a part-time apprenticeship route can deter women with children.

3. Breadth, depth and duration

Recent apprenticeship reforms have seen employers being put 'in the driving seat'. Unions strongly support employers taking far more responsibility for apprenticeships, but entirely employer-led design risks leading to narrow training and short-term apprenticeships that will only meet the immediate needs of some employers for particular jobs. Young people need the chance to move between employers and train for an occupation, profession, career or trade, not just a job. So there is of course a tension – which many employers realise.

Good employers see beyond short-term needs to their own (or their sector's) longer-term needs; and recognise the value of working with unions. Mobility between employers is a general economic good, across a sector or beyond. As the manufacturers' organisation the EEF said in a submission to the 2013 Labour Skills Taskforce, 'it is important for [employers] to work closely with unions, colleges and quality training providers to ensure that the partnership works for both the employer and the learner'.

In many other countries, employee representatives have ensured that apprenticeships are broad qualifications that include the underpinning academic subjects that enable learners to gain broad theoretical understanding and also underpin mobility and progression in the labour market[23]. Almost all other systems recognise that apprenticeship training should provide the grounding for a career, not just an entry-level job. In principle the union view is that apprenticeships should be broad and deep, allowing a wide range of skills to be developed and equipping working people with the foundations of both a substantial career and a justified pride in their own occupational identity and professionalism.

There is an inevitable tension between that long-term aspiration (to which many good employers would subscribe) and short-term cost pressures. That is why all the key stakeholders: employers, government and trade unions/professional bodies should have greater involvement in apprenticeship design, as in Germany and most other developed countries around the world.

This joint approach is the key to developing a smaller number of apprenticeship frameworks of higher quality, breadth and duration. [25] Professor Hilary Steedman notes that 'in all apprenticeship countries except Australia and England most apprenticeship programmes take three years to complete or, in the case of Ireland, four years'[26]. In England, the

---

[23] ETUC and Unionlearn. (2013)Towards a European quality framework for apprenticeships and work based learning.

[25] EEF Submission to Labour Skills Taskforce on Apprenticeships: www.yourbritain.org.uk, page 5.

[26] Steedman H. (2010) The State of Apprenticeship 2010: International Comparisons - Australia, Austria, England, France, Germany, Ireland, Sweden, Switzerland. A Report for the Apprenticeship Ambassadors Network, London: Centre for Economic Performance, LSE.

average for all apprenticeships is between one and two years (usually nearer one year) and too many are simply adopting the new minimum duration as their standard. Highly influential work by Lorna Unwin and Alison Fuller on Expansive versus Restrictive Apprenticeships has for several years underlined the importance of longer apprenticeships[27]. The design of the levy should encourage much longer duration and more expansive frameworks.

4. Apprenticeship pay

The 2014 BIS Apprenticeship pay survey[28] depressingly revealed that one in seven apprentices are still paid below the (hardly generous) Apprentice National Minimum Wage (NMW) rate. Women and young people were disproportionately adversely affected.

- The survey showed [29] that 24 per cent of apprentices aged 16 to 18 and learning at levels 2 and 3 were paid less than the apprentice minimum wage, which was £2.68 an hour at the time of the survey.
- Non-compliant pay was more common among apprentices in hairdressing (42 per cent), children's care (26 per cent) and construction (26 per cent).

Although there has been some progress since 2014, the extent of underpayment is still widespread and concentrated in the same sectors, disproportionately affecting women. As the NUS have pointed out in *Progress*[30], the effects of low Apprenticeship pay are most keenly felt by the poorest students and their families. Some families will lose their child benefit payment and child tax credit when their son or daughter starts an apprenticeship. These apprentices will face the double whammy of low pay and loss of some of their family's financial benefits.

---

[27] See Alison Fuller and Lorna Unwin. (2008) Towards Expansive Apprenticeships, TLRP.

[28] BIS. (2015) Apprenticeship Pay Survey 2014. www.gov.uk.

[29] BIS. (2015) Apprenticeship Pay Survey 2014. www.gov.uk.

[30] NUS. (March, 2015) Progress: Labour's Progressives: www.progressonline.org.uk.

The extent of non-compliance is slowly reducing. According to the BIS Apprenticeship Pay Survey 2012[31], official figures revealed that 29 per cent of apprentices were paid less than the legal minimum wage in 2012[32]. But the pace of improvement is far too slow. Unions would like to see far more resources made available to educate employers, fine offenders and prosecute employers who flout the law.

But, beyond the legal minima, how much should apprentices be paid? For many employers this is a key question. Across the developed world, unions see a trade-off: at the outset when apprentices are not contributing a great deal they are paid far less than towards the end of their apprenticeship when they will be far more productive. This trade-off is usually negotiated. In many countries, apprentices' pay is generally covered by collective agreements which trade unions have negotiated, sitting above the minimum legal wage rate for apprentices. However, in some countries, these rates are very low, where collective bargaining is in decline. In some countries, the evidence suggests that these rates are partially or completely ignored as being outdated and too low.

Germany has a well-developed system. Apprentices' pay is covered by sectoral and/or regional wage agreements. The rates vary accordingly. For example in 2012, according to figures from the Bundesinstitut für Berufsbildung, (BIBB, the tripartite national skills bureau) the highest wage in West Germany was for an apprentice bricklayer (an average of €986 per month) and the lowest wage in West Germany was for an apprentice hairdresser (€454). In East Germany the highest wage was for an apprentice media technologist (€905 − the same as in West Germany) and the lowest, also for an apprentice hairdresser (€269). [33] In the metalworking sector in Berlin in 2013, for example, an apprentice is paid €819 per month in the first year, €868 per month in the second, €925 per

[31] BIS. (October 2013) Apprenticeship Pay Survey 2012, BIS: www.gov.uk.

[32] TUC. 'Apprentices Paid Less Legal Minimum'. Workplace Issues www.tuc.org.uk.

[33] BIBB. (2013) 'Ausbildungsverguetungen-2012', Bundesinstitut für Berufsbildung: www.bibb.de.

month in the third and €961 per month in the fourth. Workplaces that are not covered by wage agreements can pay less, but by tacit agreement only up to 20 per cent less. Conversely, in the retail sector some employers (Aldi and Lidl, for example) pay more than the national rate – €730 per month.

In Ireland, the construction industry agreement, signed in 2011 and annually updated and ratified, stipulates that 1st year apprentices are paid 33 per cent of the national rate for craftsmen; 2nd years 50 per cent; 3rd years 75 per cent; and 4th years 90 per cent. In Italy, apprentices are covered by collective agreements, and according to the 2012 Consolidated Text for Apprentices Act (2012) may be promoted up the pay scale to within two levels of the maximum decided by the collective agreement for that pay grade, if they are carrying out the duties or functions of a skilled worker. As an alternative they may be paid on a percentage basis.

In the Netherlands, the construction industry national agreement as of January 2012, for example, stipulates wages for apprentices running from €196.00 a week to €567.20, according to age and seniority. Interestingly, in order to offset the difficulties experienced by the construction industry in the wake of the financial crisis, the trade unions and employers agreed to a small temporary reduction in the pay of apprentice carpenters. Not all companies are covered by collective agreements however, and in these circumstances apprentices should receive the minimum legal wage.

By comparison, the UK apprenticeship pay system is much more mixed. For example, in 2013 the national minimum rates were £2.68 per hour for under-19s and for over-19s in their first year of apprenticeship. Over-19s (not in their first year) should have received more – the rate that applies to their age group; £5.03 for 18 to 20-year-olds and £6.31 for 21-year-olds and over. But those were just the minimum rates (often ignored as noted above) and collectively agreed rates were significantly higher. The 2011 DBIS survey of apprentices' pay had a sample size of 11,020 apprentices and found a median rate of hourly pay of £5.87. Median gross weekly pay was £200.

The highest rates were for apprentices in team leadership and management (£8.33 per hour), and the lowest for apprentice hairdressers (£2.64 per hour). Year 1 apprentices had a median pay level of £5.33 per hour, Year 2 £ 5.93 and Year 3 £ 6.76. So, although the survey showed that four out of five received on or above the minimum amount they should get based on their year of study and/or age, many would have received well above the minima. Since 20 per cent received less, it is likely the range is even wider. [34] Even so, it is apparent that the UK has far more short-term (one year) apprenticeships than is the norm, and that pay rates are closer to the minima.

In most other countries, the rate of progression is much more substantial – pay may be lower than the UK median in the first year of an apprenticeship but that is felt justifiable by the much higher pay at the end of a typically much longer apprenticeship. This kind of differential is often embedded in collectively negotiated agreements. In effect there is an agreement between employers, unions and government that in return for high quality apprenticeships which typically last for 3 years and lead to decent pay and careers, apprentices' pay is lower in the early years. By contrast, the UK may have slightly higher pay in the early years (in some sectors, certainly not in those with average low pay such as hair and beauty) but far lower quality, shorter duration and worse prospects.

Of course the picture also varies across Europe. The UK may be closer to Estonia where apprentices are entitled to the minimum legal wage, as are other workers, but the rate is very low (€320 per month). Pay is also low in Cyprus, where apprentices are in sectors which are covered by collective agreements. They are paid by employers for the days spent in the workplace (not at college) according to the rate stipulated in the agreement. Cleaners, security guards, caretakers, nursery assistants, shop assistants and office workers are not covered by collective agreements and so are paid a minimum wage. However there is considerable anecdotal evidence to suggest that some apprentices are paid even less than the agreed rates. Should the UK apprenticeship system be close to Estonia or Cyprus than Germany? To which economic model do we aspire? Which labour market is closer to that of the UK?

Summing up this picture, there are three key points.

---

[34] BIS. (2011) Apprenticeship Pay Survey 2011, BIS: www.gov.uk..

First, the TUC proposes that the apprentice NMW rate might be raised to the level of the young workers' NMW rate, and that the use of the apprentice rate should be limited to younger workers in the first year of their apprenticeship. The prevalence of lower level, shorter duration, poorly paid apprenticeships means that there is a genuine, realistic fear that some employers will use 'apprenticeships' to substitute for existing or future job vacancies, including those of young workers, simply because apprentices are so cheap to employ and because there is no legal obligation to ensure the young person receives high quality training over a significant period of time. This concern is heightened by the government's decision to exclude under-25-year-olds from the minimum wage supplement. Young people starting out in the workplace, enthused by the opportunities that an apprenticeship is supposed to bring may be confronted with the reality of a poor quality, poorly paid apprenticeship. This exploitative experience is likely to be scarring for young people and provide a real setback for those apprentices who are looking to develop the skills that will help them throughout their career.

Second, high quality apprenticeships should lead to an occupation with significant breadth and depth. Without the promise of progression there is no justification for a low starting rate. The TUC believes that all apprenticeships should, over time, last for at least two years and lead to at least a level 3 qualification.[35] Of course there are many young (and older) people today who have completed a one-year level 2 apprenticeship, it would be wrong to reduce the status of such qualifications overnight. For some young people it may well be a major achievement to reach level 2. But it is not acceptable for employers to argue that some jobs should hit a ceiling at level 2 and hence a one year apprenticeship is all that can ever be offered. Every apprentice should have the opportunity, if they wish and have the aptitude, to progress to level 3 or beyond.

Third, the appropriate rate of pay on a two or three year or longer apprenticeship should reflect the productivity of the typical apprentice. As

---

[35] Some unions, such as Unite which is the main union for engineering, would argue that all apprentices should be at least level 2 and last a minimum of 2 years, either immediately or far more quickly. Others, such as USDAW the shop workers union, would argue for a much slower timetable.

in most European countries, there should be a steep gradient of increased pay. In a typical three year apprenticeship this might, as a tentative example, be 40 per cent, 70 per cent and 90 per cent of the skilled rate upon completion. These rates cannot be established centrally by government: they should be collectively bargained, a system which can adjust to the changing nature of the work, changes in design of the apprenticeship framework and which provides apprentices with a voice.

5. Advice and guidance

Careers guidance provision for young people who might consider an apprenticeship needs to be significantly improved. A TUC survey[36] shows that around 40 per cent of young people are not receiving any information about apprenticeships. A recent Ofsted report[37] into careers provision stated that the statutory duty for schools to provide careers guidance was not working well and that the National Careers Service does not focus sufficiently on supporting young people up to the age of 18. Sir Michael Wilshaw, Ofsted Chief Inspector has said: 'The fact we've only got 6 per cent of youngsters going into apprenticeships is a disaster, and it's really important that schools are fair on their youngsters and make sure that all the options are put to them'[38]

This Ofsted emphasis is particularly important as the teacher unions report that their members are often under pressure from school leaders to steer young people away from apprenticeships for fear it will harm their Ofsted or league table ratings. There is strong support for a UCAS-style system to guide young people to the appropriate apprenticeship. [39] Employers should play their part (as many unions do) by going into their local schools and talking about apprenticeship opportunities.

---

[36] Commissioned by the TUC from Believe in Young People, a charity which helps young people transition to work.

[37] OFSTED (October 2015) Apprenticeships: Developing skills for future prosperity; see also 'Going in the Right Direction?', Careers guidance in schools from 2012, OFSTED: www.ofsted.gov.uk.

[38] Evidence to the Education Select Committee, Sept 16th 2015.

[39] See for example City & Guilds (October 2015) Report on Apprenticeships, supported by TUC, AELP, AoC and several leading employers.

Careers guidance for young people is a major topic in itself which needs a separate discussion. Suffice it to say here that the model of a careers adviser (however expert) delivering a few sessions of advice at age 15 or 16 in every school or college is not the answer. Guidance needs to start far earlier; an understanding of the world of work should be embedded in lesson planning from at least early secondary school if not before; young people should be taught the skills of reflecting on their own desires and aptitudes; and they should be given far greater opportunities, well before the age of 15 or 16, to test out different kinds of occupations.

Employers, teachers, parents and young people would welcome these changes which would mean that young people entering apprenticeships (or going down any other route) would be much better prepared and more likely to succeed. Strong partnerships between schools and employers, coupled with greater use of the web should be the core of a new careers system. Intermediary organisations such as Believe in Young People [40] already provide a cost effective solution for schools and employers based on the above principles.

6. Funding

Funding is the crucial strategic issue. It shapes almost all the behaviours of the stakeholders. The TUC has long supported a levy on all employers. Some kind of levy system is very widespread around the world. It is recognition that training is a shared cost which employers should bear equally but will be unlikely to do so in a free market. The incoming Conservative government in the UK announced the apprenticeship levy in July 2015, recognising that there was a clear example of market failure. The levy consultation paper showed the relentless decline of employer funded training over several decades.

Successive government initiatives aimed at exhorting employers (the Skills Pledge) or incentivising (Train to Gain) or directly funding them (Employer Ownership) had not made a significant difference; although the work of the UKCES in developing Employer Ownership through Industrial Partnerships should have been given more time to bear fruit. The risk to UK productivity of continued under-training and poor skills

---

[40] See Believe in Young People: www.believeinyoungpeople.com. One of the authors, Tom Wilson, is Chair of this non-profit charity.

was too great to allow the slide to continue. The BIS consultation paper[41] contains detailed evidence and a remarkably candid assessment of the failure of successive previous governmental attempts to persuade or incentivise employers to invest in apprenticeships, or indeed training. A killer graph on page 7 shows the total number of employees who worked fewer hours because they attended a training course falling from 150,000 in the mid-nineties to just 20,000 in 2014. It drew on a stunning analysis by Francis Green et al charting the halving of employer training in the UK between 1997 and 2015[42].

However TUC and union support for the levy is not unconditional. It is essential that unions are fully involved in the design, governance and implementation of the levy. Equally, the levy must result in much higher quality; simply reaching the government target of three million apprentice starts between 2015 and 2020 will not be enough unless the majority are at level 3 and above.

Only employers who pay the apprenticeship levy should receive funding from the levy apprenticeship fund. Once the levy system has been implemented a two-tier system for apprenticeship funding should develop with the size threshold (i.e., the £15,000 exemption) being reduced to encompass almost all employers as rapidly as possible. The levy should not be used to support government funding for smaller employers or employers will simply see it as government taxing them to fund cuts in the skills budget. Of course many employers may continue to say it is just a tax. The better the levy is designed the harder it will be to make that simplistic accusation.

The aim of the levy must be to effect a fundamental change in employer attitudes and behaviour such that, over time, the levy ceases to be necessary except as a reserve lever to use in extremis. In Germany it is a common culture which supports the system and motivates employers, not the compulsory levy operated by the chambers of commerce - though that still exists in the background. Germany is not alone. Numerous studies

---

[41] BIS. (August 2015) Apprenticeship Levy: www.gov.uk.
[42] Francis Green, Alan Felstead, Duncan Gallie, Hande Inanc, and Nick Jewson. What has been happening to the training of workers in Britain? LLAKES Research paper 43: www.llakes.ac.uk

have shown that levy systems are successful where there is employer buy-in and support for the levy system [43].

This is why it is legitimate to extend the scope of the levy to as many employers as possible. The greater the proportion of employers who pay into the levy means greater uniformity across the apprenticeship funding system. Safeguards should be implemented to ensure that smaller employers excluded from the scope of the levy do not end up paying more for apprentices. Small employers who recruit apprentices outside the levy should not have to invest any more than large employers. Policy changes that result in small employers ending up paying more pro rata for each apprentice could drive down apprentice recruitment and or quality amongst small employers.

Payment via the tax system will utilise existing mechanisms with which employers are familiar. However, research into comparable levy systems across Europe has highlighted that levies which use the tax system do not necessarily lead to increased employer awareness of or participation in skills training. Some employers would be paying the levy, unaware of the funding they could receive from the levy fund. Promotional work and campaigning work will be needed.

Inevitably some employers will try to avoid payment or 'game' the system. 'Piercing the corporate veil' of complex company structures will be needed. The pensions legislation offers a tried and tested model, for example ensuring that all franchisees are included in an overall headcount. Some employers may try to avoid the levy by breaking up their business into smaller 'undertakings'.

Another problem could be employers setting up their own provider company (which is of course normally fine, Rolls-Royce is an excellent example) but then contributing to the costs of other levy paying employers or making excessive profits. A more effective approach might be to encourage employers to employ and train more apprentices than their business needs. This 'overtraining' works well where apprentices are seconded throughout the supply chain, picking up the skills which would

---

[43] BIS. (2014) Employer Routed Funding. BIS Research Paper Number 161: www.gov.uk.

ultimately benefit the host employers as well as benefitting the wider skills base of that industry/sector. This would also ensure that levy-paying employers take responsibility and ownership for taking on apprentices, rather than shifting the responsibility onto other employers.

It would reasonable to expect that the number of employers accessing the apprenticeship levy fund will increase over the first few years of the levy fund as awareness of the levy increases alongside the capacity of providers to supply high quality training. In the early years of the levy there may be a large surplus due to employers paying into the levy exceeding the number of employers accessing funds from the levy. This surplus can be used to fund overtraining though there should be a cap on the 'top up', rather than looking to utilise the entire yearly surplus. This carried forward fund will ensure that in future years, as the levy system is utilised by a greater number of employers, there remains an adequate source of funding to provide top ups to overtraining employers.

Governance, via the proposed Institute for Apprenticeships,[44] should be modelled along tripartite lines like the Low Pay Commission with unions, employers and independent stakeholders. It must be simple, clear and transparent. Feeding into the national board should be sectoral advice on eligible costs and tariffs. Although these should not be part of the formal levy governance, sectoral bodies should remain the responsibility of employers, working with unions, but recognised and encouraged by the new Institute.

Quality is crucial. Fair pay should be a key determinant of good quality apprenticeships. HMRC (i.e., the tax authority) should monitor the levy, and government has made it clear that any examples of deliberate evasion will be treated as fraud. It will be a statutory offence to misuse the term 'apprenticeship' in relation to poor quality training which does not comply with relevant standards. Sectoral bodies,[45] involving unions should have a

---

[44] BIS response to the consultation on the levy, published 25 November 2015.

[45] These sectoral bodies should arise organically and reflect whatever is the prevailing pattern of employer organisation in a sector. In some they may be a Trade Association, in others an SSC. There should be as little national proscription as possible. As the economy changes it is

role in advising the national board on these standards which will need to be as simple and clear as possible.

Training plans and apprenticeship agreements should be used to plan and agree high quality apprenticeships and to stipulate the standards that will be met on a particular apprenticeship. It is welcome that these kinds of agreements are highlighted in the BIS response to the levy consultation[46]. The training plan would set out what would be included in the apprenticeship and highlight the rights and responsibilities of each stakeholder. Importantly, the agreement or plan would also be signed off by all the key stakeholders, including a trade union where they are recognised. Before an employer could access levy funding a training plan should have to be submitted for approval by the new Institute.

Trade unions are the only stakeholder in the system which represent apprentices in the workplace. Too often in the apprenticeship system, the focus is simply on 'putting employers in the driving seat'. Those who have first-hand experience of apprenticeships and their representatives should be given an opportunity to help shape the programme and ensure that high quality apprenticeships lead to secure, decent employment. As in most other levy systems, trade unions should:

- Be represented on both the national Institute and sectoral levy boards
- Be represented on any governance boards which deal with oversight, monitoring/compliance etc.
- Play a role in setting any standards relating to apprenticeships
- Where recognised, unions should sign off training plans.

Lessons should be learned from the recent trailblazer evaluation [47]. Trailblazers have seen an increase in employer involvement and in some cases an improvement in quality; the key lesson is the need to involve an

---

inevitable that such bodies will change and evolve. They should have the right to advise the national levy Board, employers will have a strong incentive to make these boards representative and fit for purpose.

[46] And may well reflect TUC urging, alongside that of others

[47] BIS. (2015) Evaluation of the Apprenticeship Trailblazers: Interim Report. BIS: www.gov.uk.

array of key stakeholders, who can bring different expertise to the table, rather than a narrow band of just a few (albeit often large) employers. The levy system should involve all stakeholders with an interest and expertise in the apprenticeship system.

Unions and Unionlearn play a key role engaging employers; raising awareness and highlighting both the necessity of, and the route to, high quality apprenticeships.

The SFA provider register should be used as an additional check on quality. It should be developed to ensure that only high quality providers (including of course employers who are also providers) can be involved in the apprenticeship system. The register should be developed to ensure providers must adhere to rigorous standards.

France, for example operates a levy model where providers of skills training can be randomly inspected. Of course there would need to be proper safeguards for provider/employer staff involved in delivering training but a supportive (not punitive) skills inspectorate should be established. It is frankly unlikely that Ofsted could fill this role.

Eligible expenditure should include provision for additional support for young people such as mentoring/coaching, and designated apprenticeship officers. It should not include apprenticeship wages, for time in-work or doing off-the-job training. Clear guidance will be needed to avoid non-apprenticeship costs from being included.

The levy is long overdue but will be a major challenge to many employers. It is the price being paid for years of neglect. Significant work will need to be done to inspire new employers to develop apprenticeship schemes to match the increasing demand for apprenticeships. It will take time for FE colleges and private providers to meet this new demand. Statutory underpinning will help. The new public sector duty to take apprentices and the use of procurement will also help drive up the number and quality of apprenticeships. [48] The elements of a better system are slowly emerging.

7. Summary

---

[48] BIS. (2010) Skills for Sustainable Growth, BIS: www.gov.uk.

To sum up, the fundamental issue is raising the status of apprenticeships. Entry to apprenticeship programmes should be expanded and improved to involve many more employers and young people. The levy will boost that process. However there are contradictory drags on the system that undermine the drive to higher quality. Temporary fixes such as specifying that a level 2 is needed for entry to a level 2 apprenticeship is plainly absurd.

The answer is to invest in a systematic, expanded and improved apprenticeship system (embracing a reformed traineeship programme) which is designed to take all young people, both the very able and those with few or no qualifications, and prepare them for entry to an apprenticeship.

The government should consider making the following changes:

- Underpayment of apprentices needs to be taken seriously. The National Minimum Wage Enforcement team should be adequately resourced. The rate of the Apprenticeship NMW rate should be raised to the level of the young workers' NMW rate.

- Trainees should be paid or receive adequate financial support whilst they are on a work experience placement. Similarly for apprentices themselves. This is the only effective safeguard against exploitation and would increase participation rates amongst disadvantaged young people. If the young people remain on benefit the financial support payments need to be flexible and not affect their (or their families') benefit entitlement. In the case of trainees their work experience placements may be for a short duration.

- Progression to an apprenticeship or employment should be integral to any traineeship or apprenticeship. Progression opportunities and data on, for example, pay and careers should be far more prominent. This would improve the motivation and participation of young people and their parents, substantially raising the status of apprenticeships. Alison Wolf was right, sadly, to say that apprenticeships are viewed as worthy 'but for other people's children'[49]

- Underpinning the success and completion of apprenticeships and traineeships is an effective careers guidance service, starting at an early age and embedded in the school curriculum, making full use of professional advice, ICT, and strong partnerships with employers.

- Data collection and monitoring systems need to be improved for those participating in both apprenticeships and traineeships. To ensure that under-represented groups are offered equal opportunities to participate, it is necessary to understand the scale and nature of the problem. A portion of levy funding should be used to support research and analysis.

- Over time, every apprenticeship should reach at least level 3 and last for at least two years.

- Trade Unions should be involved in the design, delivery and oversight of the system, nationally and sectorally. This will give apprentices (and those working alongside them) a real voice in their training; it will enable unions' vast experience of how apprenticeships work in practice to feed into discussion; and it will thereby strengthen the legitimacy and currency of the emerging apprenticeship system.

- And finally, trade unions play an important role in helping young apprentices enforce their workplace rights. Therefore it is essential that young people are better informed about their workplace rights, especially the right to join a union.

---

[49] See The Wolf Review, BIS (March 2011).

21.
OPPORTUNITIES
IN INFRASTRUCTURE PROJECTS AND PROCUREMENT

Neil Robertson
CEO, National Skills Academy for Rail

———

This chapter looks at realising the growth opportunities that are there for apprenticeships at the heart of the government's plans for a multi-billion pound investment in infrastructure projects. How can these benefits be secured for local businesses and young people? It looks at how industry has been taking a lead in major projects such as Crossrail and proposes that the right approach to procurement can be one of the key levers to secure the much-needed additional investment in skills and apprenticeships.

The UK has embarked upon an ambitious, £400bn, plan to upgrade its primary infrastructure over the next decade or so. This means new power stations, transmission lines, rail lines and trains as well as upgrades to existing rail, water, telecoms, road and energy infrastructure. As the 2015 treasury productivity plan notes, a modern high capacity infrastructure drives economic growth. Few would dispute this, but, as in many other areas of life, it's how these exciting and shiny new assets are built, commissioned, maintained and operated that matters.

As well as helping other sectors grow, investing in infrastructure can create growth in its own right. The present Chancellor clearly agrees and his 2015 Conservative party conference speech suggested he is at least partly hitching the prospects of strong economic growth to the success of the infrastructure plans. These are therefore high stakes for him and for the country.

The setting up of an Infrastructure Commission and the appointment of Labour peer Lord Adonis will help drive the investment and enable the opportunities for apprenticeships to be delivered sooner if the tricky planning issues can be resolved.

This intrinsic growth however is substantially predicated on one assumption - a high degree of local 'content'. This means 'kit' is built locally and the workforce, who need not be local, must pay their taxes and spend their cash in the UK. Additionally, the corporate benefits would ideally be local too.

Considering how to ensure this local growth dividend is my main task in this chapter. Before we jump in, however, let us try and give some overly simplistic context by crudely assuming that:

- Half the spend is on 'kit' and half on labour (£200bn each)
- A further third of the 'kit' cost is labour (£65bn)
- A physical inanimate asset lasts for forty years
- Once commissioned, over this period there is a further labour cost in maintenance/operations (i.e., it's not all run by robots yet)
- There is an ROI of £4 for every £1 invested, if built optimally
- That ROI becomes £2 if built less optimally
- Wage inflation bad case is 6 per cent per annum and inflation is costed in at 2 per cent average over the next decade
- Money is cheap and comes substantially from the government
- Government has a clear leadership and regulatory role
- Profit margins are relatively low on construction
- Skills shortages already exist in relevant disciplines
- The companies that own and run infrastructure assets do not currently enjoy excellent PR, except of course Crossrail.

If these assumptions are broadly in the right area, then three things naturally follow and a fourth is never far away.

Firstly, who builds and commissions the asset is a bit more important than who and where it is assembled. Whomsoever operates it is even more important.

Secondly, if built and run less optimally (from a UK economic perspective), the Treasury will have no shortage of projects with a better ROI than £2. So if the infrastructure plan is poorly delivered, government might well take their ball elsewhere.

Thirdly, there is a £106bn bill for wage inflation risk (40 per cent over ten years x the labour component.)

Fourthly, there may be a growing feeling that these projects, funded by UK taxpayers, should benefit UK workers if at all possible and therefore create frustration if they appear to benefit mostly migrant labour. This could be seen as a big missed opportunity.

Taken together, there is an almost inescapable conclusion that those responsible for the exciting Infrastructure Plan need a good plan for ensuring a plentiful supply of locally-based labour.

Happily, the Treasury agree and have published their first thoughts on this in the National Infrastructure Plan for Skills. This, somewhat gloomily, sets out the scale of the challenge, reminds us of the well documented skills shortages and unhelpful age profile of the current workforce, and challenges leadership to address these. There are a number of strands to this challenge which I will set out before I look at potential solutions and focus specifically and optimistically (as if Captain Mainwaring to Private Fraser) on 'procuring for apprenticeships' as a key part of the answer.

So firstly what are the barriers to the infrastructure subsectors employing lots of skilled people?

Many. They are in short supply. They are in global demand. Their own workforce is ageing. The sector attractiveness is relatively low. There is a paucity of graduates in relevant fields. The specialist nature of the work means training is often best done on the job, making it demanding and expensive. There is a relatively poor supply at vocational level due to the historically low volumes, shortage of trainers, and cost of the necessary equipment. There are safety concerns too. In short, a perfect storm.

There are two further structural problems. Firstly, with uncertain workloads, short lead-in times, low margins and time-limited contracts, the supply chains have historically under trained, preferring to poach and therefore drive wage inflation. At asset owner level, infrastructure sectors are typically economically regulated and growth is limited so executives are incentivised to make additional profit by beating the cost of capital and finding efficiencies. Training has historically been an easy efficiency. The net effect is that around half of the skilled people will leave over the next ten years and replacement rates are at less than half the required levels.

Government has sought to increase the volume of apprenticeships by paying the sector the top tariffs, but £10k, £12k or even £15k are not much against an all-in cost of £121k per apprenticeship.

The energy regulator, Ofgem, to their credit recognised this and made an allowance in their settlement for training for workforce renewal. This, together with some inspirational leadership by energy and utility asset owning CEOs, has increased numbers a little at the top of the supply chains, but not significantly lower down, where the perception is still of a first mover disadvantage. New strategies published by the Department for Transport (DfT) and the rail supply chain are hoping to drive similarly positive leadership.

It remains to be seen whether the planned apprenticeship levy will drive demand – early predictions are that it won't, mainly due to the above cost ratios remaining prohibitive. The new Trailblazer apprenticeships, whilst an improvement on the previous options, are not sufficiently different to drive demand in the way some imagine. Essentially, the previous programmes were viewed more positively in these sectors than many outside the industry perceived.

What other levers are there? One significant approach is to mandate/incentivise apprenticeship development through the procurement process. This will be the focus for the remainder of this chapter.

The science, or art, of using procurement to drive skills is in its infancy. Treasury published a Crown Commercial Services note stating that public contracts will drive apprenticeship numbers. This has been confirmed in the 2015 budget, autumn skills statement, DfT-transport skills strategy, and the National Infrastructure Plan for Skills. Government clearly believe there is mileage here.

Infrastructure UK and BIS ministers (Nick Boles and Sajid Javid) have supported energy and utility efforts to establish a voluntary protocol, while the rail industry has been active in this area since 2007. What is the experience, and what prospects are there for procurement to really drive demand?

There are many examples of enlightened procurers communicating clearly, but often informally, their expectations of supply chain people development activity. Scottish and Southern Energy on the Beauly/Denny transmission line upgrade, or National Grid, led by Steve Holliday, on a careers programme. These are extremely welcome and they show doubters that the experience need not be painful. However, to achieve significant culture shift down the supply chain levels, legally mandating training offers the best hope.

One of the first meaningful attempts to formally link infrastructure procurement to skills development was in the Transport for London tube upgrade in 2007. TfL made it clear that it wanted successful contractors to engage and train specific numbers of young apprentices from particular postcodes. I was a central government official at that time and encountered the key challenges. Lawyers argued that there was a risk of challenge on grounds of a) age discrimination, b) postcode specifics and c) non-UK bidders would be disadvantaged in not having knowledge of apprenticeship delivery.

The legal debate that followed was, perhaps inevitably, not totally conclusive but the upshot was that the condition was implemented. It was concluded that it was broadly a reasonable thing to do as a) there were proven skills shortages, b) there was proven disproportionate unemployment in the target postcodes and c) apprenticeships are a tried and tested UK policy with analogues elsewhere.

The only showstopper seemed to be age discrimination, so the age specific clause was dropped – not a great hardship as most employers associate apprenticeships with young people anyway. I would observe, however, that it still required some strong leadership within TfL and also in the Department for Innovation, Universities and Skills (DIUS.) The programme delivered substantially on its promises, without challenge, and also enhanced TfL's profile with its customers, who valued what looked to be a very sensible approach. A key milestone had been reached which didn't go unnoticed in the rail sector, government and infrastructure more generally. And, let's face it, the risk of challenge was always going to be low as the accompanying dismay would not be good for that contractor's brand values.

Crossrail, the new line across London, was built on this approach. Led by Terry Morgan (now advising DfT on these questions), the Crossrail procurement process specified ratios of expected apprenticeship growth to value of contract spend. This was not done quietly; it has been identified as an important feature of the new line from the outset. With confident assertions of apprenticeship growth as a core message, the Crossrail public relations have been overwhelmingly positive, further cementing boardroom understandings of the value of a good skills story as well as creating a skills legacy

Before we leave rail, it is clear that Simon Kirby, who leads HS2, the new fast line from London to the north, is thinking along similar lines. Can other sectors follow this lead?

Before we address that, we should briefly consider what we have learnt about how to mandate with the necessary precision.

It is clear that specifying a ratio-to-contract spend e.g., one apprentice per £3m of spend, is workable. It has the advantage of being simple and, crucially, relatively easily measurable. It is an output measure that doesn't require a track record so perhaps removes some of the non-UK contractor concern. However, different infrastructure contracts have different proportions of expensive plant/equipment and technology (we made a crude average above). For example, £3m doesn't go very far when building a nuclear power station. So we need to vary the ratio (no reason why not of course) or find another method.

The energy and utilities sector, as one strand of their ambitious skills plan, identified procurement and set up a working group of lead asset owners and contractors who spent a year testing different approaches with their procurement specialists. They investigated setting expected levels of training at the outset of tendering – an input measure if you like. So a contractor, to make progress in the commercial tender stages, would demonstrate a certain percentage of apprenticeships and training in their workforce. This was more measurable as, like other construction-related sectors, there are licence to practise schemes and good visibility through the sector skills partner.

A figure of 5 per cent of workforce in training and or apprenticeships was tested and found to be stretching but not entirely unreasonable. There was a wider 5 per cent in training campaign running in service sectors at the time. The advantages of this approach are that it benefits those who have demonstrated the right behaviours on a voluntary basis in the past, and it is more likely to be rolled down the supply chain. The obvious disadvantage is that of apparently creating a barrier to entry to non-UK based companies.

The energy and utilities process is still underway and is having to negotiate some utility specific EU regulation, but I am confident that it will be successful both formally and informally, as the associated debate is contributing to a culture shift that only the most insensitive will miss.

The government has both absorbed and encouraged this paradigm shift and we can expect more targets and examples as the Infrastructure Plan unfolds; although still early days, I believe the genie is almost out of the bottle.

Such confidence cannot be exhibited in the prospects for natural commercial processes to drive training in other less regulated sectors. Whilst the minimum/living wage sectors will be influenced by the forthcoming apprenticeship levy, it's harder to see how sectors such as finance or IT could use this lever to create a more level playing field.

Voluntary agreements such as the 'The 5% Club', with their accompanying peer pressure and corporate and individual pride, may have more traction. Sectors not characterised by relatively low numbers of high value contracts will inevitably see less in this approach directly for them. Perhaps the way forward here is to develop the argument and evidence base where sustainability of capability becomes a strategic commercial necessity rather than an operational inconvenience or a social responsibility.

The early pioneers, especially in the rail sector, have certainly shown that, with leadership, it is possible to mandate skills in infrastructure procurements. The methodologies are gaining in credibility and sophistication. No longer is the dreaded first mover disadvantage scenario inevitable – prowess and a track record in renewing one's workforce can and should be a core tenet of commercial decision making. Captain Mainwaring would be proud.

22.
PRODUCTIVE WORKPLACES

Andy Smyth
Development Manager, Vocational Learning, TUI

———

Do apprenticeships really impact growth, productivity and profit? If yes, is it just a case of taking them on and the results will speak for themselves? Clearly there is more to this than a simple 'yes' or 'no'.

The straight answer to the first question is 'yes', but we need to be very clear, the impact can be either positive or negative. There is a range of factors that need to be fully understood and then correctly addressed in order to achieve the positive impact that is desired. In the next few pages, I will explain some of what can happen, good and bad, even with the best of intentions. I will also make some observations that I feel are critical to whether the impact of apprenticeships is 'positive' or 'negative'.

TUI Group has been formally involved with vocational education since 1988. During this time, the business has changed significantly. The roots of the programmes stretch back to days when Lunn Poly was still on the high street. Young people joined the business, were introduced to it by their manager and learnt how to become a travel agent. All learners were employed across the UK-wide network of travel agencies on fixed two-year contracts and were additional to the normal headcount. The business used the same on-the-job training that they would use to support the introduction of any new person but also made additional special provisions over and above this for the learners.

Private training providers were used to manage the assessment, objective setting and planning for the achievement of the qualifications that were part of the formal programme. In addition to the initial sign up of the learners, the training providers worked with them and their manager to confirm a formal plan of actions that would need to be completed. The business introduced formal off-the-job training days to support specific knowledge learning and also the development of English and maths skills as required for each individual learner. Learners progressed very quickly in both becoming a productive contributor to the business targets and in the achievement of their programme objectives. Learners were required to complete the level 3 programme over two years, completing level 2 qualifications before progressing on to the level 3.

From 1988 to 2002, the business was incentivised to take on more and more learners through the provision of public funding, and regularly had cohorts in excess of 1,200 at any one time. Overall achievement rates were often in excess of 50 per cent of all those who started, with completion of the level 2 first element being over 70 per cent. Both were good for the time. As a result of the learner being employed in addition to head counts on fixed term contracts, there was an inevitable churn of personnel. During the last six months of each contract, attention was turned to whether or not there would be 'permanent' vacancies available and which learners were to be offered roles. Those who were not successful left the business. This process also had the unintended impact of some learners leaving prior to completion to secure roles elsewhere.

Overall, the business was very happy with the programmes and enjoyed a very healthy flow of people completing the programme and moving on to permanent roles. Many of the former learners also rapidly progressed to become assistant managers and branch managers. The business was subject to inspection and it was common for the results of inspection to return a 'good' outcome, and positive observations on learner productivity, motivation and career progression.

In 2005, a new manager was appointed to lead the 'apprenticeship programmes' and a full review was carried out to determine how the programme was performing, what value it added to the business, and how the business valued the programme. The findings were very clear and presented a number of key challenges for the business.

Aims and objectives for learners were unclear. Were they apprentices or normal employees with normal targets? Achievement rates had dropped significantly over the previous three years and the learner experience was inconsistent.

The focus for the programmes had shifted towards the process of recruiting large volumes of learners to the business, and some managers were more focused on meeting their objective of employing an apprentice than always getting the right person for the role.

The scale of the programme and increased bureaucracy had caused the central team to become focused on managing the training providers, processing payments for activity and ensuring that the business was compliant with the various requirements, regulations and rules associated with public funding.

The business was unclear of the value of the apprentices and how they contributed to the business objectives.

The report led to a complete repositioning of the programmes. The whole business case was challenged and a number of questions asked, including:

1. Why is the business operating an apprenticeship programme?
2. What objectives should it have?
3. What will the business gain?
4. Why would a young person want to join the programme?
5. How do we support them?
6. What roles should the apprentices do?
7. What type of person should we be looking for, and how do we find them?
8. How should we manage the whole process?

It is in response to these questions that we can unpick how to ensure that apprenticeships create a positive impact on growth, productivity and profit rather than a negative one.

Why is the business operating an apprenticeship programme?
There must be very clear and articulated business needs identified before any business establishes an apprenticeship programme. To blindly engage an apprentice without this clarity of purpose will mean that you can never fully understand how well they are meeting your needs and it will not be possible to measure the return on your investment. It was clear for TUI that we needed a supply of energetic, talented and self-motivated people

to join our business. This can however be achieved through a targeted and focused recruitment process, so why use an apprenticeship programme? Apprenticeships provide clear structure, external recognition for what has been achieved, contain a mechanism to ensure that the learner is supported and managed through the programme, and are monitored and measured externally to ensure that standards for delivery are met.

Employers that are confident enough to say that they can do all of this over a sustained period of time without apprenticeships or a similarly structured model are very few and far between. Through the natural cycles of business and changes in the economic environment there are many pressures that have impact on the development of people, some of it through necessity and some through the gradual erosion and change to the purpose that time brings.

What objectives should it have?
Apprenticeships have been around for hundreds of years and although many people have attempted to improve or update the model, it is quite simply a case of an individual committing to a process of learning whilst working, in order to be eventually recognised as being fully independently capable in the given area.

In TUI, apprenticeship programmes have very clear purpose and objectives. We identify the skills areas we need to fill for the future to support our business growth plans. This is of course an on-going activity. As with most businesses, we have many functions and it takes time to fully establish apprenticeship programmes. Rushing the process has significant risks so it is far better to establish the process and scale up than aim too high too fast.

We have focused on the entry-level roles first but then considered what next? If you want the best people to join your business, it is likely that they will have ambition. There is no point in gathering pools of ambitious people and then providing no next steps or opportunities. Indeed, if you fail to do this, the process will be damaging all round.

TUI programmes set out to ensure that apprentices become fully capable within the job roles that they are employed in and to grow them into our culture so that they become effective in supporting themselves and deciding on where they want to go next. Our programmes provide one of our talent pipelines for the rest of the business, full of people with new ideas and new thinking for today and tomorrow. It is not an expectation that all apprentices will progress up the career ladder, but most do. It is a perfectly valid outcome that the individual finds a role that suits them, their life situation and of course our business.

What will the business gain?
The TUI programme is set up and positioned in such a way that it forms part of our talent development process. As a business, we have many entry routes and development opportunities but the apprenticeship programme impacts the business in a number of specific ways.

Taking a structured approach to recruitment and selection enables the business to become more efficient, and significant savings can be made through this. The development of clear information and processes that new people can use to access the business means that all the activity is channelled and managed efficiently. It is easy to overlook the added benefit that is derived through saving the time and effort of those in the wider business.

TUI has always had the advantage of being in the travel business. This means that we have not struggled to attract people to core business roles. However this does bring its own challenges in that candidate selection becomes a huge task where individual roles are oversubscribed. We attract great candidates and that is good for the business. The disadvantage is that we need to manage the expectations and feelings of those who are not able to join the programmes.

We have worked hard to ensure that we provide as much information as possible to all applicants for our roles so that they get a genuine understanding of what the job actually is, including the apprenticeship requirements. In addition to this, we clearly indicate the characteristics and capabilities that we are looking for. It is not about qualifications in most instances. For those who review our details and decide that it is not for them, and those who want to join us but are unsuccessful, we provide information about alternative options, personal development opportunities, and education channels that may prove helpful to them for the future.

This work has the positive effect of ensuring that only those who really want to join us are applying, and those who don't succeed feel good about the process. As well as reducing the overall amount of work required to select our candidates, this communication also enables TUI to show itself as a trustworthy employer that cares for everyone, not just the people that we employ. As a brand we are very clear that everyone is a potential customer and that people applying to join us today may also apply in the future.

In 2010, TUI conducted a formal ROI study of the apprenticeship programme and has been able to formally measure the return on investment in a number of ways. Specific areas that we were able to show are as follows:

- Apprentices stay two years longer with the business on average
- Apprentices climb the career ladder more quickly
- Fully completed Advanced Apprentices' sales revenues are 17 per cent higher
- Fully completed Advanced Apprentices' net sales profit is 11 per cent higher

There are additional areas where we know the business has benefitted, although formal ROI analysis has proved difficult.

- We know that other staff members have grown their own knowledge as a result of the apprenticeship programme being delivered in their site.
- The allocation of a mentor to our apprentices has been very beneficial for the apprentices, but has also had an enormously positive impact on the mentors themselves, who have gained opportunities to support others and demonstrate their supervisory and management capability at an early stage.
- Bringing young people into our business has led to a greater uptake of technology and has helped in our process of modernisation and change.
- The energy, enthusiasm and motivation that apprentices have brought to our business have both helped their own performance and that of others.

- Through the delivery of high quality apprenticeships, the TUI brand is enhanced.
- Every applicant and successful candidate that has a positive experience becomes a TUI brand advocate and potential customer.

Why would a young person want to join the programme?

The apprenticeship process of working and learning in TUI provides every learner with the opportunity to apply and build upon what they have learned. Based on the normal TUI performance review process, apprentices receive continuous feedback on their performance. This means that they are always aware of the standards and levels that they need to achieve and what they need to do to further develop and improve. Following the structured pathway of learning within our apprenticeship ensures that apprentices have all the knowledge and understanding that they will need when they encounter new challenges and situations. The everyday realities of working in a team and dealing with customers means that apprentices are provided with the ideal combination of regular practice that enables them to master the routine activities that they are expected to excel in whilst also placing them in unfamiliar situations on a regular basis, helping them to develop and extend their problem-solving capability.

When deciding whether to start an apprenticeship at TUI, apprentices are choosing to join an energetic and dynamic business where they will earn above the minimum wage but also have the opportunity to enhance their earnings through incentives and bonuses. In order to realise these additional benefits, the apprentices need to fully engage with the learning and development that is available to them. This is also supported and driven by the team that is in place to help them. Travel is also a very desirable and attractive sector and TUI Group is the leader in Europe meaning that apprentices who join TUI have a world of opportunities ahead of them.

There is a well-trodden path from the apprenticeship entry route through to supervisory, management and leadership positions in various occupational areas that TUI has to offer with former apprentices clearly in evidence at every stage. From initial entry to TUI, there is a vast range of diverse opportunities. They could climb the ladder within the retail environment that they started in or move over to roles in our overseas operations, or take on a role in one of our customer contact centres. Alternatively they could follow a career in our airline or perhaps move to our head office and take on the opportunities that are available in our IT, finance, trading, HR or marketing teams.

In addition to the specific benefits that TUI offers, there are also a number of other drivers that make choosing an apprenticeship a good idea. It is clear when looking at the achievements of the people who participate in apprenticeships that the model, processes and experience of being an apprentice works very well for most people. This is not to put apprenticeships into competition with other routes but there is clear evidence that those following this pathway do go on to enjoy considerable success. Indeed the journey enables the development of career networks and experience from the beginning and importantly eliminates the need for student debt that other pathways bring.

How do we support them?
All apprentices in TUI join the business through the same 'on boarding' process as everyone else. We feel that making the apprentice part of the normal team helps their development and accelerates their understanding of the business and its objectives. Managers play a key role in the recruitment process and are able to form a bond with their apprentice very early on. Once the apprentice arrives for work we quickly get them involved and introduce them to everyone who they will be working with including those who will be supporting them through their apprenticeship journey.

We buddy up our apprentices with former apprentices wherever possible as this builds a strong relationship with someone who has done the programme and is close to their age. Following the review in 2005, we chose to move all activity in-house. This gives us more control over the way in which assessors and quality assurers work. The people who surround the apprentice make a huge difference, and ensuring that they all fully understand the purpose and objectives of the programme is vital.

As part of our teams, the apprentices are also included in the usual business measures that cover team and individual performance. We are always delighted to note in our monthly shop, regional or national performance tables that apprentices have done well. This clearly shows what apprentices can do given the opportunity. It is important that apprentices understand the targets that they need to achieve. At TUI, we have found that once they get going and have established themselves, apprentices are not satisfied to just achieve targets. They want to be the best. TUI apprentices featuring in internal regional or national awards are common but they also excel externally. We are very proud of our track record of high achieving apprentices and we make sure that they know this.

As well as reporting on the good news, it is also very important to be aware when things are not going as planned or when something unexpected happens, either inside or outside the workplace. Dealing with these issues well can be the defining moments for the individual, and can be the point where the business unlocks the greatest value from its people. At TUI, we feel that we should support all our people and enable them to shine and show what they can do. We clearly need to deal with issues effectively but we want the apprentices to be the best that they can be and for them to go on to enjoy success both at work and at home.

What roles should the apprentices do?
We decided that the fixed-term contract approach was no longer suitable for our retail travel agent business so now every apprentice is recruited to a permanent role from day one. By doing this we have removed the issues that we had previously experienced at the end of the two-year contract terms. We also reviewed our intake profile and ensured that it was reflective of the business need rather than being driven by external incentives. The net result is that we now have a rolling intake of apprentices that is driven by vacancies rather than targets. Where we find that we have a number of vacancies, we will run a campaign to coincide with the educational cycles for school and college leavers. But this is involves far lower volumes than before and causes significantly less impact on our business.

Over time there has been a reduction in our overall numbers for this programme but the quality of the experience has improved as a result. We have been able to work with other TUI business areas based on the success of the retail travel agent programmes and now have a more diverse set of entry routes through which apprentices can join our business. We now look at all roles to consider if it would be suitable to take on an apprentice. We are however sometimes frustrated by the lack of a suitable programme to support the roles. In these cases we have taken the initiative, got involved, and are trying to drive developments and policy changes where we think they prohibit us from doing more.

What type of person should we be looking for, and how do we find them? As a business, we decided that our stated values and behaviours should be the driver behind who we should bring into the business. We have a very strong culture in TUI and together with our values, behaviours and clearly defined role profiles we are able to communicate to prospective candidates what jobs we have available, what is involved in those roles, and the type of personality and capability sets that would help. TUI Group is a motivated, energetic and focused business that has a great understanding of what the customer wants. As a result of this we are always looking for people who buy into our culture and have the potential to succeed.

People who have strong interpersonal skills always do well in the travel sector. The ability to communicate and build networks is essential. Determination, self-motivation, personal resilience and passion are vital when working in a holiday business. This is a sector where the phrase 'work hard, play hard' is truly deserved but for the apprentices, the rewards and opportunities on offer are huge.

We search for our new candidates in all the normal places and use the same channels and process as most others. It is not just the fact that you turn up at careers fairs or open evenings for schools, sixth-forms, colleges or universities, it's all about what you have to say…and who is saying it. Apprentices do not generally buy from people who look like businessmen or women. They like to hear messages from people who are close to their age, people who they can associate with. The preparation of clear, simple and informative materials, communicated by people with a passion for their area is all you need. The formal structures such as careers websites and recruitment systems need to be right, simple and interactive but they are all about the process, the candidate wants to be sold a vision that makes them feel confident and happy. One where they feel that they can learn, develop and prosper.

How should we manage the whole process?

TUI manages all the apprenticeship programme processes internally. As well as direct funding arrangements, we are also the Approved Centre for all the qualifications that we offer. We specifically use City & Guilds and the Institute of Leadership & Management (ILM) for our programmes as our people clearly associate these brands with quality and transferability of what they have achieved. There was a significant change following the 2005 review, initially to a new training provider but then later to a fully in-house model.

Today we have a number of business areas involved in apprenticeship delivery. As an employer, we have made decisions based on what we feel will deliver the best results for the learners and the business. It is clear that, as a business, we have people who are experts in their fields who could train our apprentices. It is also true that their time is best spent doing their day jobs. We spend time with these experts to ensure that our training is covering all the latest developments and thinking. Our solution is to have a specialist team at the centre to manage the processes and to ensure that TUI is always operating to the required standards. There is also a dedicated team of assessors and internal quality assurers with the occupational competence to support our learners in partnership with their managers. A clear principle that must be in place is that the central and field based teams support the business to develop their apprentices...and not the other way round. Ownership must sit in the business at all times.

TUI has modernised the way in which apprenticeships are delivered and indeed has worked hard externally to bring about changes in policy and qualifications that allow more people to take part. The ability to report on programme performance with a single set of data is an essential element of knowing when things are going well and where improvements may be required. Online learning and remote support are key features of the TUI programmes. We need to be efficient with our resources and effective in the work we do. The objective is to ensure that the learner has a great experience; that they are continuously learning and building new capabilities through on-going contact with the customer; and encountering new challenges. By doing this, the business is also benefitting. Win-Win!.

So, to return to the question 'Do apprenticeships really impact growth, productivity and profit?'

I said the answer is 'Yes', and they do. I believe that by careful consideration of the points that I have made it is possible to make the impact positive. It is far easier to get the parts of the process wrong and deliver a negative impact than it is to get it right and enjoy positive gains. It is also very easy to not notice that a good performing programme is failing if you do not continuously monitor it and respond to change.

Since 2005, the TUI apprenticeship programmes have changed significantly. Our retention rates have significantly improved as a result of providing more and better information to the learner at the beginning. Having recruited the right people, we ensured that managers looked after and supported them from the moment they started. We changed the offer to the individual and they became permanent from the very start. We communicated the objectives to everyone and had every member of the teams around the apprentices helping and contributing to the journey. The result? We moved from a low of 28% percent-overall achievement in 2006 to in excess of 85 per cent for our first year level 2 programmes and over 90 per cent for the advanced level 3 programmes. You could argue that mostly we have just done the basics right, and you would be right.

Finally, I would add that from an employer perspective you can create apprenticeship programmes that really add value to your business. By taking some time and care, you will be able to put all of the elements together, ensuring they link and add value. Once you have created your programme, it is then down to you to decide how you will respond to external influences. Incentives will be offered but they generally come with strings attached, so take care not to change or expand your programme if it means a net reduction in the win-win for you and the learners. Government has its objectives ,and when they coincide with yours this may bring you additional benefit, but always remember that it is your apprenticeship programme, to support your business.

23.
ENGAGING EMPLOYERS IN DEVELOPING SKILLS LESSONS FROM THE EMPLOYER OWNERSHIP PILOTS

Simon Perryman
Formerly Executive Director, UKCES

Introduction

The UK has a long history of building engagement between employers and government in helping to shape skills policy and apprenticeship design. The aim is to ensure programmes are relevant to business, up to date, provide skills for sustainable employment and support business success.

Many countries, outside the circle of northern European industrialised states, continue to struggle to build effective professional and technical education systems. Their biggest challenge remains how to draw employers into a meaningful relationship to set standards, design high quality apprenticeships and provide meaningful work experience to build practical, technical and 'soft' skills.

They envy our practical and pragmatic approach to employer engagement and skills development and seek to replicate it, particularly our approach to occupational standards development and our Sector Skills Councils, which have over the last twelve or thirteen years been the brokers in bringing a collective employer voice to the table.

Sometimes, in the UK, we forget just how far we have come on the employer engagement journey and how innovative we have been. We should, perhaps, spend more time reflecting on the lessons from all that we have done.

This chapter attempts to set the record straight. It offers a short history of employer engagement in skills since the 1980s, discusses the most recent approach through the development of Employer Ownership of Skills, and reflects on lessons learned, including some pointers for the future development of apprenticeship policy. But first, and before we become complacent about our successes, it may be useful to set some context and reflect on the continuing challenge ahead of us in tackling skills in the 21st century.

Context

The economy is recovering and recent employment growth has been impressive, but there remain deep-rooted challenges that must be addressed if growth is to be sustained and competitiveness improved.

Growth in productivity is the most important factor for sustaining the recovery and improving pay and social mobility. But productivity has remained subdued since the recession, and is 2 per cent below where it was in 2008 and 14 per cent below its trend rate at the start of the recession in 2008. We know skills play a critical part in turning this round. Skills have become the global currency of the twenty-first century. Without proper investment in skills, people struggle to find employment, and without clear linkage between the overall industrial strategy and the skills needed to deliver it, business opportunity will not translate into business growth.

Pressures on public and private finances, and a changing world of work, mean that traditional models of skills investment are unsustainable. Government funding for skills development is likely to continue to reduce and employers will need to make a proportionately larger contribution.

The approach adopted by government is also changing. Market-led models are now seen to provide a more effective way forward than government-driven and planning-led models of the past. Allied with this, there is less confidence in those traditional institutions supported by government that have historically been tasked with tackling skills problems on behalf of employers.

Government is looking for market-led solutions and to work directly with business, levering a larger share of the cost directly from them. The announcement of an apprenticeship levy on all employers from 2017, together with a digital account model into which employers must pay if

251

they wish to access government support in the form of a voucher, is entirely in line with this new approach.

The roots of these changes arguably go deeper than a change in political philosophy and a need to make cuts to public expenditure. They are underpinned by the concept of Employer Ownership of Skills, developed by the UK Commission for Employment and Skills and adopted by government in 2011. Employer Ownership was clearly highly attractive to a Conservative-led coalition. Indeed Matthew Hancock, the Skills Minister at the time, said publically that he would have invented it himself if the UK Commission had not got there first. The question remains whether government ever really grasped the wider strategic intent of Employer Ownership or the complexity and time required to implement it effectively.

But, before describing Employer Ownership in more detail and reviewing how much progress has been made in putting the policy into practice, it is useful to look back at previous models of employer engagement in skills in England. This enables us to see the origins of the current policy; to understand that the path has been complex and at times difficult; and that we are now seeing a changing balance between sector and locally based initiatives.

A short history of employer engagement

In 1964, the government passed fundamentally important training legislation that has acted as the basis for much subsequent action on skills. It paved the way for setting up a network of Industrial Training Boards, which raised levies from their industries and re-distributed this funding through grants to companies who undertook training.

Industrial Training Boards, and the state-driven model of skills development that underpinned them, fell out of fashion in the early 1980s. Industrial Training Boards were seen as bureaucratic and were abolished in 1981, except in the construction and the engineering construction sectors where employers asked for them to be retained. Both are still in place. The Construction Industry Training Board (CITB) for example remains a large and powerful body, managing substantial levy funding and its own National Construction College.

252

In other industries, the Industrial Training Boards were replaced by a proliferation of small voluntary industrial training bodies that grew up alongside the more substantial organisations that remained in, for example, the engineering and IT sectors. Their primary role was to work with employers to establish National Occupational Standards or NOS. These define what a person needs to know and do to be competent in a specific job. NOS became the bedrock on which the UK built publicly funded vocational qualifications and apprenticeships.

In the early 1990s, the Conservative government further developed this new privatised and voluntary approach by setting up a network of more than eighty Training and Enterprise Councils or TECs, based on the US Private Industry Council model, to combine the delivery of skills and enterprise services at the local level. These were employer-led private organisations managed by their business community. They had considerable freedom to develop services to support their local area, but in practice their work was dominated, at a time of high unemployment, by the delivery of job skills programmes for adults and apprenticeships.

In parallel, the Department of Industry established a small business advisory service called Business Link with offices in each region staffed by advisors who reached out into smaller companies to help them develop and grow.

A third national service, called Investors in People, was established to recognise companies that adopted effective 'people practices' and were good employers to work for. Over time, the three services became increasingly co-ordinated or integrated at the local level.

In 1997, the Labour government, under Tony Blair, came to power determined to strengthen the economic power of regions and reduce imbalances between the north and the south of the country. The initial focus was to establish new economic development bodies in each English region. These Regional Development Agencies were given a wide remit and a high degree of discretion in driving forward work on inward investment, business development and skills.

The RDAs had employer-led boards and substantial executive teams with large budgets drawn together from the various regional activities of government including industrial support and the regeneration of land and communities.

Training and Enterprise Councils came in for close scrutiny in terms of their effectiveness and value for money. They were replaced in 2001 by the Learning and Skills Council, an agency of government that operated through a network of sub-regional offices, and against a narrower remit to contract for the delivery of national skills programmes and manage the funding of Further Education colleges.

In 2002, the government, increasingly recognising the importance of developing employer demand for skills alongside supply side reform, decided to rationalise and strengthen the confusing and complex web of small sector-based voluntary organisations developing occupational standards for each industry.

A prospectus was launched inviting business to come forward with proposals to set up licensed and employer-led Sector Skills Councils (SSCs) to replace more than seventy existing National Training Organisations. Each of the twenty-five new SSCs received grant funding to cover their basic costs, giving them the capacity to concentrate on developing a more strategic approach to skills development with their employers, together with a promise that the boards of licensed SSCs would be able to exert influence on government skills policy.

Employers embraced this new model, with strong SSCs emerging in a number of sectors including engineering, IT, the creative industries, chemicals and pharmaceuticals, automotive repair, energy, and hospitality. The model was later extended with the creation of National Skills Academies. These were mainly subsidiary companies of the SSCs, providing a brokerage role between employers and the public education and skills system.

In 2006, the government commissioned a business-led review of skills policy by Sandy Leitch. His conclusions included that a UK Commission of Employment and Skills, (UK Commission), should be established to provide oversight of the skills system across the UK, to undertake high quality research of the labour market in support of an 'evidence based' approach to ministerial decision making, to advise the government on skills policy and to reform the SSCs.

Leitch pressed for a further push towards a demand-led approach to skills, arguing that one of the biggest challenges facing the UK was the lack of demand for skills by employers, too many small firms not engaging, and inadequate infrastructure to stimulate demand. He pressed for the role of SSCs to be enhanced, but was concerned about 'patchy' performance by SSCs, with some strong organisations being held back by weaker SSCs. He recommended that the UK Commission should review and re-licence the SSCs as one of its first tasks.

The outcome was to verify that about a third of the network were strong, a third needed improvement, and a third were less effective. A small number of closures and mergers followed, but more significant rationalisation occurred as a result of the move from grant aid to competitive funding, as described below, after the coalition government came to power in 2010. There had been a reduction from twenty-five to eighteen SSCs by the time the current government came to power in 2015.

Developments since 2010
The arrival of the coalition government in 2010 brought with it a fundamentally different attitude and approach. This was a government that believed in the power of markets, had little confidence in Non - Governmental Organisations and intermediary bodies, and wanted to build a skills and economic development system that was driven directly by individual employers.

Action at the regional and local level
An early decision was to close the nine Regional Development Agencies and to replace them with thirty-seven Local Enterprise Partnerships (LEPs) set up and run by employers on a voluntary basis. The funds of the RDAs were replaced with a Regional Grant Fund, managed nationally and available on a competitive basis. More recently, the LEPs have begun to have influence over local colleges of Further Education, (which provide vocational education below degree level), and the eight largest cities in the UK have begun to seek greater devolution of power from central government, including over skills, education and employment issues. This has been spurred on by the Scottish independence referendum where

Scotland has chosen to remain in the UK but with maximum devolution of powers from London.

The UK Commission's Investment Programme
The UK Commission agreed with government a similar radical approach in relation to sector-based skills support. This was to move to a market-based model where only those SSCs that were valued by their employers would be successful and where funding from government should be provided on a competitive basis.

This new approach required support from the four nations of the UK, secured through agreement that funding for core work on National Occupational Standards development would continue to be protected. The programmes the UK Commission established to deliver this new approach were called the Employer Investment Fund (EIF) and the Growth and Innovation Fund (GIF.)

The first was for SSCs only. The second was open to all employer organisations. Both were designed to support the development of new qualifications and delivery models, led by employers and tailored to meet their needs.

Essentially they were research and development programmes to design and test skills solutions in the market with the intention that these solutions become sustainable. They included the creation of business networks, apprenticeship training agencies, industry linkages with universities to improve undergraduate programmes, advice and brokerage services to support small companies and a range of online services to support employers. The criteria for accessing these funds included the quality of employer leadership of the proposal; a clear logic chain showing the relationship between the problem and the solution; whether the solution was breaking any new ground; balancing risk and reward; realistic plans for sustainability; and an expectation of matched employer support in cash and in kind.

The UK Commission invested £100m over three years in about 100 projects with thirty-five organisations including eighteen SSCs. The programme closed in 2015 and evaluation results are starting to emerge. Employers welcomed the programme. They understood and accepted a market-based approach, saw this as preferable to the regulatory model that preceded it, and had begun to see business benefits.

Perceptions about SSCs improved significantly with employers now seeing them as their skills organisation rather than as an arm of government. There was significant fallout amongst those SSCs who were unsuccessful in winning funding.

Employer Ownership of Skills

Learning the lessons from creating the EIF and GIF programmes, the UK Commission was also developing a more comprehensive analysis of the UK system, and published a paper in 2011 advocating a new approach called Employer Ownership of Skills.

Five main aims were set out for this proposed policy:

- Employers should have the space to own the skills agenda and take responsibility for the skills challenges facing their sector, with government stepping back;
- The development of a single market for skills, bringing together private and public investment;
- Skills solutions should be designed by employer led partnerships to reach more people and businesses;
- The public contribution for vocational training should move to employer incentives and investments and away from supply side subsidies;
- Transactions should be transparent, simpler and less bureaucratic, to facilitate choice and empower customers to drive quality, innovation and value for money. Accountability should be based on outcomes that achieve jobs and growth.

The government fully backed this new approach. Their first step was to announce a £350m Employer Ownership Pilot where funding would be offered, on a competitive basis, directly to companies that came forward with the strongest ideas to improve skills. The funding was to be made available on a flexible basis to support young people and adults, initial training and re-skilling. Employers were allowed to develop new forms of qualification and apprenticeship outside the normal rules including, for the first time, to deliver apprenticeships through their supply chains.

The UK Commission was asked to produce the prospectus and to actively promote the programme across the country. This attracted strong support from business, with over 300 applications. The UK Commission chaired the panel that selected the best bids and made recommendations to ministers. More than seventy projects were supported across the two rounds of the pilot.

Projects included increasing the number of apprentices in a large aerospace company; reskilling workers at an automotive factory and its suppliers to help prepare for the launch of a hybrid vehicle; a small business skills network in Cornwall; supporting the employability of young people for jobs at a major airport; and supporting the expansion of a major hotel by helping them fundamentally change their pattern of recruitment to give a large number of work placements and job opportunities to young people who lacked formal qualifications when leaving school.

Industrial Partnerships
The second round included a call for a number of much larger pilots to test the opportunity to develop industry-wide partnerships of employers able to take 'end to end' responsibility for skills in their sector.

Eight Industrial Partnerships were established in this second phase. These were in the science industries, IT, the creative media sector, automotive, aerospace, nuclear, tunnelling and energy. Their purpose was to test the ambition of employers to take collective action on skills at an industry wide level and at a scale of shared investment that could make a contribution to growth and productivity in sectors of economic significance.

These eight large projects were contracted to deliver £350m investment in skills over three years, include government funding of £130m, matched by £70m of employer cash and £150m of in-kind support from more than 600 companies. Sector Skills Councils were closely involved in supporting their industries to develop many of these pilots and in managing their delivery.

Early developments included a programme designed to help unemployed young people enter the power and water industries, new style apprenticeships in life-sciences, an undergraduate apprenticeship for the

IT industry, a social network to support skills development in the media sector, and extensive work to underpin skills development across aerospace and vehicle manufacturing.

While industry was highly committed to maintaining and further developing this approach, government became increasingly sceptical that collective action on skills would yield sufficient value. They wished to focus funding almost exclusively on apprenticeship participation rather than on development, infrastructure and intermediary activity.

Following the Scottish independence referendum in 2014, they were also increasingly being pressed to devolve funding to local areas. As a result, sharp reductions were made to Employer Ownership pilot funding and to the UK Commission's budget following the 2015 election.

Trailblazer apprenticeships
A further major development running in parallel with, and at points intertwined with, the Employer Ownership pilots was a review of the apprenticeship model in England.

While apprenticeship numbers had been increasing rapidly, increasing concern was being expressed that growth was too heavily focussed on adults undertaking low level programmes rather than on higher skills. There was also an ambition to achieve greater parity of esteem between apprenticeships and higher education and develop clearer career pathways for young people who follow a vocational rather than an academic route.

One main outcome was a decision by government to reshape apprenticeships around new, simpler standards defined directly by groups of employers. This 'Trailblazer' process attracted considerable employer interest and has arguably revitalised interest in the definition of competence standards.

Other proposed changes were to the minimum length of apprenticeships and the process of assessment. It is not yet clear what the 'steady state' position will be in terms of the overall number of new trailblazer standards, how they will fit together into families, or whether it is intended to achieve coverage across all occupations over time. The relationship between Trailblazer Standards and the National Occupational Standards, which will continue to be used in the other three nations of the UK, remains to be resolved.

In line with Employer Ownership principles, government funding for apprenticeships will flow through employers from 2017 using a voucher model and must be matched by employer cash. The proposed mechanism will be a levy on employers, discounted for smaller firms and managed through an online account with the government.

Policy development lessons.
What then are the main lessons we might draw from the long history of engaging employers in publicly funded skills development? What in particular have we learned from the move to a market led and more commercial approach since 2010 that might help us to navigate forwards in meeting the latest apprenticeship target while improving the quality and relevance of provision?

Firstly, employers have shown how willing they are to engage in skills development when they are comfortable with the proposed model of intervention. The chair of the Science Industrial Partnership said that this was the first initiative in a long time that he felt confident to back because it had sufficient scale, put business in the driving seat and provided the flexibility to allow real change to occur.

Employers understand and welcome market-based approaches provided they can exert leadership. They seek longer term, sustainable partnership with government and particularly dislike short term, top down led programmes that are subject to continual change.

The second lesson is the value of investing in leadership and 'running with the fastest'. Working directly with senior business leaders makes a huge difference to the pace of getting things done and ensures a strategic link between industrial strategy and the skills being developed. It allows scale to be achieved quickly and for obstacles to be more easily overcome.

Thirdly, employer ownership is about a fundamental cultural shift in behaviour by employers and government and will take longer than some initial pilots to set in place. It is about employers in each sector of the economy individually and collectively taking direct responsibility for the skills and talent pipeline they need for future success, with government creating the conditions and incentives to allow this to happen, while stepping away from the detail.

The pilots showed just how difficult it is to achieve this in practice. Employers needed considerable support from Sector Skills Councils and National Skills Academies to bring coherence to a collective approach, and government officials found it very difficult to let go of short-term volume-based metrics of performance.

The pilots were instrumental in highlighting the importance of co-investment backed by real employer cash contributions, which became seen by government as the most powerful symbol that business was prepared to take ownership of skills. It was also seen as an important re-assurance that government money would be spent wisely and that risk will be more evenly spread. The frustration from employers was that their very considerable 'in-kind' contribution of time and facilities was under-recognised and that the public funding made available was so quickly curtailed with no recognition of trying to build for the longer term.

Fourthly, while large companies have the resources to manage their relationship with government, the biggest challenge remains reaching the long tail of smaller companies who are 'too busy' to train people and see apprenticeships as too complex to engage with, and may be sold sub-optimal solutions by private providers and colleges. There is no easy solution to this problem.

A number of new models emerged from the pilots, including local business networks, online market places, and sector-based and sub-regional apprenticeship brokerage and diagnostic services. Each of these was about improving information and providing the conditions for higher quality business and skills support to smaller firms. The new digital apprenticeship model in England has no provision for services of this kind. Local Enterprise Partnerships will need to reflect carefully whether and how to use devolution funding to reach and support smaller firms.

Fifthly, business places a great deal of importance on the quality and relevance of the training and apprenticeships available to them and continues to expect training to be underpinned by a proper assessment of competence. Trailblazer standards have been welcomed as an opportunity to refresh and update these competences, but industrial sectors see them as a helpful 'front end' to National Occupational Standards rather than as their replacement.

What is clear is that it would be a terrible backwards step to lose the labour market research of the UK Commission in shaping policy thinking on skills, to allow the Sector Skills Councils to wither as the catalysts for standards development at industry-wide level, or to abandon National Occupational Standards as the bedrock on which the UK skills system should be built.

Conclusions
In conclusion, this chapter has attempted to briefly chart the history of employer engagement in skills in England since the 1980s, has looked at the policy of employer ownership of skills developed by the UK Commission for Employment and Skills and adopted by government in 2011, and has drawn some preliminary lessons from the employer ownership pilots.

Despite a great deal of change over time in the specific structures and organisations underpinning the skills system, it is encouraging that government has continued to place a high level of importance on engaging employers in the process of skills development and its delivery, and has increasingly emphasised the need for 'demand side' measures alongside policy measures to improve skills supply. Other countries, struggling to build this quality and depth of engagement, have established a competence-led approach to skills development based on our National Occupational Standards, supported by similar organisations to our Sector Skills Councils.

The Employer Ownership pilots will probably be seen as a mixed success. It did not help that government turned to new priorities before they were fully delivered. But some important lessons were learned and re-enforced. These included placing responsibility squarely with employers to address their skills needs, the importance of leadership, the value of co-funding and expecting real cash contributions from employers, as well as the need for brokerage in providing smaller businesses with high quality skills support. The pilots may have had their challenges but the policy intent of employer ownership should not be forgotten as the new approach to apprenticeships is rolled out.

It will be particularly important that the new Institute for Apprenticeships is employer-led, builds high quality and enduring relationships with industry at the most senior level, and establishes a programme that is both clear, flexible and not subject to endless tinkering and change. New apprenticeship programmes should be based on firm quality standards. An early task of the Institute will be to reconcile the current confusion between Trailblazer and National Occupational Standards to ensure a UK-wide model remains in place.

The Institute will also need to build a close understanding of labour market intelligence, sector by sector, with a particular emphasis given to supporting higher-level apprenticeships in economically important sectors and occupations such as STEM, construction and digital skills. We should expect them to work with industrial partnerships and SSCs to build talent pipelines and career pathways for each key industry in support of growth and productivity. The proposed digital delivery and levy model currently offers little opportunity for this kind of collective action on skills, being designed around a purely transactional model. It is hoped that the new Institute will have sufficient autonomy and employer leadership to grasp this more strategic approach.

24.

CLEARING ARRANGEMENTS FOR RECRUITING
AND RETAINING TALENTED YOUNG PEOPLE

Martin Hottass, Siemens

———————

There is considerable inequality in the attraction that certain employers offer through their apprenticeships. The applicants to companies like Rolls Royce, BAe, Jaguar Land Rover and many others far exceed the number of apprenticeships on offer at each company.

A growing number of businesses recognise that it makes little business sense to simply turn these keen and often very talented young people away when there is the opportunity to link them up with other similar opportunities in their supply chain or with others in the sector.

Government attempts to persuade employers to overtrain in order to supply more apprentices than they needed proved too expensive at a time when public finances were biting hard. New initiatives, developed by and for businesses that pass on applicants they are unable to employ themselves, may be part of the answer to increasing the supply of young people securing apprenticeships.

This chapter looks at how such arrangements can work. What were the reasons why employers are acting more collaboratively in this way? Is it proving a useful initiative? Which are the companies that are benefiting? Is this approach one that could work elsewhere?

Engineering skills in the UK – the context
Engineering UK's 2015 report summarises the UK's macro-economic climate:

> Engineering employers have the potential to generate an additional £27 billion per year from 2022, which is equivalent to the cost of

building 1,800 secondary schools or 110 new hospitals. If the UK is to benefit economically from this, we will need to meet the forecast demand for 257,000 new vacancies in engineering enterprises in the same timescale. Achieving this will take persistent and collaborative delivery. There is cross-party endorsement of the importance of science, technology, engineering and maths (STEM) education and skills and careers advice. The Industrial Strategy, Eight Great Technologies, and Growth Plan are all moving in the right direction and the new devolved local agenda and role of LEPs promises to focus growth at a local level. Youth unemployment is falling and there is a gradual change in perception of engineering as a both a viable and a desirable career.

Despite these encouraging signs, the rate of change in the growth of supply is far too slow to meet the forecast UK demand for engineering skills.

The latest Engineering UK figures show that between 2012 and 2020 engineering employers will need to recruit 2.56 million people, 257,000 of whom for new vacancies. Overall, 1.82 million of these workers will need engineering skills: pro rata, that is an average of 182,000 people per year. Within the engineering-related demand, 56,000 jobs per year will be needed at level 3 (advanced apprenticeships) and 107,000 per year at level 4+. Yet current figures show that only 26,000 people are entering engineering occupations with level 3 advanced apprenticeships and only 82,000 at level 4+.

Last but not least, there will be some significant population challenges in the UK in the coming years that will affect the pool of GCSE level and equivalent pupils and the pool available for progressing into higher education.

The number of 14-year-olds is set to fluctuate significantly, falling by 7.3 per cent between 2012 and 2017 before jumping by 15.9 per cent five years later. The number of 18-year-olds will also decrease by 8.9 per cent between 2012 and 2022. To further compound the population issue, if we assume that currently people start work at eighteen and stay in employment until sixty-five, the average number of years of employment the average person in the UK has is twenty-four years; this points to the fact that over the next twenty-four years, half of the current working population will retire. (Engineering UK 2015 Synopsis)

Coupled with the latest research from the UK Commission for Employment & Skills which points at the fundamental and profound change that digitization will play in the world of work, attraction of new talent to engineering is essential.

What could the UK's future of work look like in alternative scenarios?	1. Forced Flexibility (BAU) Greater business flexibility and incremental innovation lead to modest growth in the economy, but this flexibility often results in fewer opportunities and weakened job security for the low-skilled.	3. Skills Activism Technological innovation drives the automation of white-collar work and brings large-scale job losses and political pressure, leading to an extensive government-led skills programme.
	2. The Great Divide Despite robust growth driven by strong high-tech industries, a two-tiered, divided society has emerged, reinforcing the divergence in the economic positions of the 'haves' and 'have nots.'	4. Innovation Adaptation In a stagnant economy, improved productivity is achieved through rigorous implementation of ICT solutions.

The Future of Work: Jobs and Skills in 2030

50

Proposed Solution

Big engineering employers have benefitted from the change in perception of engineering as a viable career and are now able to attract large numbers of suitable applicants. At Siemens plc., the average number of vacancies for engineering apprenticeships is approximately 125 per year leading to a career in engineering. Year on year, over 4,000 people apply to Siemens for one of these roles. Being trained as an apprentice at Siemens means a substantial investment by the company – Siemens invests more than £100,000 in an apprentice over a four-year period. It is therefore essential to attract the best-suited candidates to the roles to ensure the company's investment bears a return (and it does – the average length of service for an apprentice at Siemens is 26 years). This means that the recruitment

---

[50] Table used with the permission of the UKCES from The Future of Work Jobs and Skills in 2030.

process selects the best candidates for each role. On average, the short list for each role is five applicants – to get on this short list, candidates go through a stringent recruitment process. In summary though, of the 625 interviewed candidates, only 125 will be offered a Siemens apprenticeship.

In 2010, Siemens recognized that while our own recruitment activities were successful, a large number of our supply chain companies struggled to attract apprentices to their schemes. Since then Siemens has encouraged those applicants who did not succeed at interview to apply for apprenticeship roles in their supply chain.

This has proven to be very successful and every participant benefits from it, but in my view the introduction of an apprenticeship clearing-house would be the only way to systematically ensure that suitable candidates are not lost to this important sector of the UK economy. Ultimately, large businesses can only thrive if the skills sets in their supply chain evolve in the same way as their own.

The fourth industrial revolution, i.e., the move to digitization, will accelerate the need for new skills sets, and the attraction of candidates with the right STEM base skills will become ever more important.

Siemens is not alone in its approach – a number of companies in the energy & utility sector have come together and implemented this model via the National Skills Academy for Power (NSAP) under the Talent Bank brand name. Member companies of NSAP have understood that there is a common gain for every business in the sector if there are more entrants into the industry – as a result, companies in this sector do not compete against one another for entry level talent, but compete as an entity to attract the best school leavers to that sector. This industry sector has also understood the importance of collaboration in the development of common skills levels across entry-level talent qualifications and beyond to remove barriers for entry.

The automotive, aerospace and rail sectors are on a similar journey and have engaged, via the Employer Ownership of Skills pilots, to raise the bar. It is commonly accepted now that employers need to own the skills landscape, and having collaborated on the development of common apprenticeship standards, the next natural step in the evolution is to share the details of candidates who have not been successful at final interview.

Government has understood the inherent inefficiency of not being able to signpost good candidates who have not managed to secure one of the coveted apprenticeships with large blue chip employers, but who have the right aptitude.

The new administration has made a number of big policy changes and is in the process of introducing them over the lifetime of the parliament. There is a fundamental shift in policy for vocational learning from government awarding grants for apprenticeships to government imposing a levy on businesses, which can only be recovered for the training of apprentices. This will be introduced in April 2017, together with the new Institute for Apprenticeships which will provide governance for apprenticeships and apprenticeship standards in England.

One of the tasks of the Institute could be to provide a clearing-house service to ensure good candidates only have to apply once for an apprenticeship. UCAS are going to offer this clearing-house service for degree apprenticeships for the first time next year, and the Institute for Apprenticeships could offer this from April 2017. While there is still a substantial amount of uncertainty about what can be reclaimed, provision of 'surplus candidates' to the clearing-house could be one way of recovering the levy for businesses that signpost unsuccessful candidates.

In terms of process, the new Digital Account Service (DAS) will be a one-stop-shop for employers. It will have the functionality to attract apprentices. Here is an opportunity to create an UCAS-style clearing-house to simplify the process for all parties, learners and employers alike. The Institute for Apprenticeships could add value to applicants, stakeholders and employers by looking into adopting the UCAS tool that is going to be used for degree apprenticeships.

The application process for candidates would be much easier, and their chances of securing an apprenticeship much higher, if they knew that their application would be forwarded to other employers offering the same apprenticeship in the same location if they were unsuccessful with their original application.

Small and medium sized employers would benefit too, as they would gain access to much better recruitment data with regards to the candidates' respective aptitude for a particular apprenticeship.

If levy paying employers could offset their recruitment costs against the new apprenticeship levy, they would certainly not mind sharing their 'surplus' candidates with local employers as those employers are likely to be in their supply chain. It would be in the interest of the large employer to have a competent supply chain with a well-distributed age profile.

This way the UK's future growth ambitions could be put on a much stronger footing and the needs of all employers for entry level talent could be satisfied.

Reference
Engineering UK. (2015) *Synopsis: Recommendations and Calls for Action,* Engineering UK: www.engineeringuk.com.

25.
A NEW APPROACH TO RECRUITING TECHNICIANS:
A CASE STUDY

Martyn Butlin
External Communications Manager, EDF Energy

EDF Energy is one of the UK's largest energy companies and, as the biggest producer of low carbon electricity, operates nuclear, coal and gas power stations along with twenty-eight wind farms.

As a safety-critical employer, the safe operation and maintenance of EDF Energy facilities is a top priority. The company focuses on its people, developing its existing team and ensuring that new employees are offered the best training.

EDF Energy is committed to providing an outstanding apprenticeship scheme to train its future maintenance technicians to the highest standard, as evidenced by the current Ofsted inspection report which judged the training programme to be 'outstanding in all respects'.

The scheme started in 2008 for an initial contract of five annual intakes and due to its outstanding success has now been extended to 2021.

The EDF Energy apprenticeship scheme receives wide exposure across the engineering community, with applications to the scheme coming from across the UK. Gender equality is strongly promoted by EDF Energy and currently 8 per cent of apprentices in training are female.

EDF Energy strongly supports the ethos of growing its own generation maintenance technicians who are trained to a very high technical standard whilst adopting the safety-critical ethos required by the generation industry.

Like many similar organisations which rely on technical and engineering skills, the company was facing a potential 'pinch point' in replacing an ageing workforce with people equipped with nuclear maintenance specific skills. The decision was to take a new approach to recruitment.

This was the inspiration for the 2008 introduction of the innovative EDF Energy apprenticeship programme, which is designed to attract talented young people into the business, able to address future skills shortages by joining the highly-skilled workforce necessary to operate and maintain EDF Energy's power stations.

EDF Energy has invested £12 million in Babcock International to deliver the first two years of the four-year apprenticeship as a residential programme based at HMS Sultan and HMS Collingwood, Hampshire. Recent additional procurement of £770,000 of training resources – bespoke trainings rigs and equipment – is further enhancing the training experience for the apprentices.

The whole quality programme, which is systematically designed, ensures training meets EDF Energy's exacting business needs while achieving the very high professional standards demanded by EDF Energy.

This unique apprenticeship scheme delivers outstanding training for all of EDF Energy's mechanical, electrical and instrumentation technicians for its eight nuclear power stations in the UK.

Throughout the four-year programme EDF Energy's corporate values are introduced and continuously reinforced; this starts at the induction week held in late summer at a Lake District outward-bound centre. During this week the new apprentices are given a range of activities to tackle, including raft building and orienteering. The tasks help the new apprentices bond into a team, ahead of the first two years of the programme.

After completing the week in the Lakes, the apprentices then start their two-year residential programme undertaking their level 3 technical certificate alongside a level 2 NVQ in Performing Engineering Operations.

Years three and four are spent at the apprentices' allocated power station where they undertake their level 3 NVQ in engineering maintenance which is completed alongside EDF Energy's discipline-specific maintenance qualification manual.

The first four intakes, some 231 apprentices, have graduated from the scheme with 100 per cent of graduating apprentices now working in full-time maintenance, operations, or engineering positions within the company.

The 2015 entry included apprentices from EDF Energy's coal and gas power stations for the first time, further demonstrating the company's comprehensive commitment to apprenticeships.

The outcomes from the EDF Energy apprenticeship go well beyond the requirements of the UK national apprenticeship framework.

It provides opportunities for personal and life skills development alongside a comprehensive technical programme which equips young people with the skills, knowledge, attitudes and behaviours required for working within the nuclear and power generation industry where an absolute safety and quality culture is critical.

These outcomes are recognised by Ofsted's 'outstanding' judgement of the training programme, and by nuclear industry recognition through the award of Nuclear Apprenticeship of the year (Nuclear Industry Association) and the 2015 Nuclear Apprentice of the Year (National Skills Academy for Nuclear (NSAN) award).

EDF Energy's key suppliers send their apprentices to the programme to train alongside their EDF Energy peers in order to gain the same technical and behavioural standards necessary for maintaining equipment. This helps deliver EDF Energy's targets as well as having a positive effect on its supply chain's workforce.

The scheme has made a strong impact at all of the eight EDF Energy nuclear stations across the UK. Graduating apprentices join their maintenance teams with an inquiring and challenging attitude which changes the culture of nuclear maintenance. The graduates join with the ethos of 'do the right thing all the time, take time to do the job in a first-class manner; question whenever you do not know'.

Retention on the EDF Energy apprenticeship is outstanding; timely success rates exceed 95 per cent each year. Graduating apprentices are offered full-time positions with EDF Energy on successful completion of their apprenticeship with very few exceptions, and thereafter job security is excellent.

The support given to apprentices on the scheme is outstanding. During the first two residential years at the Babcock Engineering Academy, there is 24 hour/7 day staff cover for all technical delivery and residential activities.

In years three and four, the apprentices are managed by a dedicated training coordinator at each of the eight nuclear stations. The coordinators plan the year three and four programme to ensure successful completion of the level 3 NVQ with apprentices fully integrated into the stations' nuclear maintenance teams.

The apprenticeship programme delivers first-class technical training using bespoke, contextualised training resources which develop high level academic and craft skills.

Additionally, year one and two incorporate an 'innovate, progressive life skills' programme which provides activities that stimulate the apprentices. These include art, health, languages and environmental and global policy issues. It is delivered through a carefully planned syllabus which allows apprentices to gain communication, team working, independence and leadership skills in a supportive peer learning environment.

It complements the vocational training that the apprentices undertake and underpins the cultural values and behavioural conduct that EDF Energy expects from its nuclear professionals.

Apprentices at HMS Sultan are given other engineering challenges during their time on base, as EDF Energy has a partnership with the nearby Submarine Museum in Gosport. The apprentices spend time on HMS Alliance, which is sited at the museum, producing maintenance schedules and instructions and also health and safety regimes within the submarine. The group then present their findings to managers from their base locations and also take them on guided tours of HMS Alliance.

Spending time working at the museum gives the apprentices a chance to expand their engineering experiences. They interview ex-submariners, examine an historic submarine, and also experience the conditions in which the submariners would have to work and live. The work also allows the HMS Sultan instructors to discuss new engineering techniques and also the health and safety challenges the submarines pose to its crew and maintenance teams. The final presentations to the broader EDF Energy management also allow the apprentice to develop important presentation skills.

The quality of the EDF Energy apprenticeship is exceptional; this is evidenced by grade 1 Ofsted inspection, Beacon status for the Babcock Engineering Academy, Semta (Science, Engineering, Manufacturing and Technologies Alliance) training provider of the year 2014, Semta skills champion of the year 2015 for the EDF Life Skills Manager, and the Nuclear Industry Apprenticeship of the Year 2015.

The Life Skills Programme includes three European trips for first year apprentices visiting Budapest, Berlin and Lyon. In addition, second year apprentices work together in groups of five to plan and organise their own itinerary for the 'Crossing Borders' trip which involves travelling independently around five European countries in six days within a predetermined budget.

After the first two years are complete, the apprentices return for the final two years' training to the base stations where they are integrated into the site's teams and gradually take on more and more responsibility within their chosen disciplines, electrical, mechanical and control and instrumentation.

The third and fourth years work closely with the sites' apprentice co-ordinators to ensure they continue the formal training towards an Advanced Modern Apprenticeship Certificate in Engineering as well as BTEC and an NVQ level 3.

Some apprentices will also move on to take HNCs, and at the end of the four-year programme it is expected that all apprentices who complete to the required standards will be appointed to a full time position at their base station.

This success is marked by a graduation ceremony held for all those completing their four-year scheme.

Graduating apprentices are achieving high positions within the company including acting team leaders, team leaders, plant engineers, health and safety representatives, and for the first time an ex-electrical apprentice has progressed to a fully authorised reactor desk engineer post at Heysham 2 power station in Lancashire.

Each autumn the exhaustive recruitment process starts with the apprentice co-ordinators visiting schools and colleges, talking to careers' advisers and pupils.

During this recruitment process, it was noted that potentially suitable technicians would not make the right academic grades and that they needed a separate route to EDF Energy or other companies' apprentice programmes.

Since 2011, EDF Energy has sponsored the 'Access to Apprenticeship' programme at colleges near its sites at Hinkley Point B and Heysham power stations. The programmes, at Bridgwater College and Lancaster and Morecambe College, are aimed at young people without the minimum qualifications to apply for an advanced apprenticeship scheme.

The course lasts for an academic year and those on the course, around twenty at each college, work towards a BTEC level 2 qualification in Engineering and Applied Science which will develop the students' core engineering skills and knowledge. The core units cover Working Safely and Effectively in Engineering, Using Engineering, Information, Mathematics for Engineering Technicians, Engineering Maintenance Procedures and Using Computer Aided Drawing Techniques in Engineering.

In addition to the core engineering units, those on the course also study Specialist Applied Science at level 2 which will include Energy and Our Universe, Biology and Our Environment, and The Application of Physical Science.

Those on the course are also offered a chance to improve their general life and job-seeking skills, with an overnight residential and one-to-one learning with a personal tutor.

The courses also offer regular visits to either Heysham or Hinkley Point power stations. In addition, the colleges invite in guest speakers to give valuable insights into the workings of the industry.

On successful completion of the course, students can enter the aptitude-testing phase of the EDF recruitment process. To date seven students have obtained EDF Energy apprenticeships, with many others successfully joining other local area engineering companies.

The apprentice programme is also attracting more A level qualified applicants who see apprenticeships as a viable alternative to university. From discussions the EDF Energy apprentice co-ordinators have had with the older applicants, many cite costs as a key driver away from degree courses and towards the apprentice schemes.

There is also an understanding that the apprentice schemes open a route into a potentially wider engineering career. EDF Energy does offer its employees continuous development opportunities to potentially take degree and postgraduate courses. The construction and operation of Hinkley Point C will create thousands of employment and apprenticeship opportunities in a broad range of occupations and careers. Opportunities will include construction, civil engineering, electrical installation, hospitality, catering, logistics, security, site services, support roles and others over the coming years.

When complete, Hinkley Point C will also have an expected workforce of nearly 1,000 people who will be needed to run the power station throughout its sixty-year operation. The construction and operation of Hinkley Point C will create 25,000 employment opportunities and aims to create 1,000 apprenticeships. Over 60 per cent of the project's construction value is predicted to go to UK companies.

Later projects will benefit from the skills gained from the construction of Hinkley Point C. To help ensure the flow of potential engineers EDF Energy uses its Visitor Centres, which have seen 100,000 visitors across the eight sites in just two years, and also experienced engineers supporting science and technology projects in schools. Working with schools and universities has proved an important element in EDF Energy's goal to ensure it has a supply of apprentices for its power stations.

26.
THE BUSINESS CASE FOR APPRENTICESHIPS: THE SWISS EXPERIENCE

Dr Dominik Furgler
Swiss Ambassador to the United Kingdom

———

Summary
This chapter looks at key aspects of the Swiss vocational training system and how employer engagement ensures that young people receive high-quality opportunities to enter the labour market.

It highlights a system in which apprenticeships are the standard route into employment for 15 to 18-year-olds, and where the voice of employers and apprentices is canvassed to coordinate the transition into work. Employer ownership of vocational training is key to its success in delivering the right skills, while the commitment of employers to offer sufficient apprenticeship places is key to a well-functioning labour market. It is a system that is clearly driven by employers but in which government has a key role in assuring quality.

There remains a strong desire by Switzerland to work with the countries of the UK to learn from each other and to inspire apprentices of the future.

The business case for apprenticeships
Swiss businesses operate in a small, landlocked country without natural resources and against a background of high wages and, often, unfavourable exchange rates. To thrive with all these economic odds stacked against them, Swiss businesses have only one route to success: they need to offer the best possible products and services in their given sector. Ask any Swiss company how they maintain the excellence and innovative power to operate at the highest level and they will tell you that a skilled workforce is one of the most important factors in their business success.

But while for some countries 'skilled workforce' refers mainly to university graduates, Swiss businesses know that they are dependent on the highest skills right across their workforce. That's why employers, business associations and other stakeholders in the economy have developed a 'gold standard' for entry level vocational training: apprenticeships.

The Swiss experience provides an excellent business case for apprenticeships because they succeed at many levels: they produce exactly the skills that employers need; they benefit both businesses and learners economically; and they are an efficient tool to combat youth unemployment and boost social cohesion. Employer ownership of vocational training is key to its success in delivering the right skills, while the commitment of employers to offer sufficient apprenticeship places, often in excess of their own immediate demand, is key to a well-functioning labour market.

Apprenticeships are at the heart of the Swiss economy and Swiss society Apprenticeships are at the heart of the education system. The default 'stream' of the Swiss school system runs - in UK terminology - up to around GCSE level, and continues in the form of apprenticeships which are based on work-based learning but also include a general education part. They are taken by around two thirds of Swiss school leavers. During the apprenticeship, the general education component and the theory of the vocation are usually taught one or two days per week at the regional vocational college, while the remaining days are spent in the company. With the standard three or four-year apprenticeship (depending on the complexity of the occupation), this stream delivers 18 to20-year-olds who have completed their general schooling and gained the 'gold standard' national qualification in their chosen profession.

Schools in the second (academic) stream only account for around a quarter of all students. They are selective state schools (grammar schools) that lead directly to A-levels and most of their school leavers then opt for university study.

Apprenticeships are offered in over 200 occupations, but crucially there is only one nationally recognized qualification for each of these occupations (apprenticeship frameworks). The ultimate quality control agency is a Swiss government department. This is intended to allow for mobility of skilled employees between companies since the competences and skills required for each national qualification are commonly understood. Ultimately, this principle ensures that apprentices are not trained exclusively for their host company but effectively for the entire labour market.

A three or four-year apprenticeship carries significant costs – who pays? While the vocational college is funded by the public sector, the employer does pay direct costs like the (comparatively low) training wage as well as indirect costs because the apprentice only spends three or four days per week at work and is supervised by a designated trainer within the company. But Swiss businesses are happy to pay much of the cost of apprenticeships because they work as a 'gold standard' which brings back a return on investment to firms who train while also benefitting school leavers entering the labour market and the economy in general.

What do I mean by 'gold standard'?
First: They are the standard route into employment. Apprenticeships can be considered the basis of the Swiss economy because they are chosen by two thirds of young people as the starting points of their career. Crucially, they are NOT a subsidized way of up-skilling existing employees.

Second: The apprenticeship is the national quality standard for entry into the non-graduate labour market. It is largely standardized across Switzerland and the frameworks are transparent. The skills and competences of apprentices in each occupation are defined by employers (not colleges or government) and they are commonly understood, so companies who don't train themselves will also hire employees who completed apprenticeships.

Third: The classic Swiss format of apprenticeships works. The mix of in-company training and one or two days per week at the local vocational college provides both the specific skills that employers need and the final chapter of a young person's general education. And it works in terms of financial benefits, both for the employer and for the apprentice.

Fourth: The apprenticeship system works because it is at the heart of not only the Swiss economy but also Swiss society. In addition to delivering skills for business, they also deliver youth employment, social coherence and, at a personal level, an early sense of achievement and career progression.

In my opinion there is one underlying reason why apprenticeships meet the four criteria above and why they have real buy-in from employers: it is employers themselves who design, run and administer most of the training, ensuring that it delivers the skills that are needed in any given sector at any given time.

Why are employers engaged?

As much as my colleagues and I like to talk about the Swiss apprenticeship system (and believe me, we do...), nothing brings home the reasons why the system works so well as actually speaking first hand with Swiss apprentices and employers. That is why we invited a range of stakeholders from the British skills sector, including Jason Holt and David Way, for a fact-finding tour in Switzerland. Of course the programme included the usual meetings with government officials and vocational training experts, but the most important visits were those to businesses. It was on the shop floor and in the company boardroom that the passion about vocational education really came across.

The main points that employers emphasized were their responsibility as an important actor within the education system, and their social responsibility at the heart of local communities. Given that around two-thirds of youngsters take the vocational route, they are right - not only because of the high number of apprentices but because the vast majority of them are school leavers aged 15 or 16 when they start their training. This means that employers are guiding new generations of youngsters into being responsible adults as well as producing skilled employees - a huge responsibility but also a source of real pride which really shone through in our discussions with trainers and managers. Usually, the response to the question 'why do you take on apprentices?' was along the lines of: 'this is the next generation of young people. If we don't take them under our wing, give them a career start and help them develop – who will?' An excellent question!

Overall, businesses appear to be well aware of their important role within local communities, not just as wealth creators but also as responsible for social cohesion. The readiness to provide high quality training is a natural consequence of this. It helps that most CEOs and trainers will have apprentices or former apprentices in their families or have been apprentices themselves.

Crucially, Swiss employers back up their commitment with significant investment of resources. Employers basically 'own' apprenticeship training: most of it is delivered on the shop floor (or in the office) and senior employees are involved in training, curriculum design and assessment. Employer groups or industry associations are tasked with coordinating and updating the curriculum to meet industry needs, a process that again involves managers and trainers from individual member companies. Importantly, this involvement is seen with pride by both the trainers involved and the company that 'donates' their employee's time.

Employers are also clearly aware of the benefit that well-trained apprentices represent for their business. Systematic research backs this up and shows that, for a Swiss business that trains apprentices, there is a significant return on investment. In a range of economic studies based on large-scale business surveys, Professor Stefan Wolter and colleagues calculated both the total costs that arise for Swiss companies due to apprenticeship training, and the financial benefits derived from the work of these apprentices during their training periods. The research shows that Swiss businesses, on average, benefit from training apprentices (Muehlemann and Wolter, 2014).

This is not only true for big companies but for SMEs as well. As might be expected, the extent of economic benefit varies between different occupations and the duration of the apprenticeship. It is important to note that this benefit already occurs at the end of the training period and that additional beneficial aspects associated with retaining apprentices (such as reduced recruitment costs or higher productivity compared to skilled staff joining from other businesses) will drive up the overall return on investment for training.

Given that business can play its role in education and social cohesion while also reaping economic rewards, a strong and active commitment towards vocational training makes sense for Swiss employers.

A concerted effort to get young people into apprenticeships
The discussions with employers and apprentices on the study trip made it very clear that the vocational route into employment is well established and its advantages are relatively easy to communicate. It offers not only smooth entry into the labour market but also, importantly, good and prestigious career progression opportunities, a well-developed system of continuing professional education at the tertiary level, and recognized commercial rewards. Because of its significance as the main route into employment and on to career progression, schools, business and government – both regional and national - are all actively involved to ensure a successful progression of young people from school into apprenticeships.

Choosing the right career option is facilitated by support from a range of stakeholders. Of course, the family plays a crucial role and, since most families have personal experience of the apprenticeship route, this is an option that will usually be discussed. Schools offer a substantial amount of career counselling from the pupils' early teens onwards and as part of the curriculum over several years. The regional and national education authorities run information and advice centres with a wide range of reference materials and online resources. Good coordination of activities between schools and local advice centres ensure that youngsters can get a consistent and comprehensive overview of career options. The analysis of the learner's personality and his or her preferences are always at the core of advice, although of course the availability of apprenticeship places within host companies is often a factor that influences students' choices.

Employers have an immediate interest in attracting apprenticeship applicants and therefore develop recruitment activities to make sure they are a potential choice for good school leavers. Important elements of companies' recruitment strategies are business open days and, especially, 'taster apprenticeships' - short work experience placements (typically only two to five days' duration) which are a good way for both youngsters and employers to find out about each other. Companies who are hiring apprentices may work with schools and their regional authorities to advertise such placements alongside the full apprenticeships but pupils are also encouraged to use their own initiative to contact businesses with a placement request. The taster experiences will then take the prospective applicant through the work processes that are covered by his or her chosen occupation. This allows the student to better understand both their career choice and the host company, while it offers the company the opportunity to choose the best applicants. Such placements are not a structured part of school or training and need to be arranged individually between applicant and host company, often during school holidays.

Matching supply and demand in the apprenticeship market
It is important to note that apprenticeships are advertised and awarded like jobs, rather than training courses with free enrolment of anyone interested (or anyone able to pay). Therefore, businesses have an interest in attracting the best-suited learners for their apprenticeship. By contrast to the UK system, vocational colleges play only a minimal role at this stage because they simply deliver the education and training content that was developed by employers and accepted by the government agency as the national standard, much like a state school in the UK that delivers the national curriculum.

Since the matching of supply and demand is critical for a functioning system, a range of tools have been developed for the transition into vocational training. An 'apprenticeship barometer' is based on a business survey carried out twice a year as well as telephone surveys of young people between the ages of 14 and 20. Conducted on behalf of the relevant national government department, it gives a nationwide indication of both apprenticeship places on offer in the various occupations and the number of school leavers looking for such places. Its aim is to monitor that there is a functioning market for apprenticeships rather than to attempt actual matchmaking, so the barometer only triggers activity by government or other stakeholders if it indicates an acute imbalance of supply (apprenticeship places) and demand (interested school leavers). In addition, the cantons carry out a monthly survey of supply and demand in the apprenticeship market. Following graduation, a 'graduate employment barometer' monitors first-time job prospects of apprentices after graduation. Again, it is a concerted effort by employers, regional and local government that ensures the right level of information to facilitate the labour market entry of each new cohort of apprentices.

It is important to emphasize, however, that this system is driven not by second guessing the number of training courses or learner interest, but ultimately by the genuine demand of employers. Because while the 'barometers' give an indication of interests and opportunities, it is the contract between a business and the selected applicants it hires that constitutes the basis of an apprenticeship, not the take-up of vocational courses by interested learners.

When there is an acute lack of apprenticeship positions, the corrective measures of government are initially focused on stepping up the dialogue with business to create a sufficient number of training places. But employers actually have shown responsibility to boost places when needed and regularly take the initiative to train as many youngsters as possible. In fact, many companies routinely train more apprentices than they themselves require, effectively supplying a pool of gold standard-trained employees for the general labour market. While this is partly in their own interest (being able to select the best apprentices to retain, for example), this type of 'overtraining' means that companies are not restricted by their own demand and can react to any danger of rising youth unemployment due to lack of apprenticeship places. During a severe economic crisis in the 1990s with high levels of youth unemployment, business worked with government to reform and update the vocational training system and to provide a sufficient number of apprenticeships.

This intricate system works well as the low youth unemployment (6.3 per cent in 2014, OECD Employment Outlook, 2015) illustrates. Sixty-six per cent of apprentices immediately entered the working world after graduation, while about 20 per cent enrolled in continuing education at tertiary level, and only about 9 per cent were unemployed or job-seeking in the year right after their graduation (State Secretariat for Education, Research and Innovation, 2012).

A recipe for success...?
The Swiss apprenticeship system works extremely well – in Switzerland! Because it is not only the basis of our economy and our education system, it is at the heart of Swiss society.

One thing that has become very clear during this study trip and in our British – Swiss Dialogue over the years so far is this: perhaps the most important feature of the Swiss skills system is not the format or content of the training as such. Instead, it is how it is shaped by - and in fact shapes - many other aspects of Swiss society. The feeling of 'social responsibility' of employers is one example of this. Another example is the school system with its standard 'GCSE / vocational stream' and a smaller 'A level stream' - to put it bluntly you could say that the school system has developed around apprenticeships. Apprenticeships are also at the heart of business: employer associations dedicate effort to setting both the content and the standards of vocational training. And apprenticeships are clearly a priority for both the federal and the regional governments. The cantons run vocational schools and provide many of the vital services in the transition from school into apprenticeship while the federal government actually sees accreditation of the over 200 recognised occupations as one of its most important tasks.

It is interesting to note that, in stark contrast to countries like the UK, there are only a few private companies who provide training, and there is no private sector 'market' for accreditation. In the Swiss mindset, some things are perhaps too important to be left to the markets. Part of the public funding in the apprenticeship system goes into matching supply and demand and into a raft of measures to enable the vast majority of school leavers to find a suitable apprenticeship even if they are weak learners or from troubled backgrounds. The result is the lowest youth unemployment in the OECD.

All these aspects illustrate the central role that apprenticeships play in Swiss society. And when you are in Switzerland you can feel the pride associated with this topic. The pride of families when their children find an apprenticeship, the pride that apprentices take in their work, and the pride of employers who help look after the next generation of youngsters.

If I may comment on the UK at this point: something my colleagues and myself have observed when talking to employers, apprentices or at the National Apprenticeship Awards is the growing sense of pride in the achievements of apprentices and their place in the labour market. A sentiment that I recognize and surely a great incentive for young people in the UK.

...or a source of inspiration?
It is clear to me that much of the Swiss success is tied to the traditional place of apprenticeships in our society – something that is, of course, not necessarily transferable to other countries. However, I think there may be

individual aspects that might serve as an inspiration to ongoing processes on vocational training reform in the different countries of the UK.

In my view the most important aspect would be employer engagement. It is only because business takes the lead in developing and updating the content and format of apprenticeships that they remain relevant in a fast-changing economic landscape. Although this involvement has a cost in terms of resources, business wins because of high returns in productivity. The variety of formats of the business organisations who make vocational training work in Switzerland may be of interest when developing employer engagement in the UK. They include sector-based industry associations, umbrella organisations, regional groupings, trade unions and employer associations.

The engagement in vocational training of such organisations or individual companies is not orchestrated top-down by government but the result of bottom-up activities and initiatives. For example, the State Secretariat for Education, Research and Innovation is in charge of quality assurance for recognizing apprenticeship frameworks, but it does not initiate changes to training standards or the recognition of new occupations, which happens as a result of internal discussions among the relevant organisations.

Also, sector organisations and occupations are not always aligned so that a bank, for example, may be engaged in defining training standards for a 'commercial employee' framework in one organization and an 'ICT' framework in another. Basically, this organic and quite fluid system is driven and shaped by the commitment of business to play an active part in the development and supervision of the training that will produce the skills that it needs.

Another important principle is the focus on actual work-based learning, as opposed to either full-time courses or the use of separate training facilities. The close integration not only into the host company but directly into its production or business process seems to work very well for both employer and apprentice. Actually, many of the more complex apprenticeship frameworks (like engineering or in financial services) now include full-time induction courses at separate facilities at the start of the apprenticeship, but this will always be followed by several years of training integrated into the host company.

A third aspect is career progression. In Switzerland, apprenticeships are only ever the entry route into the labour market. But an important feature of the system are the routes of progression into advanced training, a huge sector which ensures that apprentices can have exciting and lucrative career progression, and into higher education, particularly via the universities of applied science. These are universities for learners with an apprenticeship who have also completed a bolt-on academic qualification (a 'Vocational Baccalaureate') and they offer bachelor and master's degrees and opportunities for research. Given the practical work experience of the students and the universities' focus on applied higher education and research, their graduates are in high demand for management positions.

Universities of applied science are an excellent example of permeability between vocational and academic learning and career options. In fact, it has been a recent focus of developing the Swiss education system further to open pathways which allow for flexible, modular learning that can incorporate academic, vocational, public sector and private provider elements. The aspiration is 'kein Abschluss ohne Anschluss', a system where every qualification opens up follow-on routes. Perhaps this is an area where it would be worthwhile for the Swiss stakeholders to compare notes with UK counterparts.

A partner for dialogue!
While we are very proud of the Swiss system, we are also interested to look ahead and anticipate developments that may be important for Switzerland in an increasingly global environment. We are therefore observing the development of the vocational systems in the countries of the UK with interest. But most of all, we are interested in maintaining an active skills dialogue and exchange of ideas about apprenticeships which may turn into exchanges of learners and teachers. That would truly allow exchanges of best practice and would, I think, inspire both systems.

References
Muehlemann, S. and Wolter, S. (2014) 'Return on Investment of Apprenticeship Systems for Enterprises: Evidence from Cost-Benefit Analyses, *IZA Journal of Labour Policy* 3:25
OECD. (2015) *Employment Outlook*, OECD: www.oecd.org.

VET. (2013). *Graduate Employment Barometer 2012*, State Secretariat for Education. Research and Innovation: www.sbfi.admin.ch.

PART FOUR

NAVIGATING
THE APPRENTICESHIP SYSTEM

INTRODUCTION TO PART FOUR

David Way CBE

———————

There were a number of influences in choosing this theme for the book but the most persuasive is the recurring theme at employer events calling for demystification of skills as their principal 'ask' of future arrangements.

When the National Apprenticeship Service was first created, it was in response to a cluttered and confusing skills landscape. Does anyone seriously think this position has improved? Other countries seem to do this so much better.

This section of the book begins by examining why having a simple system matters; some of the factors that bring added complexity to the UK system; and how other countries manage to achieve relative simplicity. How could the UK simplify its skills system? Why is the institutional landscape so cluttered?

David Harbourne is the Director of Policy and Research at Edge, a respected charity that champions technical, practical and vocational learning. David has wide experience of skills systems across the world. He looks at the approaches in New Zealand, Switzerland and Scotland before describing the main aspects of a simplified system that would encourage greater participation in apprenticeships.

Emily Austin manages a successful and fast-growing apprenticeship programme at Lloyds Banking Group. I approached Emily because Lloyds are constantly innovating to ensure the best match between the range of programmes they offer and their skills needs. This includes the recent expansion into higher apprenticeships. They also have to manage apprenticeships across all the countries in the UK. How does managing an ambitious apprenticeship programme across the UK feel? Emily shares her thoughts on what has helped their success to date and how she has managed some of the complexity she has faced.

The newest and, for large employers, biggest challenge that employers are currently navigating is the introduction of the apprenticeship levy. It dominates current skills discussions. When the levy was announced I spotted a slightly plaintive cry from one of the organisations that knows how to run a levy scheme in the UK saying they could help and use their experience to help with this transition, arguably the most important development in apprenticeships for more than a generation.

Steve Radley is Director of Policy at the Construction Industry Training Board, which has supported training in the industry for the last fifty years. With growth returning to the construction industry, there has never been a more urgent need to attract a new generation of talent to the sector. But how will the government's plans for a levy aimed at funding more apprenticeships help an industry that already has a levy scheme for training provision in place? What are the wider lessons for the introduction of the levy?

Having looked at ways in which employers navigate the apprenticeship system, the next section of the book turns to young people - potential apprentices - and how they find their way to the right apprenticeship for them.

It is an unfortunate truth that almost all winners of apprenticeship awards who I met, when pressed on the advice they were given at school, said that they took the apprenticeship opportunity against the wishes of their advisers at school. This probably helped to make them the impressive individuals that they became.

I heard this so much that I commissioned research into how much careers advisers in schools knew about apprenticeships, and found that only about 20 per cent had any experience that helped them provide this all-rounded view of options for young people. While much has been said about schools only being interested in promoting the academic route, I found many schools willing to receive help to address this shortfall in their own knowledge.

The next chapter looks at how we can reform careers information, advice and guidance (CIAG) services and improve the support offered to young people. Deidre Hughes was Chair of the National Careers Council from 2012 to 2014 and produced influential reports to government on changes to careers provision. She now works for the Institute for Employment Research.

Deirdre analyses the current state of CIAG across all four home nations and argues that England is lagging behind the others, undermining prospects for apprenticeship growth and productivity improvements. She identifies policy measures that if adopted would underpin attempts to expand apprenticeships. Deirdre also draws on a case study in London that gives clear pointers to a better way forward.

Laura-Jane Rowling has been the driving force behind Youth Employment UK and in the next chapter she looks at the prospects of ensuring that there are three million young people ready to take up the opportunities that will be created. She explains why the voice of the apprentice needs to be at the heart of planning and delivering the expansion of apprenticeships. Apprentices understand the current barriers, the poor perceptions many young people hold about apprenticeships, and what would seriously attract their attention.

Laura-Jane also works for PlotR which was the innovative new kid on the block in giving young people digital access to apprenticeship opportunities when I ran the NAS. She is able to share her thoughts on how IT-supported systems can really make the difference in opening up opportunities for young people. She stresses the importance of information that is inspirational as well as informative.

The book then looks at the prospects for young people to make progress through the vocational route. One such opportunity that the government is exploring is how to make it easier for young people to move from vocational courses onto apprenticeships. This must be right as long as people are only starting the vocational course because this is right for them. In many cases, they should be finding apprenticeships from the start.

The theme of the next section is progression from an apprenticeship into higher education. This is vital if the UK is to achieve the benefit of apprenticeships for technical and other higher-level skills. It is also key to persuading more young people that they will not be missing out on a university experience if they choose an apprenticeship when leaving school.

Linking London is an organisation that specialises in understanding the issues that can help or hinder progression into higher education. Sue Betts and Andrew Jones share their experiences. They consider the current track record with big variations in progression between sectors; the importance of progression for social mobility; and how to break out of the vicious circle that prevents significant expansion in the number of advanced apprentices moving onwards, despite the clearly beneficial results when they do.

Finally, how do we help young people to choose the best career path for them? In 2014, I attended a presentation from William Walter who had recently completed research into the expected employment and earnings potential for apprentices and graduates. This research showed there were many career choices where an apprenticeship was comparable, or indeed better, for them to take.

While employment and earnings are not everything, the transparency of this information is important. Will reprises that research and shows the potential for improving the detailed information that is available. This approach feels so much more persuasive to me than the generic statistics on lifetime earnings mostly used at present that beg the question 'but what about me and my circumstances?' This research points to the need and possibility for much more specificity and relevance for individual choices of careers and learning routes.

27.
THE IMPORTANCE OF A SIMPLE SYSTEM:
AN INTERNATIONAL PERSPECTIVE

David Harbourne
Director, EDGE Foundation

Introduction

Everyone says apprenticeships are far too complicated. Are they right? And if so, what can be done about it?

This chapter starts with a brief review of the history of Modern Apprenticeships in England, up to the Richard Review of 2012, which proposed a radical reform programme with the avowed aim of simplifying apprenticeships. The following sections look at apprenticeships in Switzerland, New Zealand and Scotland. Lessons are drawn together in a commentary, leading to some concluding remarks and recommendations.

A short history of (Modern) Apprenticeships in England

I took a job with the National Association of Master Bakers in 1989. On the very first day, I was told that colleges were out of touch with the modern bakery trade. 'College lecturers haven't worked in a real bakehouse for years,' one baker said. 'Everything they teach is old-fashioned. Who leaves dough to prove overnight in a wooden trough these days? Yet they still teach it in college.'

'And they're so slow!' said another. 'Students attend college for two years. By the end they know how to make a dozen scones, but it takes them an hour. We need them to make four dozen in twenty minutes.'

'Mind you,' said a third, 'it's not all the fault of the lecturers. The qualifications haven't moved on since I was at college thirty years ago.'

Salvation arrived in stages. First we had National Occupational Standards (NOS), then National Vocational Qualifications (NVQ) and finally Modern Apprenticeships (MA), which were announced by the Chancellor of the Exchequer in 1993.

Bakers described what they needed trainees to learn and we turned that into an NVQ, followed later by a Modern Apprenticeship framework. It wasn't all plain sailing, because the new world of NVQs involved learning the language of units, elements, range statements and underpinning knowledge. My job was to translate between baker-speak and NVQ-speak.

But my members got more or less what they wanted. The first Modern Apprenticeship for craft bakers involved learning to check deliveries of raw ingredients; work safely and hygienically; weigh and mix ingredients; and divide, mould, prove, bake and finish a wide variety of products.

Hands-on experience at work was backed by off-job training and development linked directly to the NVQ. Assessors observed apprentices as they went about their normal work to check that they were working to the consistent standards and could apply their skills and knowledge to a range of products and varied contexts. They asked questions to test apprentices' knowledge, such as 'what's the difference between flour from a new season's harvest and flour held over from last year?'

On the plus side, Modern Apprentices were able to progress at their own pace. Some completed a craft baking MA in well under two years. Others needed more time – maybe as much as three years. Bakers generally welcomed the abandonment of traditional 'time served' apprenticeships.

There was also a lot of talk about 'transferable skills'. Having learned to mix baking ingredients, an apprentice could adapt to mixing something else later in his or her career – paint, perhaps, or cement.

Of course, we had to make sure assessments were carried out consistently in bakeries throughout the country. Assessors had to get qualifications themselves – the infamous and unpopular 'D units' – and were subject to scrutiny by external verifiers. Apprentices were required to create portfolios of evidence to support the assessment and verification process. Already, things were getting complicated.

These original MAs were all pitched at level 3 and deemed broadly equivalent to A-levels. That was always misleading because NVQs were very different from A-levels in content, delivery and method of assessment. Nevertheless, it gave them a certain status and meant they could be pitched at 16 and 17-year-olds as a direct alternative to staying at school or college full-time.

However, not all young people could access level 3 MAs either because they weren't ready to work at that level or because there weren't enough jobs at that level.

Either way, the government wanted to fix the problem of youth unemployment. Programmes like the Youth Training Scheme had a poor reputation. Critics[51] said they were just a way of giving employers access to cheap labour and did little to improve the skills or longer-term prospects of young people. The number of participants achieving level 2 qualifications was modest. In his review of qualifications for 16 to 19-year-olds, Sir Ron Dearing noted that between April 1994 and January 1995, only half of the leavers obtained a full or part qualification. (Dearing 1996)

Dearing recommended replacing Youth Training with National Traineeships, based on NVQs at levels 1 and 2, the three key skills of communication, the application of number and information technology, and (where appropriate) other units, short courses, and whole qualifications.

National Traineeships were launched in 1997, but didn't replace all other forms of publicly-subsidised work-based learning. A mere two years later, this led the National Skills Task Force to state that:

---

[51] For example: 'The Youth Training Scheme, introduced in 1983 and replaced by Youth Training in 1990, was characterised by low quality training and consequently failed to win a good reputation among either young people or their employers.' Harris M (2003), Modern Apprenticeships: An Assessment of the Government's Flagship Training Programme. London: Institute of Directors.

> After age 16, young people not going into an apprenticeship face a sometimes confusing array of other work-based training and National Traineeships, which do not form a coherent system.
>
> (National Skills Task Force 1999)

The Task Force recommended replacing National Traineeships with Foundation Apprenticeships, adding this caveat:

> It will be essential to design and implement Foundation Apprenticeships carefully, so that the credibility achieved over the past three years by Modern Apprenticeships is not compromised. It must be made clear that Foundation Apprenticeships are not seen as a desired end-point for most people (though they must of course be high quality programmes in themselves) but as a valuable and effective stepping stone to a full Modern Apprenticeship.
>
> (National Skills Task Force 1999)

We now had a two-tier apprenticeship programme. All apprenticeships were based on NVQs, which were devised by employers with support from people who spoke the language of range statements, underpinning knowledge and so on.

That should have been the end of the story. Sadly, it wasn't.

There simply isn't the space here to trace every step of the thirteen-year journey from the Task Force report in 1999 to the Richard Review of Apprenticeships in 2012. Apprenticeships were criticised as too narrow, too broad, too inflexible, inconsistent, failing to meet the needs of employers, failing to meet the needs of individuals, and bewildering to young people and employers alike.

The testing of underpinning knowledge was tightened up by adding technical certificates to apprenticeship frameworks. Later, these were replaced by 'knowledge qualifications'. Key skills were replaced by functional skills. Next, employers asked for flexible apprenticeship pathways, with variations reflected in an increasing range of supporting qualifications. Linked with this, NVQs were criticised as inflexible, and awarding organisations were empowered to develop alternative 'competence qualifications' to reflect employer demand. The much-criticised National Qualifications Framework was set to be replaced by the Qualifications and Credit Framework (QCF), but somehow managed to limp on alongside it. The QCF itself was in the firing line almost from birth and in 2015 OfQual announced that it would be replaced by a Regulated Qualifications Framework.

The design of qualifications and apprenticeship frameworks passed from lead bodies and Industry Training Organisations to National Training Organisations and then to Sector Skills Councils. Each successive network was described as patchy, bureaucratic and remote from the majority of employers.

Meanwhile, responsibility for public funding of apprenticeships passed from Training and Enterprise Councils to the Learning and Skills Council and onwards to the Skills Funding Agency, supported by the National Apprenticeship Service. Apprenticeships were rebranded as Intermediate and Advanced Level Apprenticeships.

None of this made the slightest difference to the take-up of apprenticeships by 16 and 17-year-olds. On the contrary, apprenticeships now appeal to less than 3 per cent of 16 year olds and 6 per cent of 17 year olds.

This is partly because most young people now stay on in full-time education until they are eighteen. Equally, however, employers have always been keen to recruit apprentices aged 18-plus, who are more mature and work-ready. The Labour government therefore expanded apprenticeship opportunities for the 18 to 24 age group, before cautiously opening the door to adult apprenticeships for people aged twenty-five and above. That said, most subsidised adult work-based learning came under the heading of 'Train to Gain'. The Labour government also experimented with apprenticeships at level 4 and above. After 2010, the Coalition government abolished Train to Gain and replaced it with apprenticeships open to people of any age, as well as further developing higher-level apprenticeships.

The Richard Review
In 2012, the Richard Review of Apprenticeships set out clear recommendations for simplifying apprenticeships. For a start, Richard

said that not all instances of training on a job are 'apprenticeships' ... for a genuine apprenticeship, the learner must be new to the job or role. (Richard, 2012, p32) On the other hand, he came down against restricting apprenticeships to a particular age group.

Secondly, Richard recommended focusing 'on those jobs that need substantial investment in skills, and rely on other forms of training to support individuals in lower skilled jobs' (ibid, p34)

He believed this would lead to proportionately more level 3 and fewer level 2 apprenticeships and support the further expansion of higher apprenticeships at levels 4 and above.

Third, Richard called for a set of standards setting out in brief and simple terms what apprentices should be able to do at the end of their apprenticeship. He felt employers should take direct responsibility for writing new standards. As a result, many groups of employers have come together to write so-called Trailblazer Standards: some have been assisted by Sector Skills Councils, while others haven't. At the time of writing, it is unclear who will maintain standards over time, or how this work will be co-ordinated or funded.

Fourthly, Richard recommended dropping continuous assessment and introducing a formal external assessment at the end of the apprenticeship. This would establish whether apprentices have indeed mastered the required skills and knowledge.

Finally, Richard set out plans for simplifying apprenticeship funding. He said employers were best placed to determine the training needed by their apprentices and decide who should provide it, and that they should be given control over public funding.

Summing up, Richard said he had resisted calls to create an apprenticeship system modelled on those of Germany or Switzerland:

> I cannot recommend we adopt a system built, over generations, upon a very different economy, labour market and social partnership. So we are, in this report, taking a road less travelled –

we describe innovations which, to some degree, do not yet exist in any other apprenticeship system. (Richard, 2012, p15)

Nonetheless, Richard's vision of a simpler apprenticeship system does bear some similarities with the Swiss apprenticeship model, which we will explore next.

Switzerland

In Switzerland, apprenticeships are the norm, not the exception: about 70 per cent of young people start one at the age of 15 or 16. Two-year apprenticeships lead to a Federal Vocational Education and Training (VET) Certificate in one of forty-two different occupations. Three- and four-year programmes – which are far more common – lead to a Federal VET Diploma in a range of 230 occupations. A growing number of people who have completed the Federal VET Diploma go on to achieve higher professional qualifications as they climb the career ladder.

Responsibility for vocational education and training is shared by the Confederation (i.e., the national government), cantons (local government) and professional organisations (POs). POs bring together trade associations, industry bodies, trades unions and VET providers to establish the content and objectives of VET programmes, define national qualifications and organise inter-company courses. POs raise income from voluntary subscriptions, but may also receive income from a small levy on employers to support their running costs.

The Swiss approach to apprenticeships is often described as a 'dual system', referring to the combination of workplace learning and off-the-job training and education. In fact, it could be called a 'triple system', because apprenticeships have three components:

1. Supervised workplace learning and experience
2. Weekly general education in a local vocational college
3. Inter-company training courses specific to the vocational field, which are usually residential.

General education includes maths, spoken and written work in the apprentice's native language, English, and citizenship, society and Switzerland's place in the world. Project work fosters research and both written and oral communication skills. There is a strong emphasis on social and interpersonal skills, with a particular focus on effective relationships with clients and co-workers.

Inter-company courses also help apprentices from different companies, towns, villages and social backgrounds learn to live and work together. However, their principal purpose is to add to the skills, knowledge and experience gained in the workplace. For example, apprentices in a food factory will learn laboratory skills so they can test the quality and purity of ingredients. They will also be taught food manufacturing techniques not used in their own place of work. This broadens their experience, making it easier for them to move between employers as their careers develop. Apprenticeships culminate in an end assessment, usually stretching over a week or more of tasks, reports and presentations.

Speaking to a Swiss employer in 2014, I asked why most apprenticeships have a fixed duration of three to four years. He said, 'We have the ability to create entirely new apprenticeships linked to emerging jobs and careers. How long will these new apprenticeships take? Three to four years. Why? Because that's how long apprenticeships last. Everyone knows that.'

However, the length of the apprenticeship is not based on a whim. There are two main reasons why it typically lasts three to four years. First, it is a product of the educational content of the apprenticeship, which is taught over two to three academic years: in Switzerland, apprenticeships are an extension of education, not a stand-alone exercise in vocational training.

Second, it is a way of ensuring that almost all employers see a return on their investment in apprenticeships. The economist Stefan Wolter has shown that in years one and two (and occasionally beyond), employment costs outweigh the productivity of the apprentice. As apprentices' skills increase, they contribute more to the business; productivity starts to outstrip employment costs. Most employers can expect net productivity to be greater than net cost over the full three to four years of an apprenticeship. There are exceptions to this rule: Wolter and his colleagues have calculated that employer investment is not fully repaid during apprenticeships for car mechanics, IT specialists, cooks and electronics engineers. Even in these cases, however, the balance shifts in the employer's favour within a year or two, provided apprentices remain loyal to their employer in the early stages of their careers – which most of them do. (Summarised from Wolter, 2014)

Two-year programmes are less stretching than full three or four- year apprenticeships. However, Switzerland is less concerned with qualifications and levels than we seem to be in England. What matters is not the qualification you get, but the person you become – a baker, an electrician, an accounts technician, and so on.

New Zealand

There are big social, economic and cultural differences between England and Switzerland. The differences are smaller between England and New Zealand.

New Zealand's current work-based learning programme can be traced back to 1992, when funds were made available for 'Industry Training' leading to qualifications at levels 1-4 on the New Zealand Qualifications Framework. Most industry training focuses on skills and qualifications up to level 3 (equivalent to our level 2). Modern Apprenticeships were introduced in 2000 to raise the quality of training at levels 3 and 4 (our levels 2 and 3) for young people aged 16 to 21.

The current government came to office with grave concerns about industry training, based on learners' slow progress and low completion rates: fewer than a third of trainees achieved a qualification within five years of enrolling. In 2013, the government embarked on a series of reforms.

First, funding rules were tightened to ensure that only active industry trainees were funded. At the same time, apprenticeships were extended to all age groups, and now focus exclusively on level 4 skills (equivalent to our level 3). The scheme was re-launched as New Zealand Apprenticeships.

At the end of March 2014, there were 67,419 industry trainees and 11,257 apprentices actively engaged in learning. Most apprentices were under twenty-one; most industry trainees were over twenty-one. (Tertiary Education Commission, 2014)

Training is arranged, but not directly delivered, by Industry Training Organisations. ITOs are employer-led organisations which agree investment plans with the government's Tertiary Education Commission, review the skills needs of their industries and devise occupational standards and qualifications in conjunction with the New Zealand Qualifications Agency. ITO staff contact employers to discuss their training needs, identify industry training and apprenticeship opportunities and broker training packages on their behalf. Some ITOs sub-contract this service to other bodies: for example, the Restaurant Association Education Trust has a contract with an ITO, ServiceIQ, to support businesses in this way.

As part of the government's reform programme, the number of ITOs has been reduced to eleven through a programme of mergers. In addition, employers now have the opportunity to manage industry training funds directly as an alternative to working with ITOs, in a programme similar to England's Employer Ownership pilots.

The amount of off-the-job training and development experienced by industry trainees and New Zealand apprentices varies considerably. Regular day release programmes are much less common than distance learning and self-directed learning using materials developed by ITOs. Some programmes include one or two blocks of training per year, usually limited to a week at a time. Training providers include a mix of colleges and private training organisations. Assessment is usually carried out by ITO staff, though they also offer assessor training for businesses that want to be directly involved.

Scotland
Scotland was a full partner in the original development of National Vocational Qualifications, though they were called Scottish Vocational Qualifications (SVQs) north of the border. In other respects, however, Scotland's approach to work-based training has increasingly diverged from the path followed in England.

The Skillseekers youth training programme was rolled out to all parts of Scotland between 1993 and 1996. Young people aged sixteen and seventeen who left full-time education and did not immediately get a job were guaranteed a training place. The programme was also open to people aged 18 to 24 on a discretionary basis, linked to local economic priorities. Training was meant to lead to SVQs at levels 2 and 3, but completion rates were relatively poor in the 1990s. About two-thirds of all achievements were at level 2.

The Scottish Executive introduced Modern Apprenticeships as a way of boosting take-up of level 3 qualifications, and set an initial target of 20,000 places by 2003 (which initially proved to be over-ambitious). In addition to an SVQ, Modern Apprenticeships included core skills units in Numeracy, Communication, Information Technology, Problem Solving and Working with Others.

The two programmes operated side by side until the decision was made in 2007 to phase out Skillseekers and introduce level 2 Modern Apprenticeships. This followed an evaluation of Modern Apprenticeships and Skillseekers, which concluded that:

> ...a single apprenticeship branding would facilitate [progression] by generating higher aspirations amongst those on the lower rungs of the apprenticeship ladder. Additionally, a stronger brand might help promote increased employer engagement, and increase the perceptions among young people, parents and advisors of the work-based route as a quality alternative to college. (Cambridge Policy Consultants, 2006)

The extended Modern Apprenticeship programme continued to give priority to young people and level 3 programmes. At the same time, the first higher-level MA frameworks were developed, leading to level 4 and 5 qualifications.

In 2010, the Scottish government reaffirmed its commitment to National Occupational Standards, SVQs and Modern Apprenticeships, developed and overseen by a Modern Apprenticeship Group comprising representatives from Scottish government, Skills Development Scotland, the Scottish Qualifications Authority, the Alliance of Sector Skills Councils Scotland, Scottish Training Federation, Scotland's colleges and the Scottish Trades Union Congress. (Scottish Government, 2010)

Numbers continued to grow. In 2014-15, 25,247 Modern Apprentices started training, 80 per cent of whom were aged 16 to 24. Nearly two-thirds of MAs (64 per cent) were at level 3 or above. (Scottish Government, 2015). MA frameworks are available in seventeen industries and 121 sub-sectors. And it's worth noting that the Engineering Modern Apprenticeship Framework for Engineering is thirty-seven pages long – about one tenth of the English equivalent. That said, it still contains long lists of qualifications which can count towards an MA, though there is far less competition between awarding organisations than there is in England.

The Scottish government's current aim is to grow Modern Apprenticeships to 30,000 starts per annum, supported by a new Modern Apprenticeship Supervisory Board. New Foundation Apprenticeships are being piloted, enabling young people to start apprenticeships before they leave school. At the same time, new high-level advanced apprenticeships are being developed in partnership with universities. All this – and a lot more – is set out in a seven-year plan published by the Scottish government in 2014. (Scottish Government 2014)

Commentary
In England, apprenticeships are seen as complicated, bordering on bewildering. With good reason.

To take just one example, the 2015 edition of the Apprenticeship Framework for Engineering Manufacture is 374 pages long. It sets out details of seven intermediate apprenticeship pathways and fourteen advanced apprenticeship pathways. Each pathway offers a choice of knowledge qualifications: for an advanced apprenticeship in mechanical manufacturing engineering, for example, there are no fewer than twenty-five qualifications to choose from.

Then there's the funding regime. Figures from the Skills Funding Agency[52] indicate that in 2014-15, public funding for an intermediate Apprenticeship (level 2) in licensed hospitality typically ranged from £1,675 for apprentices aged twenty-five or over who were trained and assessed entirely in-house, up to £5,375 for apprentices aged 16 to 18 whose training and assessment were supported by a third party. An

---

[52] Private communication with the author.

advanced apprenticeship (level 3) in engineering maintenance attracted between £4,723 and £16,267 in public funds. The exact rate in each case was calculated by reference to the age of the apprentice, the competence and knowledge qualifications included in the framework, the additional cost of functional skills qualifications and whether training was delivered entirely in-house or partially by a college or training provider.

Now explain all that to (a) a young person, (b) a parent and (c) an employer, and compare it with this conversation:

> 'What are you studying at university?'
> 'Mechanical engineering.'
> 'What will you be when you finish?'
> 'A mechanical engineer, obviously!'

But here's the thing: every degree programme consists of a set of courses and modules which vary by subject and institution, but hardly anyone ever asks about them. Parents might take an interest when their offspring choose from a list of options. Employers take an interest when considering job applications. But in the wider public perception, a degree is a degree, while an apprenticeship is just baffling. In one case, we've succeeded in hiding the wiring; in the other – to mix metaphors – we can't see the wood for the trees.

The question is whether we can genuinely simplify apprenticeships, or at the very least hide some of the wiring.

Doug Richard believes we can, and indeed his recommendations take England in a new direction. And although he is right that we cannot simply import the Swiss apprenticeship system 'en bloc', there remain some important lessons for us.

First, occupations are defined quite broadly. Apprenticeships do not train people for a narrowly-defined job in a specific factory; they lay the foundations for a career and for progression to higher-level qualifications. To repeat an earlier point, what matters is not the qualification you get, but the person you become – a baker, an electrician, an accounts technician, and so on. For young people and parents, this is a key point.

Second, social partners avoid constant changes to standards, assessment systems and so on. They believe continuity and confidence in the system go hand in hand. They also accept the need for compromise: meeting the specific, short-term skills needs of employers must be balanced against the long-term interests of apprentices.

Industry Training Organisations are New Zealand's equivalent to Switzerland's Professional Organisations. In recent years, the ITO system has been rationalised and made more efficient. However, there was no appetite for abolishing them altogether. ITOs have their critics: some employers feel their specific needs are not adequately catered for in traineeship and apprenticeship frameworks or individual vocational qualifications. However, ministers have accepted that ITOs cannot please all of the people, all of the time. As in Switzerland, designing stable, respected qualifications and frameworks involves compromise. The aim has been to make the system easy to understand and navigate.

From a New Zealand employer's perspective, apprenticeships appear relatively simple, thanks to the support provided by ITOs. Keeping brokerage, oversight and assessment separate from training delivery reduces conflicts of interest whilst also helping to keep apprentices motivated and on track.

Some New Zealand apprentices choose to take overseas qualifications (particularly City & Guilds qualifications) in addition to local qualifications, at their own expense. However, there has not been a proliferation of awarding organisations or vocational qualifications in New Zealand as there has been in England.

To a certain extent, the same is true in Scotland. Certainly, there is a wide choice of qualifications available to support Modern Apprenticeship frameworks, but there is less direct competition between awarding organisations than there is in England. In addition, Modern Apprenticeships have evolved relatively slowly, with a particular focus on young people and level 3 and above. As in Switzerland and New Zealand, there is a continuing role for sector bodies (Industry Training Organisations) and a sense that creating stable, respected programmes requires compromise and a commitment to long-term planning –

exemplified in the Scottish government's seven-year Youth Employment Strategy.

In my view, the Richard Review did not give enough credence to the views of employers and apprentices currently engaged in the English apprenticeship system. Independent research [53] shows high levels of satisfaction, but this was not reflected in Richard's final report. Conversely, Richard went too far in responding to complaints that existing apprenticeships were bureaucratic and inflexible. For one thing, he appears not to have considered that frameworks cannot please all the people, all the time. Getting a new group of employers together to write a standard isn't hard, but getting every other employer to buy into the new standards most surely will be. Secondly, a standard that fits on two sides of A4 sounds very attractive, but does not do away with the need for more detailed assessment specifications.

Apprenticeship frameworks should not be designed around the narrow needs of a particular job in a particular place. They need to be broad enough to support the career development of individuals. Consistency from one part of the country to another is valuable to individuals and employers alike, but this can only be achieved by having single, agreed assessment standards that apply everywhere.

A simple apprenticeship system in England will, in my view, include these features:

1. A finite number of standards ...
2. ... developed and overseen by social partnerships ...
3. ... with central funding, to ensure they can work independently over a long period of time
4. A commitment to stability: changes should be introduced cautiously and much less frequently than in the past
5. Consistent branding and marketing messages, including examples of career and learning paths open to apprentices in the years to come

---

[53] See, for example, BIS Research Paper 124 (2013), Apprenticeship Evaluation: Learners; and BIS Research Paper 204 (2014), Apprenticeships Evaluation: Employers: www.gov.uk.

6. Advice and brokerage –
   a. To help individuals navigate the system
   b. To help employers pick the right apprenticeships, the right training providers and the right apprentices
7. Consistent and transparent public subsidies …
8. … directed towards particular priorities, such as helping young people take their first steps onto the career ladder

If we get it right, we will start to overhear new conversations:

> 'What are you doing these days?'
> 'I'm an apprentice aircraft maintenance fitter.'
> 'What will you be when you've finished?'
> 'An aircraft maintenance fitter, obviously!'
> 'And after that?'
> 'I'll probably take a degree. I want to be head of aircraft maintenance for a major airline.'

Simple enough?

References

Cambridge Policy Consultants. (2006) *Evaluation of Modern Apprenticeships and Skillseekers*: www.evaluationsonline.org.uk.

Dearing, R. (1996) *Review of Qualifications for 16-19 Year Olds: Summary Report*. Hayes: School Curriculum and Assessment Authority.

Nana, G., Sanderson, K., Stikes, F., Dixon, H., Molano, W., and Dustow, K. (2011) *The Economic Costs and Benefits of Industry Training*. Wellington: Business and Economic Research Ltd.

National Skills Task Force. (1999) *Second Report: Delivering skills for all*. Sudbury: Department for Education and Employment.

Scottish Government. (2010) *Skills for Scotland: Accelerating the Recovery and Increasing Sustainable Economic Growth*. Edinburgh: The Scottish Government.

Scottish Government. (2014) *Developing the Young Workforce: Scotland's Youth Employment Strategy*. Edinburgh: The Scottish Government.

Scottish Government. (2015) *More than 25,000 new apprentices*. (news.scotland.gov.uk).

Tertiary Education Commission (2014) *Industry Trainees and Modern Apprentices by Counts.* Wellington: Tertiary Education Commission.

Wolter, S., Cattaneo, M., Denzler, S., Diem, A., Grossenbacher, S., Hof, S., and Oggenfuss, C., (2014) *Swiss Education Report 2014.* Aarau: SKBF/CSRE (Swiss Coordination Centre for Research in Education).

28.
NAVIGATING THE APPRENTICESHIP SYSTEM FOR SUSTAINABLE SUCCESS

Emily Austin
Group Lead - Apprenticeship & School to Work Programmes
Lloyds Banking Group

———————

The story so far
It had been twelve months in the planning when the first Lloyds Banking Group apprentice finally walked through the door on 21st October 2012. Three years on and we've seen over 3,000 apprentices commence on programmes right across our business and right across the UK.

And it's not just the numbers that have grown. We've evolved our offering from the basic frameworks to bespoke development journeys; from a handful of qualifications to a range of programmes in specialist areas such as IT, change management, marketing, and finance, and from intermediate to higher apprenticeships, providing real choice for school leavers wanting an alternative route into employment other than university.

We've recognised the value of investing and training apprentices – the hard commercial benefits have exceeded our initial assumptions; attrition is significantly reduced, engagement is significantly increased, and the performance of apprentices is very strong.

Lloyds Banking Group's ambition is to help Britain prosper and be the best bank for customers. In doing so, the Group believes it will help them rebuild trust across the industry but understands it will only achieve this through a highly motivated and diverse staff. Apprentices, trained to a professional standard, bring diverse thinking and help the Group serve and meet customer needs more effectively. It's this tangible contribution to strategy and successes so far that has firmly embedded apprenticeships in our entry-level talent management approach. We are building a diverse and professional talent pipeline that's providing an early return, and in some instances is outperforming and replacing other talent programmes.

My answer to those asking 'why should we have apprentices?' has always been 'why wouldn't we?'

We will continue to invest and expand our programme; we are lucky to have the support from the very highest levels in the organisation and have publicly committed within our Helping Britain Prosper plan to create 8,000 permanent apprenticeship opportunities by 2020 and are well on course to achieving this.

So far apprenticeships have been a very positive and successful experience although it's not all been straightforward. There have been a few challenges we've had to face and overcome on the way.

The hardest bits
1. Knowing what to choose
Lower, higher, degree, SASE or Standards – with so many different types of apprenticeships it's hard to know what might be right for a business and how it all fits within overall entry-level talent management practice – something that will be of increasing importance to understand as we move towards levy implementation.

There is no 'one size fits all' and neither should there be – if it's going to be effective and sustainable it's got to be aligned to real business need and understanding.

At Lloyds Banking Group, our IT function is a great example of this. An ageing population and heavy reliance on contractors meant we needed to build further technical and specialist capability right across the function. However, the focus of apprenticeships initially was very much on what frameworks were most accessible rather than what best matched to business need. This resulted in use of a lower level, less technical programme which quickly waned in endorsement from both apprentices and the business.

A review of entry-level requirements and how different talent programmes could address these, led to a revised apprenticeship proposition comprising a number of options and routes into the profession. They have attracted a diverse range of talent who can now see a clear career path, and as a result we have seen zero attrition and high performance.

Critical to success was harnessing the right expertise to fully understand the requirements, putting the right solution in place and then gaining acceptance for it. Sole reliance on a training provider doesn't always extend far enough. Business and HR contacts internally, combined with industry experts, peers from other businesses, and professional bodies helped to navigate the many options available in order to produce the new successful offering.

Having access to a bank of qualifications and understanding potential changes to qualifications was also helpful. Moving forward, helping employers to collaborate cross-industry to spot duplication of 'generic' subjects and identify gaps in readiness for the transfer to Standards will help to simplify the options available. This is key in growing apprenticeships. If we can't articulate our offering simply to those we are trying to persuade of the value of apprenticeships, whether it is young people, teachers or parents, we will not be able to achieve our apprenticeship ambitions for expansion.

2. The need for flexibility
Going from agreeing the concept i.e., apprenticeships are right for my business and we've selected one, to implementation isn't always straightforward. Things to consider include eligibility criteria, funding systems, qualification stipulations and assessment requirements to name just a few. Incorporating these factors into an apprenticeship solution, particularly at a higher apprenticeship level where it can be more difficult to align to business requirements and specialisms, is challenging.

We've definitely noticed this with our own higher apprenticeship programme. Having piloted a small cohort a couple of years ago, we've grown the programme substantially and by the end of 2016 will have offered 250 places. These are across a number of specialisms ranging from IT to project management to marketing and digital to finance and, more recently, commercial banking and wealth management – areas that traditionally have only used a graduate intake as their source of emerging talent. We've also embarked on our first degree apprenticeship in IT.

We have designed a programme to bring colleagues together from these different strands into a community and focus on developing the personal skills and capabilities needed for long term potential.

This development component is working well. The difficulties we've faced have actually been in mapping the technical/knowledge components of the apprenticeships to the roles. The current qualifications aren't flexible enough to meet requirements and often we've had to 'bolt on' solutions to keep them relevant.

Standards will go a long way to fix this, with a focus on employer-led design of relevant current and future content. Although as an employer operating across borders, we face inconsistencies with different design models elsewhere. Key to this is the ability to retain flexibility in the design system with employers, professional bodies and training providers all working together to mould qualifications for quality and relevance.

3. Playing catch-up on delivery

As a proposition, apprenticeships change quite a bit! While we always focus on evolving and improving our Emerging Talent programmes, and most are within our gift to do so, this is different because of where apprenticeships sit in relation to government, public funding, UK productivity and so on.

And because it's always evolving, the delivery structures needed to support aren't always ready to go straight away. For example, the range of higher-level apprenticeships available or training provider readiness for Trailblazer Standards. Sometimes you can find yourself somewhat 'laying the train track as you go'. That can be exciting as well as frightening especially if you are investing in a programme and want assurances that it will do what you've assumed.

We've always found that using our network and market experts to keep abreast of the apprenticeship landscape helps us to anticipate and get ready for changes. Certainly, when the initial levy consultation came out, using our industry connections really helped to prepare and communicate the scale of the impact to our business. Having a training provider who is flexible and willing to look at delivery options with you is also important, particularly in the implementation planning phases of a programme.

In addition, obtaining organisational buy-in to an evolving apprenticeship proposition, and being able to identify business areas that are willing to be test beds for new ways of working, is essential in managing expectations around uncertainty and ambiguity.

4. Managing across four skills systems
One of the biggest challenges we've faced as a UK employer is working with four different skills systems.

As we try to incorporate the different approaches, there are a number of impacts to our business, for example: having to develop additional components for programmes to create a consistent experience for apprentices working cross-borders; keeping abreast of policy and translating that into operational plans for each skills area; explaining to a line manager or business stakeholder why there are differences and how they can accommodate these.

Working for a major employer, I've found a number of things that are critical to navigating this.

Firstly is the importance of local and regional relationships in the areas in which you operate. Having a localised network with expert knowledge is useful for understanding policy and planning accordingly.

Secondly, collaboration with other employers geographically can provide valuable insight into their experience and the solutions they've put in place. I've always found other apprentice employers hugely forthcoming with advice and it's such a useful source of expertise.

Thirdly, a good training provider will provide knowledge of each skills area, particularly around priorities, funding systems and qualification differences. This is really essential in helping manage some of the impacts described above.

Finally, persuading business stakeholders and gaining their acceptance is crucial. Helping them past the challenges and getting them to see the longer term potential for apprenticeships goes someway to lessen concerns around impacts that could prevent them from taking the next apprentice.

In conclusion
Apprenticeships have proven to be hugely successful and Lloyds Banking Group will continue to invest in them for the future as a critical talent pipeline for our business. However there are challenges that we face with real impacts to our business and our apprenticeship programmes that we need to manage in order to maintain an effective proposition that's relevant and meets business needs.

This does require the right resource and supporting structure but of all the solutions I've come across, investing in a wide network of contacts is the most useful by far in navigating the system. Whether it is training providers, industry experts, employer peers or professional bodies, using these contacts to build expertise and evaluate the right solutions for your business is crucial.

29.
OPERATING SUCCESSFUL LEVY ARRANGEMENTS

Steve Radley
Director of Policy, Construction Industry Training Board (CITB)

———————

This chapter looks at the government's target to deliver three million quality apprentices by 2020 from a construction sector perspective, reflecting the CITB's experience of operating levy arrangements and the government's new apprenticeship levy.

With growth returning to the construction industry, there has never been a more urgent need to attract a new generation of talent to the sector. But how will the government's plans for a levy aimed at funding more apprenticeships help an industry that already has a levy scheme for training provision in place?

Without a doubt, apprentices are the life-blood of industry, bringing with them energy, new ideas and, importantly, a succession plan. But the economic downturn hit the sector hard, bringing with it a level of uncertainty which resulted in fewer training opportunities. Unsurprisingly, apprenticeship numbers fell dramatically. This, coupled with an ageing workforce approaching retirement age and strong post-recession growth has led to widespread skills shortages in the sector.

In short, now that business is booming again, there is plenty of work but not enough skilled people to do it. That's where the CITB has an important role to play.

Supply and demand
In August 2015, councils across the UK warned that a lack of construction skills could undermine the government's pledge to build 275,000 affordable homes by 2020.[54]

The Local Government Association said that there was 'a growing mismatch' between the construction industry's increasing demand for skills and a falling number of people gaining construction qualifications. Peter Box, chair of the LGA's housing board, declared that we had 'trained too many hairdressers and not enough bricklayers' and called on the industry to find a solution to the problem.

Of course, this was not news to the CITB, which was all too aware that there was a supply and demand mismatch in the sector. Our own research published in January 2016 estimated that 232,000 extra jobs would be created in the next five years as the economy improved.[55]

The issue however, was - and still is - how to encourage the industry to recruit, retain and re-train to fill these vacancies. It sounds simple enough. After all, while the UK unemployment rate fell to a seven-year low of 5.4 per cent in the three months to August 2015, according to the Office for National Statistics, the figures show that there are still a significant number of people out of work. [56]

But there are still a number of barriers to the construction industry investing enough in its staff.

Investing in a future
The first issue is that post-recession money is tight and many companies are still struggling to commit to apprenticeships after a number of lean years. While confidence in the economy is growing, many employers are still coming to terms with hard lessons learned during the 2008 downturn.

---

[54] Richard Johnstone. (2015) 'LGA Warns on Construction Skills Shortage', Public Finance: www.publicfinance.co.uk.
[55] Alan Tovey. (29 Jan 2014) 'Construction Industry to Create Nearly 200000 Jobs', Daily Telegraph: www.telegraph.co.uk.
[56] ONS. (October 2015) 'Labour Market Statistics', UK Labour Market: www.ons.gov.uk/on.

It can take up to three years to train an apprentice, so it stands to reason that many firms are weighing up the initial outlay required to take on an inexperienced member of the team against the potential long-term benefits of expanding. They will want to be sure that they have enough project work in the pipeline to sustain an extra wage, and get a return on their investment.

Of course it is difficult to justify any expense following financial hardship but predictions are that the industry is entering a period of solid growth. And this means now is the time to pay out to pay off.

There is help available to businesses to expand too, thanks to a levy and grants system that was put in place half a century ago through the CITB. The CITB is part-funded by this levy, paid by employers in the sector. It generates £180 million a year from industry which is then ploughed back into the sector, distributed to employers to help pay for training, qualifications or apprentices' wages.

Since 1964 the CITB levy has:

- Paid out £2.4 billion in grants
- Helped 1.3 million people to pass vocational qualifications
- Trained more than 500,000 new apprentices [57]

Helping the industry to help itself
Training is essential, but it cannot happen without new entrants to the sector. Attracting new talent is high on the industry's list of priorities, but the job opportunities in construction are still not well understood.

Recent research has revealed that four in five Brits wished they had received more advice and guidance about working in construction when considering their career options.

---

[57] CITB. (2015) 'Celebrating 50 years with CITB.' CITB: www.citb.co.uk.

To help industry address this critical need, the CITB has recently provided £5 million worth of funding over three years to improve the perception of career paths available within the sector. At the heart of this has been Go Construct – an industry initiative to get people of all ages and backgrounds excited about opportunities in construction.

More than 400 organisations, including employers, careers advisors, teachers, lecturers and construction ambassadors were involved in the design of the campaign which challenges some of the out-dated stereotypes about what working in construction is really like.

The Go Construct website now showcases the hundreds of career options within the sector, entry routes available and wage prospects through an online portal. (See www.GoConstruct.org.) A careers explorer matches users' interests and skills to a wide range of roles, from diamond drilling to quantity surveying, and activities and materials for teachers and careers advisors – such as a myth buster card game – can be downloaded from the site for use in classrooms.

Government targets
The government is also backing new entrants and has promised three million new apprenticeship starts by 2020. To help achieve this target, it is implementing a levy of its own.

On the face of it the apprenticeship levy sounds similar to the CITB levy. But there are some major differences between the two – one of which is that the apprenticeship levy concentrates on course fees for apprentices, while the CITB levy takes in a broader spectrum of work, as demonstrated by the Go Construct project.

In addition, the CITB levy is paid by all companies in construction, each contributing 0.5 per cent of their turnover to the pot, with a few exemptions for the smallest firms. In contrast, the apprenticeship levy will only be paid by firms with a wage bill over £3 million.

Other differences
The apprenticeship levy aims to deliver only apprenticeships and no other forms of training. It will also pool funding, managed by HMRC, so that

the income generated, it is anticipated, will be shared across all sectors. In comparison the CITB levy supports the construction sector exclusively, and all money is kept in the industry. It also supports a wider array of training.

These include programmes for people already in work, the provision of opportunities for up-skilling and assistance for those wishing to change field within the sector. It also helps people to enter the industry from routes other than apprenticeship schemes, such as through work experience, traineeships, or full time education.

In addition, while the apprenticeship levy is ring-fenced for training costs, the CITB levy supports other expenses connected to taking on an apprentice. These can include wages for apprentices or their mentors, or direct support to help firms plan training needs. The CITB can also connect an apprentice to two or more companies, through its Shared Apprenticeships Scheme, so costs can be spread out. This reduces the risk to employers and guarantees quality training for apprentices.

Getting funding to the right places
One of the issues the CITB levy has addressed is how to ensure funding gets to the right places. Here the CITB has long played a vital role – not only by targeting industry investment, but also by working with construction employers to build a clear understanding of the supply and demand for skills in the sector.

With a clear understanding of the shortages and gaps, the CITB works to help influence training providers to meet specific skills needs, in the right geographic areas. This includes current shortages but also 'pump priming' training to prevent shortages occurring in the first place.

This doesn't necessarily mean the CITB has to provide the training directly – although it could do so where required. Instead, in many cases the most effective solution is to seek to influence existing training, work in partnership with colleges and other training providers and, critically, to commission the most-needed courses in the right areas around the UK.

This commissioning model is already underway with a few pilot projects. For example, employers in north east Scotland are experiencing a shortage of scaffolders, and lack suitable courses to train new ones. In partnership with the existing ASET Training Academy in Aberdeen, the CITB is providing £500,000 funding to establish new training facilities and courses to meet skills demand – right where employers need it.

Support for small firms
A key challenge for construction is the prevalence of small firms – nearly 99 per cent of the industry is made up of SMEs. So while training in other sectors takes place largely in bigger employers, in construction it is often small or even micro firms doing the training. For this reason, funding from the CITB levy is available to all construction employers, regardless of size.

Even the smallest firms who do not pay into the system have the opportunity to access grants and support, if they carry out training. Funding is supplied as a cash transaction and businesses have a say in how the money is spent.

Across all sectors, it will be important that the apprenticeship levy gives employers a say on how their contributions will be spent. A new Institute for Apprenticeships will be launched to provide industry input, but more detail is needed on how this body will ensure the voice of employers is heard loud and clear.

Within construction, the CITB levy is underpinned by an industry consensus vote. This gives employers more ownership over the funding and the way that it is distributed, and explains why industry has continued to support the levy for over fifty years.

Looking ahead, government and industry have both indicated that they wish to find a way for the new apprenticeships levy and the existing CITB levy to coexist. But the question is: How will this work in practice?

Two levies for one industry
The CITB levy delivered over 20,000 apprentices in 2015 – steadily climbing back up to meet the pre-recession figure of around 28,000 a year. It has also listened to industry on how it can improve on delivery, and is leading discussions with the sector about the levy system and its future.

Large construction firms were asked whether the existing CITB levy was still needed and the majority answered yes. Not surprisingly, only a minority of firms - a quarter of those who felt able to express an opinion - are prepared to pay both the new apprenticeship levy and the CITB's at its full rate. But more interestingly, only 17 per cent want to pay only the apprenticeship levy while nine out of ten large firms (88 per cent) support the CITB's levy-grant system.

While no decision has yet been made on how the two levies can work alongside each other, the CITB is exploring the options available. The aim is that employers caught by both the levies will not be paying much more in net terms.

Whatever model is adopted, the CITB is keen to help smaller firms – which are more likely to deliver apprenticeships but won't be paying the apprenticeship levy – to continue to access funding. This is particularly important since some 60 per cent of construction apprenticeships are delivered by firms with fewer than 50 employees – and nearly half (46 per cent) are delivered by firms employing nine people or fewer.

Whatever is agreed, the system must be transparent, fair and easy for employers to use.

Transforming for the better
Having two levies on larger firms in the construction industry will have implications for the CITB and the industry it serves. While it could lead to rationalising some services, the organisation can also achieve significant savings on its traditional operating costs.

Long before the new apprenticeships levy was announced, the CITB began a major restructuring programme. This will see the organisation modernise and become more efficient and responsive to the construction industry's needs. So while the new levy will have an impact, the organisation was already on the path of change, based on industry demand for a more modern, responsive training board.

Making it work
Whatever sector you look at, when it comes to creating an economy that is to survive and thrive, we can all see the benefits of apprenticeships. But

delivering three million trainees by 2020 is ambitious – and the success of the apprenticeship levy in helping realise this target will rest on how it is designed.

First it will have to achieve buy-in. This means involving employers in decision-making, giving them a sense of ownership and demonstrating that the money is being spent effectively.

Next, the training on offer will need to be relevant to employers.

The funding model for smaller firms must be clear, fair and attractive to these employers. There must be a simple and balanced funding model for companies to take on apprentices of various age groups (16-18s, 19-plus, and over 24), particularly since construction employers often take on slightly older trainees.

And on top of this, it will have to address the two issues of quantity and quality. With such a shortage of skilled workers within the construction industry, quantity of staff is certainly an issue. But also key to the industry is quality training, something the CITB has insisted upon and the apprenticeship levy hopes to do too. The latter has taken steps to ensure that funding is directed toward higher quality apprenticeships and importantly, the new Institute for Apprenticeships, which will be independent of government and led by employers, will regulate this.

Finally, in order to have success in boosting apprenticeship schemes, the apprenticeship levy will need to address any barriers to people participating in this type of training. We know in construction that there is some way to go on this front – not least because the industry has not looked sufficiently appealing in recent years.

In summary

With the apprenticeship levy on the horizon, which is set to fund £900 million worth of apprenticeship spending, the government needs to assess not just how training is delivered – but how certain industries can encourage new starters to join the ranks.

The construction sector needs to look at boosting confidence in the industry – for employers as well as potential employees.

The problems:

- Employers were left badly bruised by the recession and are sometimes nervous about taking on more staff
- The industry has not done the best job selling itself, and is still viewed by some as not sufficiently open to diversity
- Young people are not aware of the range of career options within the industry
- People already in the sector do not always have clear visibility on opportunities to progress their careers
- Two levies might mean some firms pay twice, others will miss out on funding, and questions arise over the quality of training

Solutions:

- Ensure there is enough project work on the horizon to sustain new employees by firming up government plans for housing and infrastructure.
- Demonstrate that construction is a viable career choice by working with education providers and helping young people see the sector as a reliable industry – one with prospects.
- Overhaul recruitment campaigns to make construction an attractive modern employment choice – particularly to women.
- Look at ways to boost training options for those who want to up-skill.
- Explore how cash is allocated for training and development by ensuring small firms do not miss out.
- Work with the best training providers to create bespoke quality courses targeting specific skill shortages.
- Give employers a sense of ownership over how the levy is spent.

As a key partner for the construction industry, the CITB has been working on these solutions for some time, and is current reforming itself to be able to respond much more adeptly to industry need. Miracles do not happen overnight but we are certainly making headway, and the CITB is still well placed to gather the evidence on industry's current and future skills needs and then target industry investment accordingly.

In short, we can work with government to boost the market with new talent and address the skills shortage in construction. We are very much looking forward to working with the industry and the government to achieve great things. Together we can recruit and train a safe, professional and fully qualified workforce that will help the industry to shine on a global stage and build a better Britain.

30.
BRIDGING THE INFORMATION GAP:
ADDRESSING MARKET FAILURE?

Dr Deirdre Hughes OBE
Principal Research Fellow
Institute for Employment Research, University of Warwick

Introduction

In its election manifesto, the Conservative party set out its ambition to make this the most vibrant and dynamic country in the world.[58] Sustaining the UK's economic success and delivering on its apprenticeships' ambition depends on the skills, drive and talents of its people. There is a need to make the most of these, and to give everyone the chance to enjoy the satisfaction and rewards that come from a successful working life. With a government target now set to create three million more quality apprenticeships by 2020, there is an urgent imperative for individuals to be guided though an increasingly fast-changing and complex education and labour market landscape.

This chapter aims to stimulate discussion on the degree of attention needed by government(s) and other key players to ensure informed and impartial careers information, advice and guidance (CIAG) is fully in place to increase understanding and participation in apprenticeships.

It provides a brief oversight of issues pertaining to the CIAG landscape across the UK, located within an economic productivity context, since jobs, skills and growth are inextricably linked to our economy, businesses, and families, as well as individuals. The scale of the UK skills problem and its impact on productivity is briefly considered, alongside major concerns

---

[58] *The Conservative Party Manifesto 2015*: s3-eu-west-1.amazonaws.co, page 17.

330

about a legacy of underachievement, particularly among many young people.

The chapter highlights how the CIAG landscape in each of the four home nations is changing in response to tightening fiscal policies and differing government preferences and priorities for national careers services. A quasi-market, experimental approach remains the dominant discourse in England, in contrast to differing and complementary arrangements in Northern Ireland, Scotland and Wales. It is argued that England is lagging behind other nations, both within and outside of the UK, when it comes to ensuring informed and impartial CIAG for young people. This represents a serious threat to the future success of apprenticeships in England.

A case study from London demonstrates how local policymakers, employers and educationalists are working together to address this issue, whilst central government considers possible ways to mend a fragmented and incoherent careers and enterprise system across England. It concludes with policy measures that could be readily implemented, which may also benefit other policymakers outside of England involved in framing more quality apprenticeships by 2020, supported by an effective CIAG system.

The UK's long-standing skills problem that is holding back productivity
In many parts of the economy, employers are experiencing major skill shortages and see little prospect of improvement. For example, Akzo Nobel has recently warned the UK is facing 'a skills crunch' that could exacerbate the country's housing shortage (*The Guardian*, 2015).[59] The latest Royal Institute of Chartered Surveyors' UK Construction Survey shows that the country's skills shortage has reached its highest levels since the survey was launched eighteen years ago, with bricklayers and quantity surveyors in shortest supply (RICS, 2015).[60] In the most recent survey of British firms that employ engineers and IT staff, over half reported that

---

[59] Graham Ruddic. (28 October 2015) 'Tradesmen Shortfall Could Inflame Housing Shortage Warns Dulux', The Guardian: www.theguardian.com.

[60] RICS. (2015) RICS Construction Market Survey 2015, London: www.rics.org.

they could not find the employees they were looking for, and nearly two-thirds said that the shortage would be 'a threat to their business in the UK' (IET, 2015).[61] The mismatch of supply to demand across the broader range of STEM subjects (science, technology, engineering and mathematics) is just as bad.

There are persistent skills shortages that prevent many businesses securing the skilled people they need. The UKCES Employer Skills Survey (UKCES, 2013)[62] reported that skills shortage vacancies – where vacancies cannot be filled due to a lack of applicants with suitable skills or experience – increased by 60 per cent between 2011 and 2013. These shortages cover more than one in five of all vacancies and are particularly severe in sectors critical to growth, such as manufacturing and business services. These skill shortage problems are the product of a peculiarly British mismatch. By most international standards the UK workforce is well educated (Bosworth, 2014).[63] Yet our productivity growth remains slow compared to other international competitors (OECD, 2015).[64] Also, an ageing population means we need to stimulate the younger generation to obtain important skills needed in the labour market. This includes whetting their appetite for apprenticeship training. Their knowledge and skills determine the future of our society.

Skills shortages and skills mismatch are key challenges faced by many economies. OECD (2011)[65] evidence shows that, in far too many cases, a growing number of people are over-skilled for their current jobs, while others are under-skilled. In countries such as Austria, Finland, Germany and Switzerland they address these issues by ensuring that careers information, advice and guidance (CIAG) feature prominently in their

---

[61] IET. (2015) Annual Skills Survey, 2015, London: www.theiet.org.

[62] UKCES. (2013) Employers Skills Survey. London: UK Commission for Employment and Skills: www.gov.uk.

[63] Bosworth, D.L. (2014) UK Skill Levels & International Competitiveness 2013, University of Warwick: Institute for Employment Research.

[64] OECD. (2015) Economic Survey of the United Kingdom, Paris: Organisation for Economic Co-operation and Development.

[65] OECD. (2011) Right for the Job: Over Qualified or Under Skilled? Paris: OECD: www.oecd.org.

education and training systems (ELGPN, 2012) from an early age.[66] For example, an OCED report (2009)[67] summed up the strengths of the Swiss VET system as follows:

'Its [the Swiss VET system] many strengths include strong employer engagement within a well-functioning partnership of confederation, cantons and professional organisations. School and work-based learning are integrated; the system is well resourced, flexible and comprehensive, including a strong tertiary VET sector. VET teachers and trainers, examiners and directors are well prepared, quality control is ensured, career guidance is systematic and professional. Evidence is well developed and routinely used to support policy arguments.'

A legacy of underachievement

For apprenticeships, in policy terms a distinction is made between those aged 16-18 years, those aged 19-24 years, and those aged 25 years and over. Gambin and Hoggarth (2015)[68] highlight 'Increasingly the 16 - 18 group is viewed as part of the compulsory education and training system. The 19-24 group is also seen as an essential part of the initial vocational education and training system that without government support may be subject to market failures of one kind or another. For those aged 25 years and over the role of the state in funding this type of training is much less evident'

Despite improving apprenticeship completion rates, each year tens of thousands of young people fail to complete their apprenticeships.[69] For example, in London, participation in apprenticeships stood at half the

---

[66] Vuorinen, R., & Watts, A. eds. (2012) Lifelong Guidance Policy Development: a European Resource Kit. ELGPN Tools: www.elgpn.eu.

[67] OECD. (2009) OECD Reviews of Vocational Education and Training. A Learning for Jobs Review of Switzerland. Paris: : OECD: www.oecd.org.

[68] Gambin, L., and Hoggarth, T. (2015) Apprenticeships and Gender: Just the Statistics, University of Warwick: Institute for Employment Research, www2.warwick.ac.uk.

[69] Centre for Economic and Social Inclusion. (2015) Achievement and retention in post 16 education, London: www.local.gov.uk.

England average in 2014 and has fallen over the last twelve months (IOE, 2014).

In the election manifesto the Conservative party set the aim of abolishing long-term youth unemployment (2015).[70] There has been real progress in reducing the number of young people not in education, employment or training (NEET). In the three months to September 2015 there were 848,000 young people (aged from 16 to 24) in the UK who were NEET, down by 106,000 from a year earlier.[71] But this still means more than one in eight of all young people in the UK are NEET (11.7 per cent). Yates et al (2011)[72] have shown that young people who were uncertain or unrealistic about career ambitions at sixteen went on to be three times more likely to spend significant periods of time being NEET as an older teenager than comparable peers. And a number of other recent studies have reached similar results – the more realistic and certain initial teenage career aspirations are, the better young people do when they leave education – they are more likely to be employed and to earn better (Sabates et al, 2011).[73] Far too much talent is going to waste and too many young people are missing out on fulfilling their potential.

It is critically important that action is taken to support young people into and through apprenticeships and to stem the future flow of young people into NEET status. A more effective system that supports transitions from schooling to work, with appropriate independent and impartial advice, has a central role to play. Keep (2012)[74] argues that 'one of the most oft-repeated, but also most frequently ignored recommendations in the field

[70] The Conservative Party Manifesto 2015, p. 18
[71] ONS. (2015) 'Young People Not in Education, Employment or Training (NEET)', ONS Statistical Bulletin, www.ons.gov.uk.
[72] Yates, S., Harris, A., Sabtes, R., and Staff, J. (2011) 'Early Occupational Aspirations and Fractured Transitions: A Study of Entry into 'NEET' Status in the UK'. Journal of Social Policy, 40: 513 34.
[73] Sabates R, Harris, A. L., and Staff, J. (2011) 'Ambition Gone Awry: The Long Term Socioeconomic Consequences of Misaligned and Uncertain Ambitions in Adolescence', Social Science Quarterly 92, 4: 1-19.
[74] Keep, E. (May, 2012) 'Youth Transitions: The Labour Market and Entry into Employment for Priority Learners – Some Reflections and Questions', SKOPE Research Paper No. 108, Cardiff: Cardiff University.

of UK vocational education and training (VET) over the last 30 years has been the need for more and better careers information, advice and guidance'

A number of differences, as well as similarities, exist in the current arrangements for providing CIAG across the four home nations. These are briefly discussed below. This partly reflects the geography, culture and relative population sizes of the regions, as well as the differences in the constitutional responsibilities adopted by the devolved administrations. Centralised versus localised policy debates also raise new questions concerning the adaptation of employment, skills, enterprise and CIAG provision. This is particularly pertinent in England with the policy of devolving powers and budgets of public bodies to local authorities and combined authorities through the Cities and Local Government Devolution Bill 2015-16.[75]

The changing CIAG landscape
The government aims to achieve full employment in the UK, with the highest employment rate in the G7 group of nations.[76] Achieving this requires action on skills to bridge information gaps and to improve learner destinations and outcomes. There is widespread recognition by policymakers that apprenticeships will be the cornerstone of the new skills system - three million apprenticeships will have started by 2020 (up from 2.3 million in the last parliament).[77] This will contribute towards doubling the level of spending on apprenticeships in cash terms compared with 2010-11.

The Chancellor has announced details of the new apprenticeship levy to help fund this rapid expansion. For apprenticeships to be available to young people and adults across all sectors of the economy and at all levels,

---

[75] Cities and Local Government Devolution Bill 2015 16, England and Wales. The Bill formally extends to England and Wales, but its practical effect is in England only, with the exception of Clause 20 (covering the voting age for local government elections), which would also have effect in Wales. Parliament: www.parliament.uk.
[76] The Conservative Party Manifesto 2015, p. 18.
[77] Gov.uk. (2015) Fixing the Foundations: Creating a More Prosperous Nation: www.gov.uk.

including degree level, there will need to be a strong combination of pull factors (ensuring more apprenticeships are readily available) and push factors (ensuring individuals are fully aware of the options and likely returns available). This requires having access to informed, reliable and impartial CIAG, delivered by a triage system of suitably qualified and well-trained staff.

Across the UK, government CIAG strategies are situated within a broad policy context of education, employment, apprenticeships, skills, enterprise and economic growth. At present, CIAG policy strategies are characterised in terms of: career decision-making in Northern Ireland ([DEL & DE, 2011, p. 5)[78]; career planning in Wales (Careers Wales, 2012); [79] career management skills and coaching in Scotland (Skills Development Scotland, 2012) [80] and careers inspiration in England (Department for Business, Innovation & Skills, 2013). [81] In Northern Ireland, Scotland and Wales the publicly-funded careers services are directly accountable to the appropriate devolved administration, whereas the equivalent service in England currently operates within a quasi-market. A neo-liberal policy approach is the dominant discourse in England, in stark contrast to arrangements in Northern Ireland, Scotland and Wales. Here, school and college autonomy, competition, de-regulation and opening up the market in CIAG is reshaping the careers provider landscape with a multiplicity of providers, products and services targeting schools, colleges, VET providers, local authorities and universities (National Careers Council 2014)[82]. This has resulted in a crowded, confused and complex landscape, with severely disjointed and

---

[78] Department of Education & Department for Employment and Learning (2011) Preparing for success: Joint implementation plan. Belfast.

[79] Bimrose, J. and Hughes, D. (2012) Research Study on Costs Associated with a Revised Remit and Structure for the Careers Wales Service: Final Report. Cardiff: Welsh Assembly Government.

[80] SDS. (2012) Scotland's Careers Management Frame. SDS: www.skillsdevelopmentscotland.co.uk.

[81] HM Government. (2013) Inspiration Vision Statement, London: www.gov.uk.

[82] Hughes, D. (2014). Taking Action: Achieving a Culture Change in Careers Provision, London: National Careers Council Second Report to Government.

opaque careers provision for young people (CBI, 2015;[83] City & Guilds, 2015[84]).

At least three major policy developments have influenced the current CIAG arrangements in England. Firstly an all-age National Careers Service[85], launched in April 2012, became a predominantly adult careers service, with online rather than face-to-face careers support aimed at young people (unless in exceptional circumstances). The National Careers Service's current provision for young people is predominantly a telephone helpline, SMS, webchat and email service. There has been a steady decline in the use of the National Careers Services by young people online in recent years (National Careers Council 2014). In March 2014, National Careers Service prime contractors were given a remit to run local telephone and online careers services for young people only (not adults). As part of the governments' inspiration agenda, a new bolt-on 'local brokerage role' was also introduced to support schools and colleges in connecting employers and other training providers to local schools and colleges linked to a 'payment by results' funding regime.

Secondly, in September 2012 schools, after four decades of having available a publicly-funded service, were given a statutory duty to secure access for their pupils to 'independent and impartial careers guidance', with no dedicated government funding to commission such services. Despite this, some schools have risen to this challenge putting in place CIAG support for their students, whilst others continue to struggle to achieve this. Ofsted, in its sample survey, found that only a fifth of the sixty schools it surveyed were giving the right careers information to pupils (Ofsted, 2013).[86] Sir Michael Wilshaw has now given a strong indication that schools' careers provision will formally be inspected as part of Ofsted's framework.

---

[83] John Cridland. (2015) 'Skills Shortage Starts in the Classroom Says CBIS', Business Reporter: business reporter.co.uk.

[84] City & Guilds. (2015) Great Expectations, London: www.cityandguilds.com.

[85] SFA. (2012) The Right Advice at the Right Time, Coventry: Skills Funding Agency.

[86] OFSTED (2013) 'Careers Guidance in Schools Not Working Well Enough', OFSTED: www.gov.uk.

Thirdly, the Chancellors' Autumn Statement in October 2014 announced a new and extra £20m investment for careers advice and support for young people (paragraph 2.227).[87] In early December 2014, the Secretary of State for Education announced plans for a new independent careers and enterprise company in England.[88] The company was established to 'ensure employers are supporting young people with decision-making and career development at every stage of school life'. (Gov.uk, *New Careers and Enterprise Company for Schools*)

The new company's work is aimed at schools, to transform the provision of careers education and advice for young people and inspire them about the opportunities offered by the world of work. This is focused on young people aged twelve to eighteen and will help to broker relationships between employers on one hand and schools and colleges on the other. Its remit includes administering a £5 million investment fund to support careers innovation and stimulate good practice, as well as developing a system which motivates young people to take part in activities to build their employability, through the development of the Enterprise Passport recommended by Lord Young. The extent to which Lord Young's enterprise agenda[89] becomes the dominant theme is something to be reconciled.

The new company is employer-led and independent of government, with governance and advisory board arrangements established in March 2015.[90] It is currently seeking sustainable revenue streams. It plans to position itself at the heart of the careers and enterprise ecosystem – as a 'market maker' - to deliver transformational impact enabling young people to access opportunities to progress with their careers. The National Careers Service had a formal 'Memorandum of Understanding' with the new company. (It had a 5 per cent allocated budget from the Skills Funding

---

[87] HM Treasury. (2014) *Autumn Statement 2014.* HM Treasury: www.gov.uk.

[88] Gov.uk. (2014) *New Careers and Enterprise Company for Schools.* Gov.uk: www.gov.uk.

[89] Lord Young. (2014) *Enterprise for All: The Relevance of Enterprise in Education.* Gov.uk: www.gov.uk.

[90] Secretary of State for Education: Evidence to the House of Commons Select Committee. (2015) Parliament: data.parliament.uk.

Agency for 'brokerage services' to schools and colleges). In December 2015, the Minister for Skills formally announced:

> 'We have agreed that the National Careers Service should be re-focussed on the new priority groups, young people aged 19-23 not in touch with schools/colleges, lower skilled adults aged 24 plus and adults (25 plus) with learning difficulties and disabilities. The service should build on the digital first approach already in train, using a triage approach to help citizens make informed choices about learning and work and determine their career pathway and suitability for programmes including apprenticeships and traineeships as well as other learning and skills programmes intended to boost UK productivity' (December 2015)[91].

The Careers and Enterprise Company is viewed by some as a major part of the solution towards improving CIAG for young people. Its intention is primarily to act as 'a strategic umbrella organisation', supporting sustainable programmes, filling gaps in provision and aiming for coverage across the country. But this new arrangement is not without its own challenges.

For example, this new independent organisation has to find a way of securing future investment, competing with charities and other key bodies. The company must secure value for money from public funding, and whilst corporate support during the inception phase of the company is evident, this is unlikely to be a long-term funding strategy of the company as this risks the independence and standing of the company. Also, the reach and cost-efficiency of services to help young people (and others) into work is also being addressed elsewhere in other government departments to drive up co-location and co-partnering arrangements. For example, situating 'Improving Access to Psychological Therapies' (IAPT) therapists in over 350 Jobcentres (to provide support to claimants with common mental health conditions)[93], placing Jobcentre Plus advisers in

---

[91] Department for Business, Innovation & Skills. (2015) Funding Letter to the Skills Funding Agency (SFA) priorities and funding for 2016 2017, 15 December 2015.

[93] Gov.uk. Budget 2015, HC 1093, March 2015, Para 1.236, www.gov.uk.

food banks,[94] the intention to provide Jobcentre Plus adviser support in schools across England to 'supplement careers advice and provide routes into work experience and apprenticeships.'[95] The latter represents another additional new player (operating outside of the Careers and Enterprise Company) and local enterprise partnerships (LEPs) targeting schools in an ever increasing disjointed CIAG marketplace.

These changes are also set within a broader context of reframing skills. In addition to the forecast national apprenticeship targets, there is a particular focus on:

- The 2015 Spending Review and Autumn Statement[96] announced a cash terms protection of the current national base rate per student for 16 to 19-year-olds in school sixth-forms, sixth-form colleges and further education colleges in England for the rest of the parliament
- The government is creating five national colleges and will support a new network of institutes of technology across the country. The national colleges will train an estimated 21,000 students by 2020 in sectors that are crucial to future prosperity such as digital skills, high speed rail, onshore oil and gas, and creative and cultural industries[97]
- A series of area-based reviews of the further education sector has been launched. [98] Each review will assess the economic and educational needs of the area, and the implications for post-16 education and training provision, including school sixth-forms, sixth-form colleges, further education colleges and independent

---

[94] Kate McCann. (10 December 2010) 'Job Centre advisers will be based at food banks, Iain Duncan Smith reveals', *Daily Telegraph*: www.telegraph.co.uk.

[95] Gov.uk. (2015) 'The Queen's Speech 2015: What it Means For You', Section 1: Full Employment and Welfare Benefits Bill: www.gov.uk.

[96] Gov.uk. (November 2015) Spending Review and Autumn Statement: www.gov.uk.

[97] Gov.uk. (November 2015) Spending Review and Autumn Statement: www.gov.uk.

[98] BIS. (July, 2015) *Reviewing Post-16 Education and Training Institutions*, BIS: www.gov.uk.

providers. The reviews will then focus on the structure of further education and sixth form colleges to achieve a transition towards fewer, larger, more resilient and efficient providers, and more effective collaboration across institution types. A critical aspect will be to create greater specialisation, with the establishment of institutions that are genuine centres of expertise.

Whilst the reframing of skills in Scotland, Northern Ireland and Wales is also well underway, in contrast, national careers services in these countries have a CIAG framework and narrative more easily understood by young people and adults (including parents, teachers, training providers and employers). For example, Skills Development Scotland (SDS) is working on behalf of the Scottish government to identify and develop the skills that are important for the Scottish economy. A key performance indicator within the SDS business plan is to improve the career management skills of the people of Scotland.

The 'Curriculum for Excellence teaching and learning framework' (Scottish Government, 2008)[99] is the main driving force underpinning new pedagogical approaches in primary and secondary education. Within this, a dominant theme for learners of all ages is to build 'core competencies' and 'career management skills'. The transference of learning for individuals is taking place through the curriculum, web- and telephony-based careers service, entitled 'My World of Work'[100] with significant emphasis placed on educating and supporting teachers, parents and employers.

A dual approach in the integration of models of 'career guidance and coaching' and greater use of labour market intelligence/information is underway, supported by professional standards. Well-trained and qualified practitioners are employed as 'careers and work coaches', delivering a facilitative and empowerment approach both online and offline. Norway and others countries such as New Zealand are sharing lessons and adopting practices learned from this approach.

---

[99] *Education Scotland.* 'What is Curriculum for Excellence,' www.educationscotland.gov.uk.

[100] 'My World of Work', www.myworldofwork.co.uk.

In December 2014, a plan was unveiled by the Scottish government aimed at cutting youth unemployment in Scotland by 40 per cent by 2021.[101] The apprenticeship delivery plan is firmly rooted within a coherent CIAG storyboard that begins early, supported by centralised and localised innovative approaches. For example, local authorities have developed a Certificate of Work Readiness[102] designed to encourage young people to gain at least 190 hours' experience of work. Whilst there are challenges to develop a more co-ordinated, coherent and transparent range of services for individuals and employers, and to strengthen links with other guidance and support services, including the DWP and Jobcentre Plus, the national careers service in Scotland is performing well. There is a need for the continuation of healthy dialogue between those concerned with driving modernisation and those concerned with maintaining professional standards.

In Wales, the Deputy Minister for Skills and Technology (DMST) chairs a Wales Strategic Forum for Career Development. Five key themes provide a focus for policy dialogue including: employer engagement, labour market intelligence, understanding the World of Work, links to economic strategy, and parity of esteem of learning pathways. Career Choices Dewis Gyrfa Ltd (CCDG) is a wholly owned subsidiary of the Welsh Government. Trading as Gyrfa Cymru Careers Wales, it provides the all-age, independent and impartial careers information, advice and guidance service for Wales.

The service delivers to a remit set by the Minister for Education and Skills focusing on the Welsh government's strategic objectives as identified in the Program for Government and related Welsh government policies, such as the Youth Engagement and Progression Framework (YEPF). Gyrfa Cymru Careers Wales has recently been required by government to develop a new strategy for its future online presence. It has recently been refocused to concentrate mainly on supporting young people, parents and teachers in classroom-based and one-to-one activities, with its services to adults delivered mainly online. For Wales (and Northern Ireland) a

---

[101] Scottish Government. (2014) *Developing the Young Workforce Scotland's Youth Employment Strategy, Glasgow.* www.gov.scot.

[102] Scottish Government. (2014) *Certificate of Work Readiness.* www.skillsdevelopmentscotland.co.uk.

differing approach has been adopted in comparison to 'the English experiment' i.e., 'payment by results' for national careers service systems is not an obvious priority.

In Northern Ireland, the Minister for Employment and Learning and the Minister for Education established an independent panel to provide advice on the overall direction for careers provision responding to the recommendations stemming from the Committee for Employment and Learning's Inquiry into Careers Education, Information, Advice and Guidance.[103] A report by an independent panel of experts from education and employers on careers education and guidance in Northern Ireland (December 2014)[104] set out five key recommendations, including one related to improved access and transparency for users of the government-owned Northern Ireland Careers Service, i.e., Recommendation 2: To ensure that the career system reaches as wide an audience as possible, the online offer must be greatly improved. This includes a key action to establish a central system to support all structured work experiences, including an online portal for the advertising and application of work experience placements.

In November 2015, a new report on the skill requirements for the Northern Ireland economy up to 2025 aims to ensure any skills gaps are identified and addressed. The new Northern Ireland Skills Barometer[105] represents an investment by the Department for Employment and Learning (DEL) to build a forecasting tool to estimate the quantum of future skills needs across a range of economic scenarios (e.g., a reduction in corporation tax, or the impact of austerity). This will also be a useful tool to consider the skills implications of the forthcoming programme for government and future NI Executive economic strategies. The role of CIAG and LMI is a strong policy focus and an all-age careers service operates both online and face-to-face in a wide range of community settings. Risk and uncertainty

---

[103] Northern Ireland Assembly. (2014) *Inquiry into Careers Education Information Advice and Guidance in Northern Ireland*: www.niassembly.gov.uk.

[104] Northern Ireland Executive. *Careers Review Report*: www.northernireland.gov.uk.

[105] DEL. (2015) *Northern Ireland Skills Barometer: Skills in Demand, Belfast*: www.delni.gov.uk.

regarding future investment in the current national careers service framework presents a significant challenge, particularly in the context of departmental reorganisation and continued public sector cuts.

Young people need to be guided though an increasingly complex landscape

The landscape of training and qualifications is increasingly complex. Signals to young people (and parents) about the added-value of learning and career pathways are becoming more blurred. Their sources of careers information are often coming from distorted or unreliable sources such as TV and social media. And many of today's young people will go into jobs that did not exist when their parents left school.

Weighing up the costs and benefits of apprenticeship training versus higher education options is becoming more difficult for many young people. It is often reported that graduates typically have higher rates of employment and higher levels of earnings than non-graduates. [106] In contrast, the lifetime benefits associated with the acquisition of apprenticeships at level 2 and 3 are increasingly reported as very significant, standing at between £48,000 and £74,000 for level 2 and between £77,000 and £117,000 for level 3 apprenticeships (London Economics, 2011). [107] In certain sectors, higher apprentices could earn £150,000 more on average over their lifetime compared to those with level 3 vocational qualifications (AAT & CEBR, 2013). [108] There is significant variation in the estimated wage premiums of apprentices depending on the sector of employment as well as gender challenges. [109] Choices and investment decisions are becoming more difficult.

---

[106] Gov.uk. (2015) *Graduate Labour Market Statistics April to June 2015:* www.gov.uk.

[107] BIS. (2011) 'Returns to Intermediate and Low Level Vocational Qualifications', *London Economics (2011)*. BIS Research Paper Number 53: www.gov.uk.

[108] AAT and CEBR. (2013) *University Education: Is This the Best Route into Employment?*: /www.aat.org.uk.

[109] Alison Fuller and Lorna Unwin (2015) the Challenges Facing Young Women in Apprenticeships', *Education and Employers*: www.educationandemployers.org.

The education system requires young people to make early subject choice decisions (some from thirteen years old upwards). In England, raising of the participation age from sixteen to eighteen years old in 2015 signals further challenges for this current generation of school pupils, parents and teachers. There is also evidence that children begin to eliminate their least favoured career options between the ages of nine and thirteen.[110] By those ages it is argued that they have abandoned the 'fantasy' associated with the very young and started to become more aware of potential constraints on their occupational choice. But they are often poorly informed about career possibilities and routes into them. Researchers at King's College London investigating young people's science and career aspirations aged ten to fourteen found that most young people and parents were not aware that science can lead to diverse post-sixteen routes.[111]

There is widespread consensus on the elements that make up effective guidance to help young people into successful working lives.[112] The biggest challenge is delivery. Over the last three years, the track record of CIAG policy in England has been one of repeated confusion, with a mismatch between the scale of the challenge and the service delivered in practice.[113] England is lagging behind other nations (both within and outside of the UK) when it comes to ensuring informed and impartial CIAG for young people. This represents a serious threat to the future success of apprenticeships in England. CIAG performs an important moral, social and economic purpose. The UK has historically been viewed as having one of the strongest career development systems in the world. A

---

[110] Gottfredson, L. S. (2002). 'Gottfredson's theory of circumscription, compromise, and self-creation' In D. Brown, ed., *Career choice and development* (4th Edition: 85-148). San Francisco: Jossey Bass. Cited in Gutman and Akerman (2008), *Determinants of aspirations, Centre for Research on the Wider Benefits of Learning,* Research Report 17. IoE

[111] King's College London (2015) Aspires: Young people's science and career aspirations, age 10 14. London: Department of Education and Professional Studies: www.kcl.ac.uk.

[112] For example see *Good Career Guidance,* The Gatsby Charitable Foundation, 2014: www.gatsby.org.

[113] BIS. (2013) An Aspirational Nation: Creating a Culture Change in Careers Provision, National Careers Council, June 2013: www.gov.uk.

major international symposium (USA, 2015) [114] revealed many other countries are now challenging the UK in this role. Countries as diverse as Canada, Estonia, Japan, Norway, New Zealand, Saudi Arabia and South Korea are all investing significantly in their CIAG systems.

Policy measures that could readily be implemented
In England, the process of rebuilding and re-engineering careers provision for young people has been slow, with false starts and setbacks. There exists a crowded, confused and complex landscape, with a multiplicity of disjointed careers initiatives.

London, Bradford, Leeds, Solihull, Cornwall and many other parts of England are now forging ahead independently with implementing a local careers offer for young people. For example, the Mayor of London, as Chair of the London Enterprise Panel (LEP), launched 'London Ambitions: Reshaping a careers offer for every young Londoner' (July 2015). This provides a vision and sets out a pragmatic way forward to tackle some of the challenges that young people face when trying to make the right career choices. It highlights the importance of face-to-face discussion and independent and impartial advice, as a requirement for every young person on a regular basis.

Having access to up-to-date, user-friendly labour market intelligence/information (LMI) as part of modern schooling system is an essential part of countering the distorted or unreliable information that is so widespread. While there is plenty of open source LMI at national level,[115] making high quality LMI available in an easily accessible form that is meaningful for non-specialists is a challenge. This can be achieved by learning from good practice that already exists in differing parts of the UK system.

There is a strong case for going further, to require every secondary school and college to have in place an explicit, publicised careers policy and careers curriculum encompassing young people's experiences of the world

---

[114] International Centre for Career Development and Public Policy, International Symposium 2015, Iowa. 15th 17th June 2015: www.is2015.org.
[115] For example the 'LMI for All' resource developed by UKCES: www.lmiforall.org.uk.

of work, links with business, careers provision, and measurement of destination outcomes. That policy should be reviewed and approved by the governing body at least every three years. All schools and colleges should also report annually on delivery of the careers policy and curriculum.

The destinations of leavers are every bit as important as exam attainment. In a world where the great majority of young people are comfortable in accessing data online, there is scope to share information and resources across universities, colleges, schools and jobcentres in a locality. There is also a real opportunity to incorporate best practice delivered by primary schools[116] and to systematically build upon those activities that work from a young person, parent, teacher and employer perspective.

The status of CIAG could be better reinforced if schools and colleges are expected to have a nominated governor with special responsibility for leading on the issue, ensuring the institution supports students to relate their learning to careers and the world of work. Teachers also need curriculum support and resources to embed CIAG within the classroom and in their work with parents and employers (London Ambitions, pp. 39 - 70.)

An effective CIAG strategy and delivery plan needs to work in a diversity of settings. It also has to reach out to those young people currently not participating in education, and to others such as apprentices who are re-considering their future career direction. Experience has shown that no single organisation can do that on its own. Three macro-policy initiatives could be a helpful way forward, particularly in England for ministers planning to announce a new careers strategy in early 2016.

1. Convene a cross-departmental review of activities and funding streams that impact on careers education and CIAG to ensure a more cost effective, fair and equitable system. A useful starting point for reconsidering existing approaches could be an evidence-based review of quality and policy systems drawing on lessons learned from within and outside of England. At present a number of central government

---

[116] See for example: 'Primary school children get a taste of TV at Sky Academy Skills Studios': news.cbi.org.uk.

departments are supporting and/or funding initiatives for young people, parents, teachers, and/or employers. A multi-faceted, integrated careers strategy, based on evidence of what works, now needs leadership and co-ordination across government departments to review investments and set national priorities.

2. Put a greater focus and investment on building capacity throughout the system and within schools to support the delivery of high quality careers education and CIAG

Making regular face-to-face guidance for every young person a reality depends on building capacity in the system. At a time when resources are constrained, this must rely on a partnership of teachers, careers and enterprise advisers and other local players with skills and expertise developed through joint projects and joint training, led by experts. In this context, careers professionals trained and experienced in building capacity in the system are a resource significantly under-utilised. For example, in Scotland[117] (and elsewhere in Europe)[118] career coaches are performing a key role in supporting schools and colleges in careers system design and delivery. Having everyone know what success looks like in a 21st century careers curriculum, starting from Key Stage 2 upwards, would be a helpful starting point. There are also economies of scale to be achieved through schools and colleges being centrally supported to embed LMI within their portals, aimed at students, parents and employers, and having greater awareness of tried and tested LMI dashboards and apps for student support and curricular developments. If every school simply does its own thing in isolation, there is a real risk of unnecessary cost, duplication of effort and lack of sharing good practice.

3 Put in place incentives in the accountability framework and regulatory regime to encourage schools, colleges and universities to invest in, build on and disseminate best practice in careers education and CIAG to ensure this is widely shared

---

[117] Education Scotland. *Careers Information: Advice And Guidance Delivered by Skills Development Scotland in West Lothian.* www.educationscotland.gov.uk.

[118] ELGPN. (2015) *Strengthening the Quality Assurance and Evidence-Base for Lifelong Guidance.* www.elgpn.eu.

There should be incentives as well as challenges for schools, colleges and universities to engage in being more attentive to careers education and CIAG – moving away from 'pain share' towards 'gain share'. Whilst inspection bodies perform an essential role in accountability and performance monitoring, other gains are required at a local level. For example, Local Enterprise Partnerships (LEPs) or Enterprise Zones have access to rich local data and have the potential to bring together all parties concerned with careers provision to deliver more effectively. In some instances, partnership may be most effectively cemented through 'clusters' and 'co-location' of services. Career development professionals (qualified to level 6 or above) should be enabled to support clusters to meet statutory responsibilities. This would help deliver a multi-agency, one-stop shop of careers guidance and support in each locality.

Conclusion

Careers provision is a public good as well as a private one. It is essential to keep more young people switched on to learning, to encourage them not to close down opportunities too early, and to broaden their horizons. For apprenticeships to be valued by more people in society major work is required to ensure individuals' access to independent and impartial CIAG. In England, the current patchwork system of careers provision is far too uneven in handling these essential tasks. To what extent can a largely unregulated careers provider market in England deliver in the public interest? Lessons from the Netherlands who have adopted a similar approach show marketisation of career guidance leads to an impoverished supply of services, both in the quantitative as well as in the qualitative sense (Hughes & Meijiers, 2014).[119] There is a pressing case for change to achieve a CIAG system fit for the 21st century, but change should focus on reducing fractures in the present system rather than constantly launching wholly new ones.

---

[119] Hughes, D., Meijers, F., Kuijpers, M. (2014) 'Testing Times: careers market policies and practices in England and The Netherlands', *British Journal for Guidance and Counselling*, London: Routledge: www.tandfonline.com.

31.
ARE THREE MILLION YOUNG PEOPLE READY FOR THREE MILLION APPRENTICESHIPS?

Laura-Jane Rawlings
Founder, Youth Employment UK

As ever, following any announcement on youth employment schemes, since the government announced its commitment to three million apprenticeships, we have been inundated with the views of employers, training providers and policymakers. With plenty of arguments both in favour and critical of the measures, the debate has been loud. Yet, in what is a worryingly familiar trend, we have heard very little from the young people who are to take up the three million opportunities.

Why does this matter? Many are quick to assume that young people have neither the experience nor the maturity to know how best to structure apprenticeships or training opportunities in the world of work. We believe that is profoundly wrong. To maximise the success of such a major scheme, we believe it is vital to put the youth voice at the heart of its development.

We should be asking questions to large numbers of young people across the country: what perceptions do they have about apprenticeships? do they understand the offer? do they actually want one? what would a good opportunity look like? what are the barriers and challenges? and what needs to happen to encourage young people to actively choose this pathway?

No government or agency can simply design and grow an offer such as apprenticeships without genuine consultation and involvement by the consumer. Young people must be brought into the very heart of the design and delivery; tokenism simply won't cut it.

When a young person is invited to speak about apprenticeships, they are usually guided as to their message or they are already a converted champion. Rarely do we hear from the young people who have had to turn down an apprenticeship pathway or have refused to consider it. These young people must be encouraged to voice their opinions and insights in a way that is open, frank, and ensures that recommendations are not just heard then ignored but are acted upon.

At Youth Employment UK we regularly consult with young people and have a large network of volunteer ambassadors nationwide who support the work we do. It is through discussions with our young members and ambassadors that we gain insights into the youth employment issues that face this generation, not least apprenticeships.

The most common barriers around apprenticeships, according to our consultations, fall into one of three categories:

- Awareness
- Perceptions
- Opportunities

Awareness
In a survey conducted by us in 2015[120] we found that only 57.9 per cent of respondents were provided with an interview with a professional careers advisor whilst in school or college and only 6.7 per cent were given information about apprenticeships.

Young people are then surprised to find out about the range of career opportunities, quality and benefits of apprenticeships.

A Youth Employment UK ambassador who has a First-Class Honours degree said she wished she had known about apprenticeships when making her post-sixteen choices as she would not have gone on to

---

[120] Youth Employment UK. (2015) 'Yeuk Launches Report On Careers Education – Young People's Experience of Careers Education', *YEUK*: www.yeuk.org.uk.

university, believing that an apprenticeship would have helped to progress her career further than her degree had.

In the Nothing in Common report by the Education and Employers Taskforce[121] the career ambitions of young people aged between thirteen and eighteen were explored.

The top three occupational preferences of young people aged fifteen and sixteen were recorded as teacher, lawyer and accountant. Out of the ten top occupational preferences of 17 to 18-year-olds, nine of the choices are traditionally academic careers which require a university degree. Whilst so many young people are not aware of the apprenticeship pathways that include higher or degree based qualifications, university will remain a more obvious choice for these aspirations.

Despite the investment by government, employers and training providers, too many young people remain unaware of the opportunities through apprenticeships. To recruit three million apprentices, much more has to be done to raise the awareness of the diversity, quality and academic opportunity they offer to young people across the UK.

Perceptions
It is interesting to discuss with young people who do know something about apprenticeships what their perceptions really are. These perceptions have often been influenced by parents, teachers and peers:

- Blue-collar careers
- Cheap labour for employers
- For the less academic
- Limit of potential employers
- Limited progression
- Menial work
- Narrowing options
- Poorly paid

---

[121] Anthony Mann. (2013) *Nothing In Common: The Career Aspirations Of Young Britons Mapped Against Projected Labour Market Demand* (2010 2020): www.leeds.gov.uk.

- Removes university as an option

Reaching and informing influencers is key when it comes to growing the numbers of young people wanting to choose an apprenticeship pathway. In our consultations with young people we are often told that school teachers and careers advisers promote university as the recommended pathway to secure a good career.

Parents can also have outdated perceptions of apprenticeships; understandably for many parents the information they have about apprenticeships is biased towards blue-collar careers as this would have been the case when they had exposure to careers information.

One of our ambassadors, who is an apprentice in the Civil Service, recently spoke at a conference, telling the audience he had to actually defy his parents to take up his apprenticeship. They were adamant that he would only achieve his potential through university.

The influence of other young people is also a major factor; peer endorsement is key to many young people when it comes to making decisions. Having a large number of peers plan to attend university can have an impact on the appeal of an apprenticeship pathway.

Opportunities
If you were to type Apprenticeship Vacancies UK into Google today you will be offered about 3,640,000 results. Type in Careers Help UK and there are about 564,000,000 results. Independently navigating careers information and apprenticeship vacancies is more difficult than it has ever been, and not because of a lack of opportunity.

When young people want to apply to go to university they know that the most popular information resource is UCAS. It's also the centralised place that manages the application process for British universities that includes the Clearing process.

For potential apprentices the choice is overwhelming and the process unclear. By registering with one site you risk missing opportunities, which means the need to register with potentially upwards of twenty websites. It was a welcome measure from government to require all apprenticeship vacancies to be advertised by the .GOV Find an Apprenticeship website. This will allow young people to register and search in one place for all of the available vacancies.

However, this new Find an Apprenticeship website comes with a downside, in that it lacks the inspiration and information to help young people explore, learn and choose from a range of different options. Which means it cannot be the only go to place careers advisers and schools can recommend to students.

Other challenges that face young people are the chronically low wages and inaccessible locations of some apprenticeship vacancies. For some young people living in rural areas there is a limited selection of opportunity. This may mean the need to travel some distance to work each day, which comes at a cost.

One of our ambassadors who has just started an apprenticeship told us she had to turn down two apprenticeship roles before accepting her current job as the travel cost to get to work would have wiped out her salary.

There must be more work done to encourage SMEs in rural and coastal areas to create apprenticeship opportunities. The large majority of employers offering apprenticeships are those corporate employers who are based in cities or large towns.

Employers must be encouraged in the strongest possible terms to pay fair wages and offer support for young people who are starting out in their career. Youth Employment UK launched a report in January 2016 on the wages offered to young people, and the barriers and disincentives that low wages can mean.

Key Training, a Member of Youth Employment UK, have recently applied a higher minimum wage rule for employers they work with. They found that by paying a fair wage to young people, the attrition rates of apprenticeships fell overall and the quality of applications increased.

As retail and hospitality employers raise their hourly pay rates in line with the new national minimum wage there is a risk that young people may turn down training opportunities to earn higher salaries in low-skilled positions.

Entry requirements for some apprenticeship vacancies have also been cited as a barrier by young people. Some employers and training providers require a set number of GCSE grades as part of their application criteria. For many young people who would benefit from level 2 apprenticeships this can be a real challenge, as they have left education without the grades required to progress. In the academic year 2013/14, 37 per cent of pupils did not achieve grades A*-C in both maths and English according to a report by The Education & Training Foundation.[122]

Young people who do not obtain the required GCSE results in a formal school setting can flourish and add real value whilst achieving the functional elements of their course in a quality apprenticeship environment.

Young people want to progress and they want to have opportunities to develop; this is why a university option can appeal. Young people believe that with a degree they will have more choice and opportunity to develop their career and earning potential. To secure good apprenticeship candidates, employers must consider how they can continue to develop and support the growth of their apprentices and demonstrate the career opportunities to young people.

What do young people want?

- ✓ Information
- ✓ Opportunity
- ✓ Progression
- ✓ Reward
- ✓ Support

---

[122] Freddie Whittaker. (2015) 'Education and Training Foundation Workforce Survey Highlights Functional Skills Teacher Recruitment Problem' *FE Week*: feweek.co.uk.

Young people have told us that they would like to have comprehensive and inspiring careers information whilst they are at school. This information must include insights into the range and opportunity of apprenticeships. We have also been told that young people want to be able to explore and research the range of pathways available to them from trusted sources.

Young people have told us that they want to work and they want to be able to access opportunities. They want to be given the opportunity to learn and develop, and to also give back. They are ambitious and want to be able to progress, and they want to be successful in their lives, from getting good qualifications to having a good career and lifestyle.

They tell us that they want to be treated fairly, to be recognised for the work that they do, and to feel valued.

They tell us that they are aware they sometimes need support post-education - mentoring, or an approachable supervisor who can offer guidance and support is important to them.

Recommendations
Youth Employment UK strongly recommends that government, employers and training providers build in more consultation with young people around the design and development of apprenticeships.

We recommend that at least one young person is invited to join the panel of the trailblazers and that a cohort of young people are recruited to provide advice and insights to BIS, SFA and colleagues in the Cabinet Office with a responsibility for the apprenticeship target.

We would encourage all employers to adopt the principles of the Youth Friendly Charter [123] developed by young people through Youth Employment UK. The Charter encourages businesses to:

- Create/listen to Youth Voice

---

[123]Youth Employment UK. (2015) 'Youth Friendly Charter', *YEUK*: www.yeuk.org.uk.

- Create Opportunity
- Recognise Talent
- Offer Fair Employment
- Develop People

We recommend that more support is offered to schools to help and incentivise them to deliver quality careers information on all post-sixteen options for young people. Young people should be able to clearly cite apprenticeships as a quality choice post-secondary education.

The government should urge schools to use a small set of quality resources that have already had government funding as an example these could include:

- Careers & Enterprise Company
- Plotr
- National Careers Service
- UCAS
- .Gov Find an Apprenticeship

Investing in and building on these resources that already exist will help young people to be able to self-manage, explore and navigate the world of work.

It is also our recommendation that the current government secures the application process for future governments. The recruitment and application process for apprentices must be consistent if we are to see long-term growth.

Finally, Youth Employment UK recommends the support for our own ambassador network as a way to encourage peer-to-peer inspiration and insights into the world of work. Our ambassadors receive comprehensive training, support and mentoring so that they can confidently go and meet with and speak to other young people about their own experiences and share tips and advice.

By growing and developing our ambassador network we will be able to offer schools and colleges access to great inspirational talks from young people who are still on their own journey of development.

Are three million young people ready for three million apprenticeships? Perhaps, in fact, this is the wrong question to be asking. We believe young people are eager and ready for apprenticeships. Rather, we feel that the three million opportunities may not be quite ready for them. With a relatively small amount of targeted work and a focus on placing young people at the heart of their development, three million young people will get the training that they deserve.

32.
FROM ADVANCED APPRENTICESHIPS TO HIGHER-LEVEL LEARNING

Sue Betts
Director, Linking London
and
Andrew Jones
Deputy Director, Linking London

————

1. Introduction and context - progression to higher-level learning from advanced apprenticeships.

This chapter describes our experiences in promoting apprenticeship progression onto higher-level learning. Linking London is a collaborative partnership of forty institutions (universities, colleges and strategically important organisations) based at Birkbeck, University of London, committed to the progression of all vocational learners, including apprentices.

In 2005, Brenda Little and Helen Connor illustrated well the sometimes difficult route that vocational learners had to follow to get to higher education. (Little and Connor, 2005) There has been a vast increase in the number of BTEC learners studying at level 3 – the number has doubled since 2008 – and therefore in the number of vocational learners (particularly BTEC) progressing to higher-level learning. Our research illustrated this change. (Joslin and Smith, 2015)The development of access agreements in universities overseen by the Office of Fair Access (OFFA) has added further momentum to the recruitment of students from all parts of the vocational track.

Since the recession at the end of the last decade, there has been much focus on the potential for expanding apprenticeships and the government's emphasis on productivity sees their expansion as critical to economic recovery. This has led to a number of improvements including a long-term increase in numbers, a short-term policy emphasis on quality, the abolition of programme-led apprenticeships (PLAs), increases in the participation of minority ethnic groups, and a more even gender balance in apprenticeships, which have historically favoured men.

However, issues still remain in terms of what an apprenticeship means, who they are aimed at, their quality, and their place in progression to higher-level learning.

Context
Until recently there has been a lack of detailed data on advanced apprentice progression into higher education. Research carried out by Mark Gittoes for HEFCE identified that the progression rate from advanced apprenticeships was around 6 per cent. (Gittoes, 2009)

Research carried out for the Department for Business, Innovation and Skills (BIS) into the progression to higher-level learning of advanced level apprentices over a seven year period showed that the pattern of progression to higher-level learning is very different from that of traditional full-time school and college leavers, the majority of whom progress in the year following study at level 3. Of the advanced level apprentices who progress, 58% do so within three years of starting their apprenticeship but significantly 42% of them do so four, five, six or seven years later. In total 19.3% percent of the 2006-07 tracked apprentice cohort progressed to higher-level learning when tracked for a total of seven years. (Joslin, Smith, et al, 2915)
In terms of progression to higher-level learning by framework, there are stark differences. For example, tracked over seven years, 76 per cent of accountancy advanced apprentices enter higher-level learning compared with 5.7 per cent of vehicle maintenance and repair advanced apprentices. Higher education qualifications studied also vary. While 71.5 per cent of active leisure and learning advanced level apprentices who progressed went onto study a first degree, only 3.6 per cent of apprentice engineers went on to this level of study. Most engineering apprentices go on to other undergraduate (OUG) study, particularly Higher National Certificates.

In terms of progression onto higher apprenticeships (HAs), the number of advanced level apprentices progressing on to HAs increased from 1,130 to 1,630 between 2008-09 and 2010-11, with a progression rate for the 2010-11 cohort of 2.6 per cent, slightly higher than the 2.3 per cent rate for 2008-09 apprentices. The majority of tracked higher apprentices were on an accountancy framework, although in 2010-11 numbers on business administration, management, and health & social care frameworks increased.

Further Education colleges (FECs) deliver higher-level learning to a higher proportion of advanced level apprentices than universities but the gap has narrowed. For the early cohort in 2006-07, 63 per cent of apprentices progressed to higher-level learning in colleges but for the cohort in 2010-11 this had dropped to 52% per cent While 68.4 per cent of the 2006-07 cohort progressed to part-time learning in higher education, this had dropped to 50.3 per cent for 2010-11, an indication perhaps that more advanced level apprentices are choosing to make a life change and progress to education on a full-time basis, but the drivers for this are not investigated in this study.

However, the research also shows that the overall three year progression rate has dipped over the five cohort years from 11.2 per cent in 2006-07 to 8.8 per cent in 2009-10. This reduction is influenced by the significant increase in the numbers of apprentices who are over the age of twenty-five . The numbers of 25-plus advanced level apprentices in the cohorts increased from 0.3 per cent of the total in 2006-07 to 40 per cent of the total in 2010-11. While tracking apprentice progression into higher-level learning longitudinally has shown greater numbers going on than originally thought, there remain a number of issues in terms of progression to resolve.

2. Why greater rates of progression are important.
In order to improve social mobility and enhance economic prosperity for both the nation and the individual, apprenticeships should keep options for progression open. In some of the older established areas, routes through to higher-level learning and chartered status are well known. The same level of opportunity needs to be available for all apprentices who can benefit from it and it is important that this work-based pathway provides opportunity for further study and qualification. For some vocational learners the next step up may be into management rather than a further level of vocational learning.

Drawing on recent research 22 per cent of advanced level apprentices who entered higher-level learning were classified as coming from the most educationally disadvantaged parts of the country (POLAR2 Q1). (Joslin, Smith et al, 2015) This compares to 11 per cent for all young undergraduate entrants and 12 per cent for mature undergraduate entrants (HEFCE, 2012). Apprenticeships clearly play an important role in social mobility.

In terms of benefits to the economy and future prosperity there is considerable evidence that an individual's employability is improved by participating in an apprenticeship. The employment rate for those aged twenty-five to sixty-four with no qualifications was 48.5 per cent, but for those with apprenticeships it was 80.7 per cent (ONS, 2014)

Research undertaken by the Sutton Trust indicates that completing a HA at level 5 results in greater lifetime earnings than undergraduate degrees from non-Russell group universities. (Kirby, 2015) Research carried out by Joslin, Smith et al highlighted that 82 per cent of higher education leavers from the apprentice cohort were in employment six months following their degree, higher than the all-England higher education leaver rate of 76 per cent, and a further 12 per cent were in further study. The unemployment rate was low at 2.4 per cent and the average salary of the apprentice higher education leaver cohort was also higher than that of higher education leavers generally. (Joslin, Smith et al, 2015) Perhaps if these outcomes were more widely known by policymakers as well as by employers and higher education institutions (HEIs), more higher-level opportunities for apprentices would be made available.

4. Higher-level learning opportunities – what are the current options for employees who have completed an advanced apprenticeship?
There are several possibilities for employees who have completed an advanced apprenticeship:

Higher apprenticeships
A clear progression pathway from intermediate through advanced on to higher or degree apprenticeships is an obvious and desirable option that for many would be an attractive stepping-stone. It would enable many to continue to learn and put theory into practice while working and earning a wage.

Part-time and flexible learning

Part-time and flexible learning that enables employees to continue to work and study is also an appealing option for many. As noted further on in this chapter, the significant increase in tuition fees and for many the lack of accessibility to the loan book may be off-putting. Work based learning that draws on the experience and knowledge of the employee and supports them to apply theory to work-related practice is particularly relevant to this group.

Full-time higher-level learning

As highlighted by Joslin, Smith et al, a significant number of former advanced apprentices progress on to full time study. While this may come as a surprise in terms of the learner having to potentially give up full time employment, it may be due to a number of factors, for example, a career change, because they have to (e.g., nursing is studied predominantly on a full time basis only at undergraduate level) or perhaps because they were unaware of alternative study modes or there were a lack of them. The research therefore suggests progression into work-based routes is currently limited.

Short courses

Many short courses or even in-house training provided for employers by HEIs may be accredited. These may include short courses to degree courses but also professional qualifications. Some HEIs also offer bite-sized introductory higher-level courses. Bite-sized and incremental learning which enables employees to 'dip their toe in the water' at low risk to themselves in terms of commitment and costs is likely to appeal to some former advanced apprentices.

5. Progression onto higher-level learning – the challenges

The challenges in our experience are many and complex. They include, but not exclusively, a lack of HA vacancies and employer engagement, a largely uninformed higher-education sector (although recently developments suggest this is changing rapidly for some institutions), a lack of suitable higher-level learning provision, and funding complexities and quality issues which the TUC has long campaigned on. The TUC has also been critical of short training programmes in the guise of apprenticeships. If all of these challenges are overcome, however, there is still a lack of information in terms of informing potential higher-level learners about what is available.

a) Employer engagement and the quality issue
More employers need to be fully engaged with both the costs and opportunities in supporting young people, in particular, to get their foot on the first or second rung of the working ladder, especially in small and medium sized enterprises (SMEs). We know only 17 per cent of employers, for example, offer work experience.

This is not good enough if employers continue to say employees arrive with a lack of experience. Apprentices are apprenticed to a job. For some employers at higher levels they may see the cheaper option as recruiting already qualified individuals rather than supporting local people to progress.

The Ofsted report '*Apprenticeships: developing skills for future prosperity*' (Ofsted, October 2015) was damning in terms of both quality and who was being described as an apprentice. Inspectors found that many of the 24-plus apprenticeships at lower levels were simply a re-badging of an employee already working for that company.

b) Lack of higher apprenticeship vacancies
Apprenticeship vacancies, especially HAs, are in short supply, available in a limited number of sectors, offered in the main by larger employers rather than SMEs, and taken up predominately by older employees. Data on HA numbers (i.e., starts) shows, if we focus on the most recent data for 2013-14, that 40 per cent of HA starts were in the 16-24 age group. This figure drops to 27 per cent for 2014-15. (FE data library)

When we look at higher apprenticeships starts by sector, two areas dominate:

1. Business, administration and law
2. Health, public services and care

In 2013-14, of the 9,200 HA starts, 7,900 were in these two areas. The pattern continues in 2014-15 with 17,210 of the over 19,000 starts in these two areas.

When discussing higher apprenticeships/degree apprenticeships (HA/DAs) with potential employees there is a danger of raising false expectations. Caveats need to be made in terms of availability as well as competition levels. In 2013-14, there were 1.8 million applications for 166,000 vacancies: a ratio of eleven applications for each vacancy across all apprenticeships.

Despite the lack of information and 'poor' Information, Advice and Guidance (IAG) in general, competition levels when applying for advanced and higher apprenticeships is high. Indeed, young people accounted for the majority of applications through the official apprenticeship vacancies system, but a far smaller proportion of starts. Older people accounted for an extremely disproportionate share of starts.

> 'Under 19s made 56% of applications but only 27% of starts; 25+ made 7% of applications but 37% of starts'. (IPPR, 2015)

Where competition levels are lower (especially at intermediate and advanced level) this may be in part due to low rates of pay, for example, in the care sector, hospitality and hairdressing. It may also be due to timing, a rarely referenced issue. Feedback from Linking London college IAG practitioners has highlighted this issue of timing in terms of when apprenticeship vacancies are advertised. While employer recruitment patterns do not mirror academic cycles, promoting apprenticeship vacancies that would start mid-course present difficulties in terms of the impact on both the college and the learner. Dropping out of a course mid-way is a retention and potential funding issue for the college and an achievement issue for the learner.

Although negative perceptions of apprenticeships the 'oily rag' image - are often cited as an issue in terms of take up of apprenticeships (and it doesn't help that some policymakers tend to talk about apprenticeships as a solution to the numbers of young people not in education, employment or training or NEETS) we don't believe this applies in terms of interest in HA/DAs. Explaining that they are funded by the government and the employer and that you could end up obtaining a higher-level qualification debt free, at the same time as gaining work experience and a wage is not 'a hard sell' to potential higher or degree apprentices. The harder sell is to convince employers, especially SMEs, of the benefits of HA/DAs and for

HEIs to engage in the learning and qualification aspect of the training in consultation with employers rather than offering off-the-shelf courses.

c) Complexity of funding arrangements and excessive bureaucracy

The complexity of funding for providers has not helped, and the funding straddles two funding bodies, a potential recipe for confusion. There are also challenges around bureaucratic processes, although we appreciate that it is important to guard against fraud. As the 2015 Ofsted report notes, 'the complexity of funding arrangements has meant that employers, especially SMEs, have been slow to get involved in designing programmes and developing standards, or in taking on apprentices out of fear of becoming mired in bureaucracy'. The IPPR report notes that for employers looking for clarity about a potential employee, the system is opaque and confusing. Anecdotally in our dealings with higher education providers, many have found the terrain equally confusing.

Greater rates of progression for people on a work-based route are vital for the economy, employers and their employees. The fact that there are approximately five million people in this country qualified to level 3, offering a massive potential source of further skills development, is reason enough for freeing up progression to higher-level learning. The government target of three million has made people sit up and take notice but who knows the number of new employees learning on the job the economy requires.

We have recently researched the labour market needs of the London labour market and there does not appear to be a link-up yet. The government is putting employers in the driving seat but there is a potential conflict between the interests of the employer (not completing a full framework or including a qualification in the mix) and the future of the employee and their training which should have national transportability.

d) Lack of awareness of advanced apprentices as potential learners and of relevant higher education provision

There is an argument to be won on behalf of progression for apprentices. Some parts of the higher education sector need to be convinced that an advanced apprenticeship is a grounding for further study, but there also needs to be change in the type of provision on offer in higher education.

High quality work-based education has a pedagogy of its own and it is demanding.

One of the major barriers to progression for apprentices is the fact that many HEIs do not have appropriate provision onto which apprentices can progress. This lack of readiness for a whole section of the population equipped with the equivalent of a level 3 qualification and who may want to continue working while learning has been a major stumbling block.

The last few years has seen a complete collapse of the part-time market in higher education which has had dramatic consequences for what is now on offer. Part-time, work-based and distance learning provision is not core activity for most HEIs and there has been little attempt to market an offer to advanced apprentices and their employers. With government policy and funding that encourages HEIs to chase the 'full fat' full-time learner at the expense of developing alternative, innovative, flexible, work-based and bite-sized provision, this situation is likely to worsen unless the levy can alter behaviour.

For current or former apprentices interested in progression to higher-level learning, there is a lack of appropriate information on higher-education admissions requirements generally, whether for full-time or part-time and flexible provision. When prospective higher-level learners do identify relevant provision information specifically aimed at former advanced apprentices, entry requirements are in most cases absent. The potential applicants may therefore assume that they would not be interested and therefore de-select themselves.

It should be noted that advanced apprenticeships are not UCAS tariffed and that frameworks differ widely in terms of content and qualifications obtained. Some require the advanced apprentice to complete a BTEC level 3 qualification, which does attract tariff points, while others include level 3 qualifications, e.g., City & Guilds and NVQs, which do not. This presents challenges to higher education admissions departments in terms of setting entry requirements.

Higher education institutions as well as potential receivers of former apprentices, are also providers of curriculum content for HA/DAs.

The introduction of the nine thousand pound annual fee for full-time students in 2012 has also possibly not helped universities look more closely at this potential cohort of learners. The very nature of an apprenticeship is complex; unlike putting on a course of study for a cohort of learners, it requires the coming together of an employer, a need for an apprentice, a job and a suitable course of study which can be part of the award. This requires flexibility on the part of the provider, and a curriculum that is appropriate for a work-based learner on an apprenticeship framework. Add to this the complexity of funding arrangements and high levels of bureaucracy and these become potential barriers to further engagement.

e) Poor information advice and guidance
While websites have made strides in improving the availability of information aimed at those interested in apprenticeships, the wealth of information and advice aimed at those interested in higher-level study is still dominated by the eighteen year old A level student interested in progressing on to full time (and usually campus-based) undergraduate study. Information about progression onto higher-level opportunities including HA/DAs, is in most cases an 'add on', and is often generic and lacking detail. Detailed information and advice aimed specifically at current or former advanced apprentices on higher-level progression opportunities and identified pathways is therefore notable by its absence.

In addition, while the internet is awash with information on full time higher education study opportunities, along with the UCAS website which provides a definitive centralised admissions service for full time undergraduate study, information on alternative study modes is more difficult to obtain. For part-time and flexible study there is no equivalent comprehensive and publicly available source of information or centralised application or admissions service.

Providing effective information and advice is largely dependent on the accessibility of comprehensive and up to date information. It is hard to provide advice to someone in their absence. Current or former advanced apprentices are not a captive audience in comparison to other level 3 learners situated in schools or colleges and this poses challenges in terms of providing IAG to them.

In summary, there are not many options for advanced apprentices, information is not readily available, and so they do not enquire about higher-level learning. As a consequence, HEIs do not currently develop more opportunities for them, or specifically market the opportunities they already have. It is a vicious circle.

6. Examples of good practice in supporting progression: A Linking London perspective

Linking London has been working in this area for many years trying to address some of the issues and challenges for our partners.

a) Research

Linking London conducted a survey of entry criteria information available for advanced apprentices. We examined a sample (550) of entry criteria from thirty HEIs across England, representing (at the time) approximately 10 per cent of the UCAS membership. Apprenticeships, unsurprisingly, were least well served with information. Just 2.5 per cent provided information that suggested the qualification could be accepted as a route to higher-level learning. This compared with 88 per cent of entry profiles which provided complete information for A level applicants. (Linking London, 2010)

Linking London also commissioned research into progression for apprentices as a sub-set of national research funded by BIS. Some interesting findings emerged, progression was not always immediate (7 per cent) and the rate of progression did improve if people were tracked over a five year period (11 per cent). (Joslin and Smith, 2013)

b) Resource development

Linking London has produced a good practice guide for admissions staff, giving details of the advanced apprenticeship, how to make a meaningful offer to advanced apprentices and where to go for further information. This publication, originally developed in 2011 has recently been updated. (Linking London, 2011)

In 2011, Linking London published '*Apprenticeships and Progression to Higher Education*'. This publication argued that more was needed than guidance and information and it supported the University Vocational Awards Council (UVAC) recommendations that called for 'level 3

369

vocational qualifications and apprenticeships offered to young people should only be eligible for public funding if they outline clear progression routes and opportunities for progression to higher education'. UVAC also recommended that the government should 'invite universities to make an apprenticeship admissions pledge – where an appropriate course is available they will guarantee interviews to advanced apprenticeship framework completers'. (Chappell, 2011)

A 'Routes into HE' postcard and poster has also been produced which matches qualifications to the relevant Regulated Qualification Framework (RQF) level. Both publications have proved very popular not only with higher education admissions departments but with advisers, outreach and teaching staff in both higher and further education.

In 2012, Linking London published a detailed guide, that included information on reasons to consider higher-level study; the range of courses; qualifications and study modes, including HAs; examples of related higher-level courses by framework; the relevant higher education curriculum offer; and details about the Linking London Apprenticeship Admissions Pledge.

c) Working with partners to improve progression
Linking London has funded work with university partners, such as Middlesex University and Birkbeck, University of London, on projects to see how progression could work. Linking London also worked with partners on the suggested 'apprenticeship admissions pledge' but to date it has had limited success.

d) Engagement with higher apprenticeship developments
Through our links with UVAC we have also engaged with the Greater London Authority (GLA) on the GLA's demanding targets of 2,500 HAs in London over the next year and a half. We are now working with the thirty-two London Boroughs as Associate Members of Linking London to increase awareness and linkages with our partners. We have played a brokerage role for our partners, for example, in relation to the new HA developments at the Knowledge Quarter in Camden.

e) Staff development and networking events

Over the last nine years Linking London has held numerous events focussing on apprenticeship developments, sharing good practice, and ways in which our higher and further education partners can engage more effectively in the development particularly of higher apprenticeships.

7. What next?

Addressing the challenges, particularly in terms of more employer, especially SME, engagement in the development of apprenticeships and improving quality, will not be solved overnight and will require a firm commitment from both the government and employers to put right. The government needs to strengthen mechanisms to implement and monitor the improvement of both the quality of apprenticeship training and to cajole and incentivise employers to target their recruitment of apprentices at all levels at the younger age group. As noted in the Ofsted report, government should also hold providers to account for the value their apprenticeships add to apprentices' careers, evidenced by progression to higher-level learning and training, increased responsibility at work, and improvements in earnings.

In terms of improving the quantity of apprenticeships at higher levels, the government also needs to simplify and sustain policy approaches to enable all agencies and individuals involved to know what an apprenticeship means and to help encourage more SME take up. This needs to include reducing, as far as possible, unnecessary bureaucracy allegedly 'putting off' employers. The public sector, including education institutions and local government, should lead by example in recruiting more apprentices at all levels.

A dedicated service for individuals, providers and employers to independently encourage, broker and support progression onto higher-level learning, including the provision of up-to-date labour market information and intelligence identifying those sectors with skills shortages would also be hugely beneficial. This would enable providers to start an informed dialogue with local and other employers to plan apprenticeship provision to meet local, regional and national priorities.

More HEIs and FECs need to engage in the development and delivery of HA/DAs. This is fertile territory for a collaborative partnership such as Linking London to continue its role in brokerage and innovative

371

development. The apprenticeship levy due to be introduced in 2017 is set to raise £2.7-3 billion and most certainly will be a 'game changer'. Government priorities also have a role to play and they appear to see apprenticeship development as the main purpose of further education and the route for vocational learners. We will work with our partners to ensure they are more involved, with the caveat that apprenticeships are not the answer for everyone.

It should be noted that a number of recommendations (the pledge, the use of public money and funding to support alternatives to full-time provision) included in the UVAC report mentioned earlier still hold value and have not been addressed.

Government levers are needed to encourage the growth of flexible bite-sized, part-time and work-based higher-level learning to provide a greater range of opportunities for those who want to go further. In turn, HEIs should do more to promote their existing offer to advanced apprentices as well as the five million people in the workforce already qualified to level 3. Organisations such as UCAS and the National Apprenticeship Service could play an important role in supporting this.

HEIs should be encouraged as a starting point to map apprenticeship progression opportunities by framework/standard, and to identify potential progression routes into higher-level provision, enabling them to develop and promote admissions information for potential advanced apprenticeship applicants. In this work they could learn from successful examples of progression in the accountancy sector. HEIs also need to be more aware of what an advanced apprentice can offer in terms of skills, knowledge and experience acquired. Building on this work, HEIs could then commit to an advanced apprentice admissions pledge, guaranteeing an interview for applicants while working to clarify admissions requirements in external marketing and on their websites.

In seeking to address the issues of poor IAG we suggest that key stakeholders, including the National Apprenticeship Service, National Careers Service, Careers Development Institute and colleges, professional bodies, Unionlearn and universities should work together to review current IAG resources and support that can be accessed by apprentices (and former apprentices). Collectively they need to develop a holistic offer that is widely disseminated to schools, colleges, training providers and HEIs. Ideally all advanced apprenticeships should include an embedded module that enables apprentices to explore the full range of potential progression opportunities, and flags up further support to those who may want to access it.

Government should instruct UCAS to do more to raise awareness of alternatives to full time study, in terms of information, resources aimed at potential higher-level learners, advisers and HE staff. This would ideally include in the longer term a service for applications to HA/DAs as well as to part-time, work-based and flexible higher-level learning. Work with UCAS to develop a UCAS tariff for advanced apprenticeships, building on the success of the UCAS tariffed AAT NVQ in Accountancy would be a start.

And finally to help inform these recommendations, more data and research is needed. As a starting point we would suggest that this should include the identification of where advanced apprentices are located, current progression into and through higher-level learning (in all its forms) from advanced apprenticeships, by framework, building on the work of Joslin, Smith et al. Research also needs to be conducted to determine the level of awareness of and interest in higher-level learning amongst those undertaking, or who have recently completed, an advanced apprenticeship. Research to identify existing good practice in work to support advanced apprenticeship progression would also be helpful, as would the identification of lessons to be learnt from the high progression rates of accounting apprentices. The role here of the professional body appears key.

We are currently mapping curriculum provision in our further education college partners to see how what is on offer aligns with economic needs and progression opportunities. We will re-invigorate our pledge work with our partners. Funding, a government target, and a change of will are important if the key players in this many-sided relationship building, on behalf of apprenticeships, are to work. Our work will continue with admissions tutors, developing relevant resources, and actively brokering on behalf of potential employees and employers through our Borough links. For Linking London our work on arguing for, and raising awareness of, apprenticeship progression will continue.

References

Carter, J. (2010) *Progression from vocational and applied learning to higher education in England.* University Vocational Awards Council (UVAC).

Chappell, J. (2011) *Apprenticeships and progression to higher education.* Linking London.

Gittoes, M. (2009) *Pathways to higher education: apprenticeships.* HEFCE.

IPPR. (2015) *Learner drivers: local authorities and apprenticeships.* Institute for Public Policy Research for the Local Government Association.

Joslin, H. and Smith, S. (2013) *Progression of apprentices to higher education in London 2004-2010.* Linking London.

Joslin, H. and Smith, S. (2015) *Progression of college students in London to Higher Education 2007 to 2012.* Linking London.

Kirby, P. (2015) *Levels of success – the potential of UK apprenticeships.* Sutton Trust.

Linking London. (2010). *Quality of admissions information for applicants to full-time undergraduate study,.*

Little, B., and Connor, H. (2005) *Vocational ladders or crazy paving? Making your way to higher levels.* Learning and Skills Development Agency.

Office for National Statistics. (2014) *Qualifications and labour market participation in England and Wales.*

Ofsted. (2015) *Apprenticeships: developing skills for future prosperity.* SFA and BIS, Statistical Data Set FE Data Library: Apprenticeships.

Smith, S., Joslin, H., and Jameson, J. (2015) *BIS research paper number 240: progression of apprentices to higher education – 2nd Cohort Update'.* BIS .

33.
EMPLOYMENT AND EARNINGS PROSPECTS: IMPROVING INFORMATION ON HIGHER EDUCATION AND APPRENTICESHIP CHOICES

William Walter
Associate Director, Kreab

This chapter argues that the UK will not witness the apprenticeships revolution it needs without first seeing improved transparency in higher education returns. At present, young people are unaware of the differing prospects for employment and earnings between higher education institutions. In very many cases, apprenticeships offer better prospects but without greater transparency these benefits will not have the impact needed on young people, parents and advisers.

Policymakers in the UK regularly reaffirm they are committed to levelling the playing field between apprenticeships and university degrees. There is cross-party commitment to boost apprenticeship numbers and tackle the negative perceptions that still remain pervasive among both school leavers and their parents regarding the merits vocational training can offer young people in Britain today.

Government strategy to achieve this goal, however, remains confused and disjointed. Figures for 2014/15 show there were 499,900 apprenticeship starts in England, some 59,500 (14 per cent) more than the previous year. Indeed 2014/15 was the first year since 2011/12 in which apprenticeship numbers actually increased. While these figures are welcome, growing numbers of apprenticeship starts alone do not necessarily indicate success at improving the perception of vocational qualifications.

The real test of whether perceptions of apprenticeships are improving is to examine the take-up of higher level (level 4 and above) apprenticeships. Unfortunately, figures for this category remain modest. Examining the

375

figures above in more detail shows that only four per cent (19,800) of apprenticeship starts in 2014/15 were higher level, compared to 60 per cent for intermediate apprenticeships.

A significant proportion of parents and school leavers are wedded to the idea that university is the key to a successful and prosperous future, unaware of other avenues available. While for many, university remains the most suitable option, this is not the case for all.

A 2014 study by the Million Jobs campaign explored the strength of often quoted anecdotal evidence that vocational educational routes provide a more suitable avenue for many young people to join the workforce than more traditional academic options, such as university. (Walter and Malhotra, 2014)

The study was the first of its kind to compare the earnings, employment and taxpayer returns of graduates to those of apprentices. It was also the first to investigate, in any meaningful detail, the impact of the type of university attended.

Using data from a combination of sources, the authors examined the average earnings of the two routes in more detail, while also seeking to determine the impact of a range of variables on earnings including HEI type, debt and foregone earnings.

In terms of lifetime earnings, the authors found that over the period 2005 to 2013, the proportion of graduates earning less than the average wage of an apprenticeship completer has remained broadly constant at 28 per cent.

However, after factoring in debt, the proportion of graduates earning less than the average lifetime earnings of an apprentice increases by an average of one per cent. The extra earnings apprentices receive from earning while studying relative to their graduate counterparts increases the proportion still further by an average of 4 per cent.

Meanwhile, attending a 'new' university increases the proportion of those earning less than the average lifetime earnings of an apprenticeship completer by 6 per cent taking the proportion to 39 per cent (once debt and forgone earnings are accounted for).

Using NVQ levels 4 and 5 as a proxy for higher-level apprentices, the study investigated how graduates compared to higher-level apprentices. The authors estimate the proportion of graduates from 'new' universities earning less than the average higher apprenticeship completer increases by an average of 7 per cent to just under half (46 per cent).

The study also looked at how graduate earnings change over their working life. Their findings reveal that graduates can expect their earnings to increase well into their forties and peak at around the age of 45 at which point around 12 per cent of those from Russell Group universities and 20 per cent from 'new' institutions will still be earning less than the average annual salary of an apprenticeship completer.

After peaking in their mid-to-late forties, graduate earnings tend to decline until the age of retirement at which point over half of graduates will be earning less than the average lifetime wage of an apprentice.

In terms of the impact of subject studies, medical graduates were found to have the highest average lifetime earnings with the vast majority (98 per cent) earning more than the average apprentice.

The majority of those graduating in STEM subjects (science, technology, engineering and maths) have earnings in excess of the average for an apprenticeship completer.

Over half (54 per cent) of those graduating with a degree in the humanities from a 'new' university had lifetime earnings in excess of the average for an apprenticeship completer.

Those graduating with degrees in 'arts' or 'media & information studies' fared the worst relative to apprentices. After taking into account lost earnings and graduate debt, over two-thirds (69 per cent) of graduates in 'media & information studies' from 'new' universities earned less than the average wage for an apprenticeship completer. For arts graduates the equivalent figure was 58 per cent.

The exact proportions of graduates by degree subject earning less than the average lifetime earnings of an apprenticeship (once debt, extra wages and attendance at a 'new' university were accounted for) were as follows:

- Medicine - 10 per cent
- Engineering - 21 per cent
- Physical Sciences - 28 per cent
- Maths & Computing - 29 per cent
- Languages - 35 per cent
- Linguistics, English & Classics - 41 per cent
- Biological Sciences - 42 per cent
- Law & Social Studies - 42 per cent
- Business & Finance - 45 per cent
- Humanities - 54 per cent
- Arts - 58 per cent
- Media & Information Studies - 69 per cent

The analysis also found that the proportion of graduates with lifetime earnings below those of the average NVQ level 4 or 5 completer (used as a proxy for higher apprenticeships) is seven per cent more than for apprentices as a whole.

Meanwhile, three-quarters (74 per cent) of 'media & information studies' graduates from 'new' universities had earnings less than those of the average NVQ level 4 or 5 completer. For 'arts' graduates the figure is 64 per cent or two-thirds.

The exact proportions of graduates by degree subject earning less than the average lifetime earnings of an NVQ level 4 or 5 completer - once debt, extra wages and attendance at a 'new' university had been accounted for - were as follows:
- Medicine - 15 per cent
- Engineering - 2 per cent
- Physical Sciences - 33 per cent
- Maths & Computing - 38 per cent
- Languages - 42 per cent
- Linguistics, English & Classics - 50 per cent
- Biological Sciences - 50 per cent
- Law & Social Studies - 50 per cent
- Business & Finance - 52 per cent
- Humanities - 63 per cent
- Arts - 64 per cent
- Media & Information Studies - 74 per cent

The report clearly demonstrates that HEI type and subject studied have a significant bearing on lifetime earnings. It also suggests that contrary to popular belief - and after factoring in the high levels of debt many graduates will incur to obtain the qualification - a degree is often not the most effective route to maximise lifetime earnings for many young people.

The authors recognise that while earning potential should not be the main motivation behind making the choice between university and an apprenticeship, it is nonetheless important for many young people.

The study highlights a chief shortcoming in policymakers' approach: too often university degrees are treated by politicians as a homogenous unit. The evidence shows the concept of 'the graduate' is a falsehood and that there is an enormous range - at least in terms of earnings – between graduates, depending on a number of variables, including subject studied and institution attended.

The report also examined contrasts in employment prospects between graduates and apprentices. It found that from 2005-2013 in terms of the under-25 unemployment rate between the two cohorts, apprentices had a broadly more favourable rate than graduates. Over the period, the apprenticeship unemployment rate peaked in 2009 at 11 per cent, while the graduate rate rose more slowly peaking a year later in 2010.

As speculation surrounding a double-dip recession gripped the media in 2012, graduate unemployment returned to its 2009 peak of 11 per cent, while unemployment among apprentices continued to fall to a little over 6 per cent in 2012. Figures for 2013 showed a slight rise in apprenticeship unemployment taking it to 7.5 per cent while the graduate rate subsided slightly in 2013.

Apprenticeship completers under the age of 25 also enjoyed a more favourable rate of employment than graduates. Relative to their pre-crash levels the 2013 unemployment and employment rates showed an increasing divergence in apprentices' favour.
The authors cited graduates having to study longer to obtain additional qualifications such as master's degrees in order to gain employment as one possible explanation for this trend. Another possible explanation is

that graduates are finding it harder to secure 'graduate level' jobs, and so are delaying their entry into the jobs market.

Unfortunately, due to limitations on the data available, the study was not able to determine the effect of HEI type on graduate employment or unemployment rates (although judging by the evidence supporting the effect of HEI type on earnings, it seems reasonable to assume HEI type does play an important role).

Despite the results of the investigation, a poll of 500 British school leavers found that apprenticeships remain a minority option. Only 2 per cent said the majority of their peers are planning to become an apprentice, and just 6 per cent of those surveyed were planning an undertaking an apprenticeship themselves.

Apprenticeships are struggling to acquire the social status they deserve. When asked, more than half of school leavers feel apprenticeships are of no interest to them. Those surveyed were more than twice as likely to associate university rather than apprenticeships with providing good long-term earnings potential and job prospects. Fewer than one-in-six say they are preferred over university by their parents and friends.

The poll highlighted clear drivers to encouraging the uptake of apprenticeships. Three-in-five of those school leavers not planning on undertaking an apprenticeship said a guarantee of a job or qualification at the end of the apprenticeship would encourage them to do so, while half felt a clearer idea of the earnings potential would have this effect. Emphasising these key attributes will help to reposition apprenticeships as a desirable and practical alternative for school leavers.

Tackling such misconceptions is vital if policymakers are serious in their determination to tackle the country's youth unemployment and to equip young people with the skills they need to meaningfully participate in Britain's 21st century workforce.

The strategy to achieve the government's objective of improving the perception of vocational qualifications must take place on two fronts.

Firstly, to promote awareness among students, their parents and teachers of the opportunities modern apprenticeships offer. But secondly – and in order for apprenticeships to gain their rightful place alongside graduate degrees – we must see greater transparency and competition between higher education institutions. Young people must be made aware that not all degrees offer the same returns.

In a recent job interview for a policy position with a leading higher education membership body, a prospective candidate was asked what they deemed to be the greatest threat to the UK's higher education industry. They replied: loan repayment rates by institution.

This attitude is wrong. Increased transparency and competition among higher education institutions can only be a positive. It will help the country retain its place at the forefront of higher education and research, while giving young people the opportunity they deserve to compete in the global race.

The increase in tuition fees to £9,000 a year was a bold and much-needed step. For the first time it forced universities to compete to attract students, while affording students the power to hold institutions to account. But there is more still to be done to ensure the higher education system in the UK remains among the best in the world, and to ensure students get a sufficient return on their investment. The status quo, whereby 45 per cent of student loans are never repaid, is unacceptable.

It is crucial we move away from the idea that graduates are a single homogenous unit. They are not. We will not see the apprenticeships revolution policymakers strive for without a revolution in the higher education sector.

Reference
Walter, W. and Malhotra, S. (2014) *Varsity Blues. Time for apprenticeships to graduate?* A Million Jobs Report, sponsored by Pera Training.

PART FIVE

DELIVERING APPRENTICESHIPS

INTRODUCTION TO PART FIVE

David Way CBE

No analysis of apprenticeships, current and future, is complete without looking at those tasked with delivering them. As the preceding chapters have made clear, recent years have seen government trying hard to establish employers as the principal investors, innovators and drivers of apprenticeships.

This part of the book therefore starts appropriately with an analysis of how the opportunities to put employers in the driving seat can be seized.

Toby Peyton-Jones is an employer, and Commissioner on the UK Commission for Employment and Skills. He looks at how we can create the right conditions for employers to lead the delivery of a high quality apprenticeship system. This includes how we can become more successful in scaling up those practices that clearly work well; how Apprenticeships need to be materially different from an extended training course and provide the meta skills so needed by employers; and at ways in which the full potential of the levy can be realised. Finally, Toby considers the vital role of the Institute for Apprenticeships.

Training providers have traditionally been the crucial foot soldiers in persuading employers to take on more apprenticeships. In this rapid move to an employer led approach it is vital not to overlook the key role that most providers have performed, especially in supporting SMEs who lack the HR infrastructure of big businesses.

Unlike in most other countries, apprenticeships in the UK are mainly delivered by independent training providers, often in partnership with FE colleges. These providers are represented by the Association of Employment and Learning Providers and their former Chief Executive, Stewart Segal, shares his thoughts on the future and the innovations that will be needed to respond to employers who take up the challenge of leadership. He emphasizes the potential opportunities emerging, especially in higher apprenticeships and in the public sector; as well as the importance of ensuring that the levy does not create a two-tier system to the disadvantage of SMEs.

Growing the role of higher education in delivering high quality apprenticeships is vital for the government's ambitions for increasing the higher-level skills needed to boost the economy and the nation's productivity. HE engagement will also attract more young people and their parents who will see apprenticeships as growing in status and offering a clearer pathway to many of the attractions and benefits of a university-based education. The engagement of universities brings reputation and reassurance to young people and parents facing daunting decisions about education pathways.

Conditions seem ripe for universities to extend their positive influence on skills to more businesses. The introduction of the apprenticeship levy also presents important new opportunities for universities as and where employers choose to invest in higher-level skills.

The next chapter therefore looks at how the apprenticeship requirements of employers are moving towards the higher levels of skills training and education offered by universities; and at how this process can be better facilitated and accelerated so that more universities in England become engaged in the expansion of higher and degree apprenticeships.

Adrian Anderson is the Chief Executive of the Universities Vocational Awards Council (UVAC) and Mandy Crawford-Lee is an Associate Director there. UVAC has an unrivalled track record in the development and implementation of higher and degree apprenticeships and in working in partnership with universities and further education partners to improve higher-level technical and professional education. Adrian and Mandy reflect on past experiences for universities engaging with higher apprenticeships and consider how higher education can drive a world-class apprenticeships system.

One of the most significant recommendations in Doug Richard's influential review of apprenticeships in 2012 was a new approach to assessment. Can the apprentice demonstrate that they can do the job for which they have been trained at the end of the apprenticeship?

I found many interesting approaches to assessment around the world. In Switzerland, I saw rival companies assessing each other's apprentices. I was told that it ensured standards remained high. Neither business wanted the other to think that they were producing low-skilled people and, if their competitor was doing something that was making their apprentices stand out, they wanted to emulate it.

Judith Norrington is an acknowledged expert in assessment policy and practice having been at City & Guilds and the Association of Colleges as well as a contributor to government assessment policy reviews. She serves on the Skills Commission, which is an independent body of leading education and skills thinkers. In her chapter Judith examines the proposals for changes to assessment in the UK, and then considers what is good practice in assessment drawing on her international experience. What lessons are there for the UK if effective assessment is to underpin the expansion of high quality apprenticeships?

The final two case studies look at employers and training providers who are doing things differently.
Phil White explains how United Utilities tired of their existing training arrangements and truly took ownership of their apprenticeships from top to bottom. They had to shake up the whole system to get what they needed. Trailblazers provided the vehicle for engineering these changes.

Craig Crowther looks at what Group Training Agencies have been contributing to delivering apprenticeships, especially in STEM occupations. He looks at the strengths of the GTA model and its above-average growth performance across a varied range of employer projects and sectors. He provides compelling evidence for GTAs to have a bright future and to be in the political spotlight, with a great future as well as a great past.

34.
EMPLOYERS IN THE DRIVING SEAT:
AN EMPLOYER-LED, HIGH QUALITY APPRENTICESHIP SYSTEM

Toby Peyton-Jones
Commissioner, United Kingdom Commission for Employment and Skills

---

Summary
This chapter provides a short history of apprenticeships from the employer perspective. It describes and welcomes many of the efforts to put employers in the driving seat, but also notes where these efforts could be improved. It then proposes a way forward for a truly employer-led, high quality apprenticeship system.

A little history
As recently as the 1950s, almost every male teenager in the country entered an apprenticeship. These were similar apprenticeships to those which operate today in countries renowned for large-scale high quality apprenticeship systems, such as Germany and Switzerland. They involved a contract between employer, apprentice and the state, so each knew their roles and responsibilities. Apprentices would go on day-release or evening classes at the local technical college and receive training and on-the-job support from their employer.

However, from this period onward, apprenticeships entered several decades of decline, to the point where they were almost forgotten. This decline was hastened by the economy's shift away from sectors like manufacturing and mining which had a strong tradition of apprenticeship.

Then, in the 1990s, interest began to grow again. Modern apprenticeships were reintroduced in 1994. However, numbers participating were low until 2009/10, when starts reached 280,000 and the policy aim had become one of rapid expansion. Under skills minister John Hayes, numbers exploded, reaching half a million starts in 2011/12. Judging on scale alone, the policy aim of expanding and revitalising apprenticeships was achieved.

However throughout this period of growth there were question marks over the quality of these apprenticeships. Some of this is down to the existence of another training policy. Before the surge in apprenticeships, there was an in-work learning programme called Train to Gain. This offered free training to employees who didn't have a first full level 2 or level 3 qualification. Train to Gain had a very different approach to apprenticeships - the training was shorter and tended to be at a lower level. The focus was on certification of existing skills rather than on developing new ones. Also, employers were simply the passive recipients of free training for their staff, not the active deliverers of training they are in European apprenticeship systems. The programme closed due to concerns around deadweight and value for money.

It now appears that a significant part of the growth in apprenticeship numbers is simply a result of the changeover from Train to Gain to apprenticeships. Many providers simply switched their training from Train to Gain to apprenticeships and employers did likewise, moving from one free training programme to another and re-badging existing activity as an apprenticeship .

Unfortunately, many of the issues around quality were transferred over too. The apprenticeships that appeared during this period of growth were not like those of yesteryear, or indeed those of Switzerland or Germany. They were predominantly at level 2 (equivalent to GCSEs), where European apprenticeships tend to be at level 3 or above (equivalent to A-levels). They concentrated on assessing and accrediting existing skills instead of developing new ones, particularly among adults aged over 25 where most of the growth came from. And they were provider-led, with employers complaining that they were not getting what they wanted

Focus shifts to quality

As a result, the focus of government policy shifted to quality, encapsulated by Doug Richard's review. This was commissioned in June 2012 and reported in the November of that year. Its aims were to 'ensure that in the future the programme is meeting the needs of the changing economy, consistently delivers high quality training and the professionally recognised qualifications and skills which employers and learners need, and is maximising the impact of government investment'. (Richard, 2012)

The Richard Review concluded that many apprenticeship frameworks were not sufficiently driven by employers. Instead, the view was that awarding organisations and other intermediaries were in the driving seat. What's more, Richard pointed out that many frameworks were not providing a sufficiently deep or broad experience for the apprentice, were not always linked to a specific occupation, had low requirements in terms of guided learning hours, and no clear line of sight to a career. In some cases, frameworks were too employer specific and were therefore deemed unfit for purpose by other employers.

In response, government policy made a concerted effort to put employers in the driving seat and test some new and exciting approaches to employer led apprenticeships and vocational education. One of the conclusions of the Richard Review was that frameworks need to be replaced with employer-designed apprenticeship 'standards'. This was picked up by the government, who set up pilots, with apprenticeships being redesigned by trailblazing groups of employers to make them as relevant and as valuable as possible.

This has had some clear successes. Employers have welcomed the opportunity to collaborate and develop apprenticeships that work for them and their industries. This includes sectors that don't traditionally offer apprenticeships such as law, banking and nuclear science. As of November 2015, we have 127 Trailblazers developing 345 standards in total, of which sixty (17 per cent) are ready for delivery.

However like many of the changes made over the last twenty years, the trailblazer approach has had good intentions, but the implementation has lacked proper transparency and governance for the sign-off of standards. The resulting uncontrolled proliferation of standards, and the 'making it up as we go along' approach to implementation, has left employers and providers frustrated by something that still has the potential to be part of the solution. In addition, the poverty of end-to-end thinking regarding how the standards will be delivered, assessed and maintained is adding to the confusion.

In parallel the government has supported other employer-led approaches. 'Employer Ownership of Skills' pilots galvanised large networks of industry leading employers to collaborate on skills and recruitment issues. The 'Industrial Partnerships' that followed have underpinned many of the new apprenticeship standards and have seen some of the most exciting examples of employers getting together and taking end-to-end ownership for skills development in their sector.

Problems remain

In recent years, the government has made many of the right moves. The pilots and initiatives have shown the real appetite of employers to step up and take a lead in both reforming and owning the skills agenda for their area of business, and yet we are now three years on from the Richard review and problems with quality remain. Ofsted has recently found that apprenticeship training is too often limited and short, that employers are not sufficiently engaged, and that the impact on the careers of apprentices is poor.

In addition to Ofsted's findings, and perhaps most striking of all, over one-third of apprentices are not aware that they are, in fact, apprentices. This figure rises to around half of apprentices who already worked for their employer before becoming an apprentice. Even amongst apprentice employers themselves there is an issue. Around four in ten apprentice employers are unaware that the training they are providing for their workers is, in fact, an apprenticeship. This lack of awareness would be unthinkable in European apprenticeship systems or, indeed, any other form of education. It's hard to imagine someone attending university and graduating with a degree without ever knowing it.

These were precisely the problems that Doug Richard had set out to tackle, and yet they remain. There are three broad areas that need to be confronted. They are: stability, quality and employer engagement, particularly the engagement of small employers who make up the vast majority of businesses in the UK.

1. Stability

Despite the UK pioneering some of the most innovative approaches to the development of apprenticeships and the wider skills agenda, even when seen against international comparison, we have not been able to scale up and deploy what we have learned. We have now had twenty years of constant change from successive governments, who have changed not only the funding but also the content of policy on an epic scale, leaving behind a trail of complexity, frustration and lost opportunities.

The most critical imperative therefore is to move apprenticeships onto a stable and sustainable footing away from the whims of successive governments. Only then will we be able to stabilise the 'pipe work' in the system so all players (employers, parents, students, providers and awarding bodies) learn how to relate to each other. We should, of course, retain the demand-led approaches to ensure that what then goes down the 'pipes' is dynamic and can deliver what employers need.

To date the succession of changes have led to huge amounts of duplication, overlap and fragmentation between well-intentioned players. This applies in government as well as in the wider skills landscape. At a time when productivity is such a key issue, there is opportunity for consolidation if this is properly thought through.

There is a risk now that the latest changes are viewed with the scepticism of the past and not seen as worth investing in by either employers or colleges and training providers. We have to make sure that the new arrangements are built with the common goal of creating a sustainable landscape for technical and professional education in which apprenticeships clearly play such a central role.

2. Quality and standards

With regard to current apprenticeship provision, Ofsted says 'Because they were passing their apprenticeships, apprentices appeared to be successful, but these programmes tended to add little value to the apprentice or business.' The frameworks are so low level that they can be passed with relatively little effort by the provider, employer and apprentice. (Ofsted, 2015). This concurs with Doug Richard's view that the current frameworks are the main issue in quality and a shift to new standards is the solution.

But this is only part of the problem. There is nowhere that sets out in concrete terms what an apprenticeship actually is. By contrast a university education is well understood by all. It involves studying at a university, for three years in almost all cases, after which you receive a qualification called a degree.

At their worst, current apprenticeships are indistinguishable from a long training course during which the learner has an unconnected Saturday job or vice versa. By contrast, an apprenticeship combines earning and learning on the job and has a clear line of sight to a career. It is this combination of learning and doing, both on and off the job, that develops the vitally important 'meta skills'. These allow the apprentice to combine their technical knowledge with their team-work, commercial, problem solving and project management skills to tackle problems they have never encountered before.

While the technical knowledge is vitally important, in our industry it has a shelf life of about three years before they are out of date. By contrast, it is the meta skills that can be applied in a variety of settings, industries and occupations and truly boost the apprentice's career.

At Siemens we start to see meta skills emerge during the second year of the apprenticeship and truly flower in the third year. This is one of the reasons we advocate longer apprenticeships. It is the meta skills that we prize most as these are the skills that have the biggest pay back to Siemens as an employer and to the individual who soon finds out that it is these skills that can take them from the shop floor to the top floor. Indeed our current CEO and three of our current managing directors are ex-apprentices.

So we need a concrete definition of what an apprenticeship is and what it should deliver. An apprenticeship must be in a skilled job, based on a standard designed by industry, include exceptional teaching and learning both on and off the job, require investment from all parties (the employer, apprentice and government) and, most importantly, provide a route into a recognised and well understood career that receives decent pay. A critical outcome must be the metaskills which are valuable in every career. It's by these yardsticks that we should measure quality in apprenticeships, not simply how many there are and how many pass.

For the government, success of their policy must be measured in outcomes for industry and increased earning power for apprentices and not in terms of meeting a three million target.

3. Employer engagement

Employers, SMEs in particular, remain unengaged partly because the policy landscape changes so frequently. They have not got the bandwidth or balance sheet strength to get involved in the development and delivery of apprentices who may or may not be retained. Small employers therefore need local intermediaries of some sort to support them. But policy has shifted away from supporting intermediaries. Instead it has adopted a more organic approach of providing pots of funding and asking the market to organise around them. The result has been a high level of innovation but at a cost. The system has become more fragmented and difficult to navigate and as the funding or initiative ends the players drift away and the learning is lost.

The current system lacks an organising framework that describes clear 'docking stations' for all players and in the meantime the 'let a thousand flowers bloom' approach has left all but the largest of sectors and employers lost and without partners they can turn to for help.

Any new set up must therefore address how employers and the market will be engaged, and the model of engagement must work for small as well as the large employers.

A way forward

Having been through many learning points, we can now do something transformational.

1. A stable apprenticeship system

It's clear that the apprenticeship system needs to become more independent of government. This is necessary to ensure employer leadership and buy-in, political independence and stability. We need an employer-led national body to take 'end to end' responsibility for apprenticeship quality, with oversight of standards and assessment development; and with the authority to 'farm' the complete apprenticeship ecosystem. To do this it stands to reason that it must in time develop the wider accountability for vocational education more broadly or technical and professional education as it is sometimes referred to.

The government has recently announced the establishment of a new body by 2017 to do just this. While the details are as yet not worked out, it is clear this body and its functions will be central to achieving a large-scale high quality apprenticeship system, something that employers, government and apprentices themselves desire. In my view, it should be governed by a partnership led by employers but including trade unions, apprentices and other stakeholders to ensure collective buy-in to its purpose.

The Institute for Apprenticeships must have a strong research function to create an organising framework that avoids proliferation and poor quality apprenticeships leading to dead-end jobs. There is a wealth of information available about the labour market. It can tell us which occupations and sectors are growing, where the skills needs are, and what the employment and earnings outcomes for apprentices are. And yet this information is under-used to inform apprenticeship policy.

The Institute for Apprenticeships must make sure that both the development of new standards and the growth of apprenticeship numbers are guided by what's happening in the labour market more generally. It must use LMI on employment growth, earnings outcomes and skills gaps to inform the development of new standards and help us understand which apprenticeships are delivering for employers and apprentices.

Ofsted found that 'the growth in apprenticeships in the last eight years has not focused sufficiently on the sectors with skills shortages'. The Institute for Apprenticeships should also have the task of stabilising and structuring the engagement model with employers and providers in particular so that a stable demand led system emerges that can respond rapidly to emerging market demands.

2. High quality apprenticeship standards

Many apprentices pass their apprenticeships despite poor quality training. This is down to poorly-designed frameworks with limited content. So the government's decision to move to new Trailblazer apprenticeship standards has been driven by positive intentions.

New apprenticeship standards must continue the drive to broaden and deepen standards. They must be set so that they can only be passed with significant effort and investment by the apprentice, employer and provider alike. A focus on curricula is vital.

Critically they must relate to an occupation. Some estimate that if we continue as we are with the Trailblazer approach there could be as many as 1,600 to 1,700 apprenticeship standards in eighteen months' time let alone five years' time. Germany, Switzerland and the Netherlands have far, far less than that because they have to be written to relate to an occupation in order to ensure transferability. This again highlights the important role the Institute for Apprenticeships will play in deploying research-led governance to hold in check employers who under the current free-for-all are being funded for their narrow needs without proper market or peer challenge.

There is no doubt, however, that the degree of employer engagement and direct involvement in the Trailblazer process has been impressive, and employers have welcomed freedom from the old rules. One of the strengths of the new Trailblazer standards is that they are presented at a much higher level than the current National Occupational Standards (NOS), typically a two-page summary.

However, while welcoming the simplicity of the two-page standard, Trailblazer supplementary detail is needed to guide curriculum developers and those designing the assessment. Many of the trailblazers have made use of NOS to help provide this detail; this is certainly the case for the trailblazers that Siemens has been involved in.

There is current work going in the UKCES to reform NOS standards to dovetail with trailblazers. Combining the two approaches would deliver both the high level simplicity that trailblazers standards offer with the layered drill down for those who have to deliver the standards in the field. NOS also have another characteristic that is vital to large employers, like Siemens, who have a presence in each of the nations - they are UK-wide.

Large employers like ourselves crave consistency, and UK-wide NOS provide the necessary underpinnings for this consistency.

However, it appears from the announcement that the Institute for Apprenticeships will be an England-only organisation. This is concerning for all large employers with a presence across the nations. It is vital that the Institute for Apprenticeships has the appropriate 'docking stations' with the devolved nations and a reformed and combined version of NOS/Trailblazers would be an important common currency that would keep the wider vocational education system aligned across borders.

The Institute for Apprenticeships would have the task to convene and maintain groups of employers who coalesce around broad occupational areas for precisely this purpose. These employers could develop and maintain broad occupational standards that can underpin both apprenticeships and other vocational pathways. The latter includes the Department for Education's 'technical and professional routes' for all those who don't go down the traditional A level to HE route. The aim here is to develop these routes with the 'direct input' of employers. The Institute for Apprenticeships' convening role is ready-made for this purpose and it makes complete sense for apprenticeships and classroom-based technical and professional education to share some common underpinnings.

Getting the 14-19 'pipeline' right is vital not only for an expansion of apprenticeships but also for meeting the needs of our economy more generally. All vocational education needs a clear line of sight to a career, and this has been missing in too many cases. Another critical interface for the Institute for Apprenticeships would therefore be the Department for Education.

All of the above, combined with an LMI-based occupational map, will ensure that the right apprenticeships are in the right place at the right time. It will also help bring stability.

3. Employer engagement

The introduction of an apprenticeship levy will be a sea change for employers, apprentices and government. At the very least it ensures that large employers are aware of apprenticeships and have a strong incentive to get involved in order to 'get out more than they put in.' Employers up and down the country, including ourselves at Siemens, are now busily figuring out how the levy will affect them and how to get the most out of it.

So, in one sense the levy has had a positive impact already, raising awareness, but there is much more to do. The levy solves a couple of the government's problems: how to pay for an expansion, and how to get employers aware of apprenticeships and to consider taking them on.

 But the levy will only pay for the apprenticeships, it won't help deliver them. It needs to be designed correctly to drive the right behaviours. There is a real risk that many levy-paying employers will treat it simply as an employment tax to be minimised or, at worst, a way to exploit low cost labour by re-badging existing activity as an apprenticeship.

This means that there needs to be a pro-active strategy for engaging with employers. This would identify employers in high-growth, high-skill sectors and encourage them to develop apprenticeship standards in a systematic way. It would also be informed by who the levy paying employers are in each sector, the scale of their combined levy bill, and their current engagement with apprenticeships.

It is my view that employers are best placed to speak to others about the benefits that apprenticeships bring, but they need to know the rules of engagement. It is essential for instance that all levy money raised in a sector is spent in that sector. While digital vouchers are only allowed to be spent by the employer they must be allowed to transfer that to another employer/ party to deliver on their behalf. Here again the Institute for Apprenticeships has a major role to play in structuring the discussions and supporting collaborations. Some of these collaborations may even be a continuation of the successful dialogues started in the Industrial Partnerships of the last two years.

In the meantime, if the government uses the levy money raised as a windfall to spend in other areas then this will be hotly challenged by employers and will bring the whole positive introduction of the levy into question.

Local employer structures would also need to be engaged in driving demand and ensuring quality, and could play a vital role, together with providers, in supporting small employers. Again the Institute for Apprenticeships not only needs to manage and convene the sectoral employer voice but also the spatial / local employer voice.

There are a variety of roles that local stakeholders can fulfil in the system. Local Enterprise Partnerships and others are in a strong position to convene local networks of apprentice employers to share good practice. The most important aspect is to structure and enable these conversations so that they align and enable rather than compete and confuse.

Conclusion

Apprenticeships are jobs. They are work-based routes to qualification and so it is entirely natural that they are governed by an employer-led Institute for Apprenticeships. The introduction of the levy in 2017 makes this even more vital. There are some cornerstones for this body and the quality assurance of apprenticeships more generally:

Employer leadership - alongside the introduction of a national living wage, the introduction of a £3bn apprenticeship levy is one of the most dramatic changes to the employment and skills landscape for decades. Given the scale of their investment, it is vital that employers have the scope to genuinely lead apprenticeship policy to deliver on what will be their investment. Any attempt to have policy carry on, with employer leadership as simply a 'fig leaf' is likely to be fiercely resisted. The levy will focus employers' minds and employer leadership is the key to harnessing that interest to facilitate a demand-led approach.

Clarity of role and responsibility – the Institute for Apprenticeships must have a clear role that is well understood by all other stakeholders. Indeed, this is the case for all the actors in the system. Each must understand their own role, their relationship with everyone else guided by clear end-to-end governance that is laid down by the Institute in which they should also have a voice

Independence and quality assurance – to be truly employer led, the governance must be one step removed from government. Indeed, this is the case in successful arrangements in other countries. Like the Central Bank which was given the task of setting interest rates away from political interference, this new Institute for Apprenticeships must also have research capability to drive evidence based approaches that reflect the market not the latest fashion. One key aspect will be how success will be measured and here we must be driven by quality not quantity and by outcomes not inputs. The true test of success is that industry gets the skilled workforce it needs and the apprentice gets a clear line of sight to a career with earnings progression.

Simple, stable and sustainable – this is key. If this chapter has shown anything, it is that the history of apprenticeships is one of constant change. No system is perfect but the new independent Institute for Apprenticeships combined with the levy do have the seeds that could give the UK a chance of setting up a truly world-class and sustainable approach to apprenticeships.

History has also shown us that we don't stick with things long enough for them to deliver. It is therefore my view that the next steps need to be very thoroughly thought through. In particular the scope and terms of reference for the Institute for Apprenticeships and the associated set up needs to be co-created with the input of all stakeholders and importantly should command cross-party support. The prize is well worth it: a dynamic apprenticeship eco system that can be an engine for UK future growth and prosperity. There is a lot to play for!

References
Richard, D. (2012) *The Richard Review of Apprenticeships.* Department for Business, Innovation and Skills BIS publication, London
Ofsted. (2015) *Apprenticeships: developing skills for future prosperity.*

35.
HOW TRAINING PROVIDERS CAN RESPOND WITH INNOVATIVE SOLUTIONS

Stewart Segal
Former Chief Executive and Current Board Member
Association of Employment and Learning Providers

The Association of Employment and Learning Providers (AELP) represents the interests of a range of organisations delivering vocational learning and employment and employability support. The majority of our 750-plus member organisations are independent providers (from both the private and third sectors) holding contracts with the Skills Funding Agency (SFA), with many also delivering Department for Education (Education Funding Agency) and Department of Work and Pensions (DWP) funded provision. Many of our members are involved in the delivery of employment and skills programmes in Wales and work with employers who operate throughout the United Kingdom.

In addition to these we have a number of colleges in membership, as well as non-delivery organisations such as Sector Skills Councils and awarding bodies as Associate Members, which means that AELP offers a well-rounded and comprehensive perspective and insight on matters relating to its remit.

The apprenticeship challenge
When the Conservatives won a majority at the last election, most commentators were surprised. They were even more surprised when a few weeks after the election they announced a new levy[124] on large businesses

---

[124] After the Summer Budget announcement, the government launched a consultation on the matter. See BIS. (2015) *Apprenticeships Levy Consultation*: www.gov.uk.

to pay for the apprenticeship programme. This was a low tax, business-friendly government imposing mandatory fees on businesses. Mandatory charges on employers to pay for training have never been very successful. Employers need to be committed to training rather than forced to be at the table but we could see the logic and the benefits of a levy.

Government funding is under real pressure, so a levy creates a more secure and predictable source of funding for a programme that is growing year on year. The levy would include all public and private sector large businesses so it will have the benefit of raising the profile of apprenticeships with a number of employers that do not currently get involved in the programme.

We should also be pleased that apprenticeships have been singled out to get this sort of backing from government, because ministers were obviously of the view that introducing a new levy would be supported by most employers. Their reasoning is that the benefits of apprenticeships have been proven and the levels of satisfaction of employers involved in the programme are very high. They calculated that the government would probably get the support it needed for the introduction of a substantial new tax on business given that it is a 'hypothecated' tax that employers can use to purchase training. This is undoubtedly a very bold and positive move for a new government. We have seen some 'push back' from employers and their representatives such as the CBI.

The idea of a levy was bold, but it was probably the only way that the government could increase the number of starts on the apprenticeship programme without significantly increasing the investment by government at a time when most budgets are reducing. This is clearly a way of passing responsibility for the funding of this programme to employers.

This need for additional funding is driven by the target to deliver three million apprenticeship starts in the next five years and the likelihood that changes to the content of the apprenticeships will increase the cost per apprentice. The government has estimated that an average of about 600,000 starts a year will need to happen between now and the end of the year 2019-20. This is a significant jump from the 440,000 starts at all ages in 2013-14 and while we will probably see increases in starts before the levy begins in April 2017, the really big leap will come after the levy's introduction i.e., when the employers will be paying.

The government had to find a way of funding this expanded programme and it has probably enabled the government to maintain rather than significantly reduce the adult skills budgets that funds everything that is not apprenticeships.

The role of training providers

Training providers are now seen as an important part of the solution to engaging more employers in apprenticeships. That was not always the case. In the early stages after the Richard review, the view was that training providers were part of the problem in the system that prevented employers from taking ownership of their own training. That was far from the truth.

The government had created an apprenticeship system that was maintained by inflexible funding rules. It was training providers that had to make those rules work. Despite the inflexible rules set out in the SASE (Specification for Apprenticeship Standards for England) providers are able to create programmes that are responsive to employer needs and delivered high quality. Satisfaction and success rates are higher than ever with employers recording a satisfaction rate of 82 per cent and apprentices registering an even higher score of 89 per cent.

AELP has never supported the status quo and has promoted change that builds on the success of the current programme. The results of the current system are evidence that we do not require fundamental change to the structure of the apprenticeship programmes. One of the main recommendations made by employers in every survey about improving the system is to reduce the level of year on year changes. Employers want simplicity and consistency. Training providers are a way of delivering that consistency.

AELP has set out a series of changes that build on the fact that employers are already in control of the apprenticeship programme. Employers make the decision whether to employ an apprentice, what job they will do, which training programme they will be on and which training provider they will use. These are the key decisions that drive engagement and quality.

Employers, especially smaller employers, want a training provider to work with them and reduce the level of bureaucracy and administration they have to manage. Government is now beginning to accept this approach. It has taken a long time to convince them of the fact that it is the strength of the partnerships between employers and providers that will deliver growth and quality.

Now that a large part of the funding of the programme will come from the new apprenticeship levy it will be more important than ever that the role of the training provider is recognised. That is particularly true for SMEs because the levy is likely to apply to large employers with a payroll cost of £3m. That leaves the majority of businesses as non-levy payers. There is a real risk that the focus of the programme will now be on large employers including large public sector employers.

Training providers will be an important part of making sure SMEs views are taken into account during the next phase of reform where funding systems will be agreed. Training providers work with thousands of smaller employers and know what is needed to keep them engaged. Simplicity is essential which is why AELP has always said that 'up front' mandatory cash contributions will be a barrier to entry for SMEs. With a new levy proposed for larger employers we need a new funding system for SMEs that avoids the barrier and administration costs of cash contributions.

SMEs continue to make a huge contribution to the cost of their apprentices through wages, support in the workplace and provision of training time. We need to recognise this and move on from a narrow focus on mandatory cash contributions to training. We have recommended that the government reviews its mandatory cash contributions and cash incentives that go back to employers. Using training providers to manage those transactions will make the system work for SMEs.

Training providers also have a role to ensure that the apprenticeship programme delivers what apprentices themselves need as well as the employers. Young people in particular need a lot of support in the workplace and smaller employers cannot always provide the rounded service both in and out of work that is sometimes required. SMEs don't always have the in-house skills to ensure that the apprentice is covering all of the skills and knowledge they need to ensure that the programme is transferable. It is a difficult balance to ensure that apprentices get training that their employer wants to do the job whilst getting the broader skills that make them employable in the market place. That is the delicate balance that training providers can help deliver.

The apprenticeship levy

Some details of the levy scheme were released as part of the Spending Review in 2015 but there are still a lot of details to be agreed. Whatever form the levy takes, there are likely to be some unintended consequences of those changes. The levy will be applied to large employers only, which may mean that those large employers will want an even greater influence over the structure of the programme and how the money is spent. It will be important that the needs of smaller employers and apprentices themselves are recognised. There is a constant balance between programmes benefitting individual employers rather than the sector needs or even wider economy needs. The benefits of common standards and transferability of skills is important for apprenticeships, so there needs to be a governance structure that recognises the needs of employers, government, apprentices, providers and other stakeholders.

The government has announced the creation of an Institute for Apprenticeships which will be employer-led. This group needs to reflect the knowledge and experience of SMEs, training providers, awarding organisations and apprentices. The role of the Institute is to manage the development of the standards but it will clearly have a very influential role in the development of the whole system.

It is also not clear as to how the funding system will work for SMEs. The levy's application to large businesses means that around half of all apprentices (our estimate) will be employed by SMEs not covered by the levy. A different system applied to those SMEs needs to be flexible and responsive. We have recommended a system that recognises the many ways that employers contribute to the employment and training of an apprentice. Every apprenticeship is a job, and the cost of employment is always met by the employer. It is right that the government contributes to the training of apprentices in SMEs, especially when they are small businesses or where the apprentice is young, inexperienced or out of work. These contributions should be set by the government, with employers managing the use of the contributions through their chosen training provider.

The best apprenticeship programmes are those where a committed employer works closely with their chosen training provider to deliver a programme that meets the needs of the employer, the apprentice and makes a contribution to the economy.

With a levy system that applies only to large employers, we need to avoid the dangers of a two-tier system. The impact of a structured work based training programme is felt most in SMEs who would not be able to apply this type of programme without the infrastructure provided by government and large employers. We need to see the programme as a benefit to all employers and not allow the fact that large employers will have to pay the levy to affect the reality that apprentices can and should be employed in all sizes and types of businesses.

Apprenticeship growth
So where will the growth in the programme come from? We know already that the schools are trying very hard to retain as many pupils as possible within the school system as long as they can as part of the Raising the Participation Age framework[125].

The cohort of 16 – 18-year-olds is not noticeably growing with about only 6% of school leavers starting an apprenticeship[126]. Realistically we will not see a huge growth in apprenticeships in this age group although there are many young people working in jobs that are not apprenticeships. It is important that we address the balance within schools to present the apprenticeship route as an effective alternative to A-levels and going direct to university. Increasingly young people are seeing the alternative to an

---

[125] All young people are required to participate in education or training until their eighteenth birthday. See *Duty to Participate Until 18 from Summer 2015*: www.gov.uk.

[126] See table 6.1 of *Statistical First Release dated June 2015 for 16 18 Apprenticeship Starts*: www.gov.uk.

academic route and high debt levels when they go to university. An apprenticeship route is not an alternative to a degree but it is a different route. This should not be a debate about choosing between university and work based learning. It is more about which route people take to get the higher skills they need.

Training providers will be key to delivering this growth. Most employers become engaged in apprenticeships following discussions with a training provider. Providers have an interest in engaging employers and they therefore make a significant investment in time and resources to do just that. Governments have tried a number of initiatives over time to drive apprenticeship growth such as the National Apprenticeship Service, the large employers group, and the Ambassadors network. All of these initiatives have had a positive impact in terms of awareness and perception.

However, when it comes to the hard-nosed selling of the apprenticeship programme where someone has to convince a small business to recruit an employee and make a commitment to the cost of training which is substantial, then training providers deliver the numbers.

Training providers have always responded to the targets set out by government. Despite the constant changes in the programme, training providers have always spent the allocated budgets and more. In fact the main cap on numbers is the budget set by government every year. Providers have over delivered against that budget over the last few years.

Higher numbers of starts could have been delivered if we had had a responsive contracting system that encourages growth. That will be the case going forward. If the government is able to make the long-term investment in growth then the training providers will deliver. The current programme has grown even though the funding rates have been frozen. There have been some changes, such as the introduction of loans for those age 24-plus studying for level 3. There is a requirement for even higher quality assurance levels and employers are feeling the pinch of a recession.

What AELP has recommended is a much more consistent contract management system that provides growth budgets for those providers that have a history of delivering high quality programmes. In 2015/16, there are even fewer opportunities to get growth in contracts, so the SFA really does have to make longer-term commitments to that growth.

There are many other opportunities for growth. The introduction of higher apprenticeships is a very positive step forward where training providers, employers and universities can work together to combine the best of academic and practical delivery. What we do not want are university degrees rebadged as degree apprenticeships. This has to be a collaborative approach where training providers experienced in the integration of academic and work based learning are able to work with universities to ensure the programmes meet the needs of employers and apprentices.

The public sector has been slow to adopt the apprenticeship approach to training. They are very often focussed on graduate intakes or professional qualifications. We need new routes into these major public services such as the health service including nursing, the police and fire services. Apprenticeships can provide a very effective route way into these services to ensure that they are open to those people that have the practical skills.

We believe that substantial growth in the apprenticeship programme can be delivered through the public sector and the government has announced that it will use its Enterprise Bill to set targets in this regard[127]. In the Governments five year vision for apprentices the NHS has committed to 100,000 apprenticeship starts per year by 2020.

With a renewed focus on apprenticeships in the public sector, training providers can work with managers across a whole range of sectors to develop workforce development plans where apprenticeship programmes play a key role. We know that despite the reductions in budgets, many public services have skills shortages and face a demographic profile where a large number of their staff are coming up to retirement.

Nurses are a good case in point. We all agreed that the level of qualifications for nursing needed to be raised. Nursing became a degree level profession. However the government misunderstood the different

---

[127] For the Government briefing on the Enterprise Bill, see BIS, *Apprenticeships Target Enterprise Bill Factsheet*. www.gov.uk.

routes into the job and should have retained the apprenticeship route. They assumed that the apprenticeship route meant that nurses could not go on and do a degree level course.

It is now accepted that apprenticeships are just a different route to high level skills and for many people a better route to high level skills because they get more practical training. Many more apprentices will be going on to do degree level courses like nursing and we need to open these routes. Training providers will work with the professions to develop these essential routes into professions where there are serious skill shortages.

Flexible delivery models

The previous section highlighted the need to open up new and innovative routes to high level skills. The reforms to the way the apprenticeship programmes are structured should allow employers and training providers more flexibility in how the skills, knowledge and behaviours are delivered. This will allow training providers to adapt programmes so that they can balance the need of an individual employer with the needs of the apprentice who needs a wide range of skills that makes them employable across a sector and not by just one employer. This is a delicate balance and it is the partnership of an employer and a training provider that can make this work.

Too often we see employers not prepared to train people in skills that they don't need themselves. Some of those skills are essential to the apprentice if they are building a career and not just doing a job. It is the skilled provider acting as the training manager of the employer that can find that balance.

The market for apprenticeships will become competitive. Employers will want more for less. Providers will only be able to deliver that if they have the flexibility of delivery where they can use the new training technologies that exploit the ICT developments of recent years. Using online tracking and teaching tools will be the way to respond to the changing needs of employers and the new way of learning for many apprentices. However that will have to be combined with effective one-to-one teaching, mentoring and coaching.

The challenge and the strength of an apprenticeship programme is that providers have to adapt all of the learning techniques that make up learning on the job and off the job. It has always been the case that most of the learning, especially for young people will be on the job. Combining these new and traditional approaches will be a major challenge to retain the quality of delivery. Off-the-job training is important, but we need to drop old-fashioned ideas of how and where people learn.

Maintaining the quality of delivery will be an essential part of growing the programme. We do not see this being a choice between numbers of starts and quality of delivery. In fact it is just the opposite. We will not get the growth in the programme unless the starts are underpinned by a strong quality assurance process.

Training providers have an important check and balance role to play in the system. Employers have to be engaged and leading the programme but training providers understand what makes a quality programme and will need to have a clear role in that respect. We continue to support an external, independent quality assurance process. If this is to be Ofsted then the framework will need to reflect the growing importance of employer engagement and the impact on the apprentices rather than the current emphasis on success rates, teaching methods and protection of apprentices.

This is a major opportunity to allow employers and training providers to develop new and innovative programmes that meet the needs of the business and apprentices. Training providers are ready to respond to that challenge.

Conclusions
In conclusion, we believe that, given effective implementation, the apprenticeship levy will increase the profile of an already successful apprenticeship programme. There are some dangers associated with the new levy but if employers, providers and stakeholders are given more control of the programme then we should be able to deliver an even more focused and high quality programme that drives the productivity of the UK.

The quality of the programme can be maintained whilst expanding it to new sectors, new levels and new employers. This will require careful implementation of both the levy and the new system of funding for SMEs, giving them the opportunity to make a major contribution to improving the economy of the UK and give more individuals the opportunity they need to maximise their potential.

The opportunities will only be taken if employers, training providers and other stakeholders work closely together over the next five years. That is a lesson that has been learned since this reform process started. Providers bring both their expertise on the ground and the drive to engage employers. Developing an apprenticeship programme is not easy, and even large employers who are new to the programme when they begin to pay their apprenticeship levy will need a lot of help and support to ensure that they and their apprentices get the best from the programme.

Training providers will be key to delivering the three million target, particularly if we want to maintain the quality of the programme. It is inevitable that if larger employers raise the majority of the funding for apprenticeships they will want to see the programme focussed on their needs. SMEs, where the productivity benefits of the apprenticeship programme are felt most, will need the support of training providers to ensure that they remain engaged and can deliver the opportunities to apprentices.

36.
THE ROLE OF HIGHER EDUCATION
IN A WORLD-CLASS APPRENTICESHIP SYSTEM

Adrian Anderson
Chief Executive, University Vocational Awards Council
and
Mandy Crawford-Lee
Senior Associate, University Vocational Awards Council

———

Introduction

For many years politicians have argued that England needed a world-class apprenticeship system as an alternative to higher education[128]. An apprenticeship was frequently seen as the choice for those who weren't academic, and for the 50 per cent of young people who didn't go to university.

In this paper the authors suggest that apprenticeships should not be seen as an alternative to higher education, but that instead higher education is an essential partner in the development of a world-class apprenticeship system and the target to achieve three million apprenticeship starts by the end of this parliament in 2020.

The authors argue that the existing provider-led apprenticeship system has largely restricted apprenticeship provision to level 2 (intermediate – GCSE equivalent) and level 3 (advanced – A level equivalent) job roles.

The current apprenticeship reforms, where 'control' of apprenticeships has been placed in the hands of employers, has demonstrated that

———

[128] Definitions in vocational learning can mean different things to different audiences. Here we use the term higher education to encompass provision that leads to the award of a prescribed higher education, typically a Foundation, Bachelors' or Masters' degree.

apprenticeships should not be restricted to, or indeed largely focused on, level 2 and level 3 occupations. As at October 2015, a third of the new employer developed Apprenticeship Standards were at higher education level, with some requiring higher education qualifications for professional registration.

Apprenticeships should be a key tool in enhancing business productivity in occupations of strategic importance to the economy and, as employers are clearly demonstrating, a significant proportion of such apprenticeships will be at higher education level.

Finally, while supporting the target for three million apprenticeship starts this parliament, the authors make a plea that the achievement of this target must not be at the expense of the delivery of high cost and high quality higher and degree apprenticeships in the occupational areas that are most needed by employers and which will make the greatest impact on enhancing UK productivity.

Role of higher education in apprenticeships
A good place to start our analysis of the role of higher education in apprenticeships and the target for three million apprenticeship starts by the end of this parliament in 2020 is to consider what we mean by apprenticeship.

The UK government defines an apprenticeship as follows:

> 'An Apprenticeship is a job, in a skilled occupation, that requires substantial and sustained training, leading to the achievement of an Apprenticeship standard and the development of transferable skills to progress careers.' [129]

This definition says nothing about level, further education, higher education or indeed qualifications. New Apprenticeship Standards focus on defining, in plain English, the knowledge, skills and behaviours

---

[129] HM Government. (October 2013) *The Future of Apprenticeships in England: Implementation Plan.* page 9.

required to be competent in a defined occupation, demonstrated by passing an end test.

The UK government definition relates to the current apprenticeship reforms, but it is interesting to review the definition in the context of apprenticeship delivery in recent years. Key issues to consider are:

Why has apprenticeship development and delivery been so skewed towards lower level job roles and occupations?

Why has the emphasis on the higher-level occupations required by the economy and the engagement of higher education in apprenticeships been so marginal in the apprenticeship system?

In the academic year 2013/14, 65 per cent of apprenticeship starts in England were at intermediate i.e., level 2, akin to GCSE level. Indeed, the proportion of level 2 apprenticeships actually increased between 2012/13 and 2013/14 from 57 per cent to 65 per cent[130]. Of the remaining apprenticeship starts in 2013/14, 33 per cent were at level 3, akin to A level, and only 2 per cent at higher level (levels 4 to 6 – level 5 equates to a foundation degree and level 6 to a bachelor's degree).

Apprenticeship frameworks[131] with the highest take-up (in 2013/14) were health and social care, business administration, management (at intermediate level 2 and advanced level 3), hospitality and catering, and customer service. In 2013/14, construction skills and engineering apprenticeship frameworks each amounted to approximately 3.5 per cent of total starts[132].

Such figures demonstrate the myth of apprenticeships in 2013/14 as an alternative or equivalent to higher education. The majority of apprenticeship starts were at level 2 (GCSE level) and only 2 per cent at higher education level. The lack of STEM focus is similarly a concern. It

---

[130] Apprenticeship Statistics: England: Briefing Paper 06113, June 2015 by James Mirza Davies, House of Commons Library

[131] Apprenticeship Frameworks are currently in the process if being replaced by the new employer developed Apprenticeship Standards

[132] Ibid 3

is also worth considering the specification for apprenticeship frameworks (the system currently being replaced by the new employer developed Apprenticeship Standards) and the 'size' of apprenticeships. Take advanced apprenticeships. According to the Specification of Apprenticeship Standards in England (SASE[133]):

> 'An Advanced Level Apprenticeship framework must specify the total number of credits which an apprentice must attain for a qualification on the RQF (Regulated Qualifications Framework). This must be at a minimum of 37 credits'.

Using the further education credit system, thirty-seven credits equates to 370 learning hours. Contrast this minimum size with the size of a foundation degree, 240 credits or 2,400 notional learning hours, or a bachelor's degree, 360 credits or 3,600 notional learning hours. The SASE requirement for thirty-seven credits was, of course, a minimum requirement. Many apprenticeships were substantially larger, but some were not. Apprenticeships were, as such, in some cases relatively short training programmes, and not the 'sustained' training programmes identified in the UK government's current definition of apprenticeship. This is not to say that there were not excellent apprenticeships, even world-class apprenticeships involving world-class employers which were of equal status to higher education. The problem was that many, if not most, weren't.

Of course, we're not arguing that a short work-based programme at level 2 in business administration or customer service doesn't have a role in supporting young people enter the workforce or in tackling social inclusion. Indeed, such programmes may also help enhance business productivity. Our argument is simply that the English apprenticeship system was far too focused on work-based programmes in such areas and not sufficiently focused on higher-level skills and on raising productivity in

---

[133] Specification of Apprenticeship Standards for England (SASE), BIS September 2015 Revisions to SASE in 2013 mandated a minimum size of 90 credits for an Apprenticeship at level 4 and 5 and 120 credits for an Apprenticeship at level 6 and 7.

occupational areas and industry sectors of fundamental importance to the UK economy.

Why was it the case that the apprenticeship was so focused on level 2 job roles? Was this focus related to economic and business need? As our argument will demonstrate the answer is clearly 'no'. At a policy level an apprenticeship was often positioned as both a workforce development programme and an approach to support young people aged 16 to 24 not in education, employment or training (the NEET cohort) enter employment and engage in learning. There was also a chase to increase apprenticeship numbers often, regrettably, at the expense of quality.

Skills policy was also seen as separate from higher education policy – the 'learning and skills system', the Learning and Skills Council (LSC) and its successor the Skills Funding Agency (SFA) equated to further education. Sector Skills Councils (SSCs) following the practice of their predecessors, National Training Organisations (NTOs) and Industry Training Organisations (ITOs) largely focused on the development of apprenticeship frameworks at intermediate and advanced levels. Indeed apprenticeships, in the guise of Modern Apprenticeships , emerged in the mid-1990s as a Training and Enterprise Council/Industry Training Organisation initiative specifically and solely focused on level 3. Subsequently, the incorporation of National Traineeships, a level 2 programme, into the apprenticeship family as intermediate apprenticeships expanded apprenticeships downwards to include level 2 work-based programmes.

In information, advice and guidance (IAG) terms, apprenticeships were also positioned as for young people who didn't want to or couldn't stay on for A-levels or go to university.

The provider 'allocation system' managed by SFA, and its predecessor the Learning and Skills Council, also resulted in a supplier-led system. Providers (overwhelmingly private training providers focused on level 2 provision) and Further Education colleges were allocated funding from the SFA to deliver apprenticeships. The SFA, although it funded higher apprenticeships would, until 2015, only fund non-prescribed HE qualifications (e.g. QCF/RQF qualifications) and not prescribed higher education qualifications (e.g. foundation degrees and bachelor's degrees).

Therefore regardless of whether employers wanted or even needed to use a prescribed higher education qualification in an apprenticeship (for example it was specified as a requirement for professional membership) they couldn't – or at least not if they wanted SFA funding. Not surprisingly the growth in higher apprenticeships and particularly HEI involvement in apprenticeships up until 2015 has been disappointing.

Through the apprenticeship reforms, the government is tackling these issues head on. Placing apprenticeship development in the 'hands of employers' has been the ideal start. An apprenticeship is also being clearly positioned as a workforce development policy designed to raise business productivity – a welcome clarity of purpose. Although this is not to say that work-based learning programmes aimed at tackling NEET issues are not important, they are, however, different from apprenticeships, or certainly where employers are now positioning apprenticeships.

An analysis of the Apprenticeship Standards being developed by employers clearly demonstrates that the apprenticeship is migrating upwards in terms of education levels. At the time of writing (October 2015), approximately a third of the Apprenticeship Standards developed by the employer Trailblazers were at higher education level (level 4 – certificate of higher education to level 7 – master's degree levels). Neither are Apprenticeship Standards at higher education level just being developed in 'niche' areas. Apprenticeship Standards have been developed in key areas of higher education provision including; engineering, manufacturing, digital, construction, accountancy and law. Revealingly, employers had not rushed to propose trailblazers to develop Apprenticeship Standards in many of the lower level 'job roles' that have dominated apprenticeship framework provision in recent years, such as business administration, management (at level 2 and 3) and customer service.

Of course the proportion of Apprenticeship Standards developed at higher education level does not necessarily directly relate to the proportion of actual apprenticeship starts at higher education level. There are, however, some occupational areas where Apprenticeship Standards are being developed at higher education level where there are potentially significant numbers of apprenticeship starts.

If apprenticeship starts in such areas do not materialise, a detailed analysis of why this is the case will be needed. We would question any answer based on an absence of demand. For a young person a higher or degree apprenticeship will be an attractive offer. They are employed and earn a salary from day one. The government and their employer will pay their fees (as part of the apprenticeship) and the learner will therefore avoid student debt. Government funding and future compulsion through the apprenticeship levy and public sector procurement requirements will similarly 'attract' employer engagement. There should therefore be both individual and employer demand for higher and degree apprenticeships.

We do, however, see several risks to realising the potential of higher and degree apprenticeship:

- A Tension between Cost and Quality - higher and degree apprenticeships will, in most cases, be more expensive to deliver than intermediate or advanced apprenticeships. There is a risk that to achieve the three million apprenticeship target – numbers will be chased and cheaper but less valuable forms of apprenticeship will be prioritised through the SFA's existing supplier base. The government's decision to place control and purchasing in the hands of employers will help avoid this. But ambition in setting targets will be needed. Take for example the BIS target of 20,000 higher apprenticeship starts in 2013/14 and 2014/15. This target was comfortably exceeded with 28,000 actual starts. But before celebrating this achievement, it is worth noting that the figure of 28,000 actual starts equated to just 3 per cent of all Apprenticeships. The increase in the number of higher apprenticeship starts between 2011/12 and 2014/15 (81 per cent from a low base) can also be accounted for almost entirely by growth in one or two frameworks: the Higher Apprenticeship in Accounting Level 4, and the Higher Apprenticeship (level 5) in Care Leadership and Management (the latter accounting for 8,300 starts of the total 19,300 starts in 14/15)[134].

---

[134] BIS/SFA. (October 2015) *Statistical First Release, Provisional Full Year Data for 2014 15*: www.gov.uk.

Establishing a target for the proportion of the three million apprenticeship starts that are at higher and degree apprenticeship level – based on actual economic and business need – could be an ideal way to counter any trend to focus predominately at lower levels. A target would be based on an analysis of economic need and the potential role of apprenticeship in increasing productivity. Perhaps local enterprise partnership (LEP)/city-based targets could also be developed. If we're serious about the role of apprenticeships in raising productivity, targets should be challenging – perhaps a quarter of apprenticeship starts at higher and degree level?

- The Ability of the Skills Funding Agency to Support Higher Education Engagement in Apprenticeship – The Skills Funding Agency (SFA) is predominately a further education agency, using further education systems and processes with a further education ethos[135]. The Advisory Board of the SFA has further education representation but no higher education representation. This is, to say the least, odd when SFA is the agency responsible for the government's flagship skills programme, apprenticeships that encompass both further education and higher education. Further education will have a fundamental role in apprenticeships, but so too will higher education if the higher and degree apprenticeships that employers want are to be delivered.

- The Implementation of the Apprenticeship System should be based on the Aims of the Apprenticeship Reforms: Employer-Driven, Simplicity, Quality and Giving Employers Purchasing Power - Regrettably, our experience of working with approximately twenty-five universities has identified multiple and significant barriers to university involvement in higher and degree apprenticeships. The 2015 invitation to universities to join the SFA register of approved

---

[135] SFA describes its role as follows: 'We fund skills training for further education (FE) in England. We support over 1,000 colleges, private training organisations, and employers with £3.7 billion of funding each year.' SFA: www.gov.uk.

training organisations (ROTO) and two procurement rounds to enable HEIs to secure an 'allocation' to deliver apprenticeship proved in some cases, but not all, relatively straightforward. Moving to actual delivery has, however, been anything but.

As examples, at first there was uncertainty as to which quality assurance system universities would need to comply with. Apprenticeships have been overseen by Ofsted, whereas universities, for the type of provision concerned, operate under the auspices of QAA. It took nearly eight months to confirm that 'HEFCE-funded HEIs delivering higher and degree apprenticeships funded through the SFA will not be subject to separate Ofsted regulation, and will only come under the remit of one regulator, the HEFCE-appointed Quality Assurance Agency (QAA) during 2015/16.'

Why the delay, why the uncertainty? Similar problems may occur when HEIs are asked for management information purposes to use the further education based Individual Learner Record (ILR) system instead of higher education based HESA returns. Again – why? It is early days, but early experiences are remembered and influence future behaviour even if the system develops and evolves.

The potential bureaucratic burden to higher education of engaging in apprenticeships has the potential to be substantial – yet it need and should not be. Our plea would be that the SFA and those involved in the implementation of higher and degree apprenticeship focus on how best to support the engagement of both further education and higher education providers on the basis of employer demand and national and local skills requirements. Perhaps this could be achieved if the SFA was remodelled to become not just a further education agency, but as its name implies a 'skills' funding agency. Alternatively, the SFA could be remodelled as an apprenticeship agency focused on achieving the government's apprenticeship target making the system work for the employer as outlined in the Apprenticeship Reforms.

- A Misunderstanding of Higher Education - We should also confront head on the absurd notion that higher education isn't about skills or occupational competence. England has a world-class higher education sector and several universities have world class reputations in the delivery of higher education programmes that develop and accredit work-based learning and occupational competence. UVAC's peer reviewed journal *Higher Education, Skills and Work-based Learning* published quarterly by Emerald showcases many such examples.

   Greater capacity is needed in higher education, but the apprenticeship reforms could stimulate the development of such capacity on the basis of employer demand – if an appropriate implementation approach is adopted. Of course, further education colleges and private and alternative training providers will also have a fundamental role to play in both higher and potentially degree apprenticeship delivery. New and innovative delivery partnerships will be needed particularly at the higher technical level (level 4) where we are seeing a large proportion of Apprenticeship Standards developed. Our argument is simple, if employer demand for higher-level skills/apprenticeship provision is to be satisfied it will require the expertise of further education and higher education, awarding organisations, private and alternative providers and professional bodies. UVAC working with our HEI, further education and awarding organisation members is focusing on encouraging and supporting collaboration in this area.

Of fundamental importance to the engagement of higher education in apprenticeships has been the decision to introduce the apprenticeship levy which will be paid by all large public and private sector employers. Large employers in England will be able to spend the levy to support all their post-16 apprenticeships, including higher and degree apprenticeships. The levy will act as a 'game changer' for higher education and could fundamentally alter the market for vocational higher education provision. In recent decades the expansion of higher education was funded by the state and more recently by the individual underpinned by state backed student loans. As such, the purchaser and customer for most higher education provision is currently the individual and state.

An analysis by the Edge Foundation provides an interesting perspective of the expansion of higher education in the decade between 2002 and 2012. The relative growth of different subject areas reflects student choice and subjects studied at age 16 to 19. Many employers may be critical of the relative growth of subject areas. It was, however, the learner who took on the burden of fees and student debt – with employers making no systematic contribution other than through general taxation.

Subject focus of Recent Growth in First Degrees 2002 – 2012
>Humanities – 81.3%
>Business/Administration – 77%
>Creative Arts/Design – 76.3%
>Physical Sciences – 23.7%
>Engineering/Technology – 16.3%
>Computer Science – 6.5%
(The Edge Foundation – compiled from data from HESA)

For higher and degree apprenticeships the employer (and not the student) through the apprenticeship voucher system will be the purchaser and customer of the training provision offered by an HEI or other provider. The individual learner will remain the customer, if not purchaser, for the high education qualification associated with the apprenticeship.

It seems probable that more industrial planning and devolution at city level could also incentivise and support higher education provision which is linked to skills and labour market requirements.

The new apprenticeship system government wants encompasses higher-level occupations. The Apprenticeship Standards being developed are clearly demonstrating that employers want apprenticeship to be a programme that encompasses higher-level occupations.

As at October 2015, the following number of Apprenticeship Standards by level had been developed:
>Level 4 – 31
>Level 5 – 8

Level 6 – 19
Level 7 – 4
Total Degree Apprenticeships – 17
Total in Regulated Professions – 35

Rightly, the apprenticeship system is not now prescriptive on which type of provider an employer uses to support delivery of the training to support the individual learner pass the apprenticeship end test. The employer will decide on which organisation to use for the training required by the apprenticeship be they a private training provider, alternative provider, further education college or university or a combination of providers. If the new apprenticeship system is to be a success and to deliver the number of starts the government wants higher education will need to play a fundamental role for the following reasons:

- Without HEI involvement it will be difficult to develop capacity and expertise to support delivery of apprenticeships at higher levels and in the occupational areas demanded by employers – Existing apprenticeship 'providers' (the majority of whom are private training providers) are mostly focused on level 2 and 3 provision. Current providers will need to adapt and new providers will need to enter the apprenticeship market if the potential demand for higher and degree apprenticeships is to be met. Of course a degree apprenticeship will by definition also require the involvement of a HEI as the awarding body for the bachelor's degree or master's degree.

- Several Apprenticeship Standards require the use of HE qualifications particularly as a route to professional membership and registration - Apprenticeships, particularly at level 6 and 7 will, in some occupations, require an individual to acquire a prescribed higher education qualification (a bachelor's degree or master's degree) in order to secure professional registration and membership.

- Employers and learners want HEIs to be involved in apprenticeships – Many employers want to involve HEIs in the development and delivery of apprenticeships as demonstrated by the interest in degree apprenticeships, which by definition must involve HEIs. Individual learners are attracted by the 'best of both worlds' by undertaking an apprenticeship and gaining a degree as part of their apprenticeship. The message that an apprenticeship can lead to a degree or actually incorporate a degree will be a very important factor in promoting the apprenticeship brand, particularly to parents and schools.

- Progression from advanced apprenticeship to higher education will continue to be important in terms of social mobility and widening access – Apprenticeships (based on standards) are not focused on qualifications. Apprenticeship standards do not give any indication of credit value or guided learning hours. HEIs may therefore find it difficult to support progression from advanced apprenticeship to higher education programmes.

HEIs will need to work with further education colleges, independent training providers and assessment organisations to ensure progression opportunities remain, and indeed are extended, for individuals completing advanced apprenticeships who have the aspiration and ability to pursue a higher-level learning programmes. The development of higher and degree apprenticeships – if suitable work based craft, technician, profession and associate profession career pathways are developed - should also help grow demand for advanced apprenticeships by demonstrating progression routes to professional and managerial occupations and higher education to 16 – 19-year-old learners and equally importantly their teachers and parents. Appropriate development in this area will also help widen access to higher education.

To the vast majority of HEIs, the apprenticeship is a new product. As has been noted earlier until very recently, prescribed HE qualifications (i.e. foundation degrees and bachelor's degrees) were not eligible for SFA funding when used in an apprenticeship. HEIs will therefore not tend to exhibit reluctance to move from apprenticeship frameworks to delivery of apprenticeships based on standards – because most haven't to date been involved in the delivery of higher and degree apprenticeships. If the right conditions are established then employer and learner demand for higher and degree apprenticeships could see HEIs leading the changeover from apprenticeships based on frameworks to apprenticeships based on standards.

The rationale for HEI engagement in higher and degree apprenticeships is very strong and makes a compelling case. The rationale – which UVAC outlined in a guide released in March 2015, encouraging HEI engagement in apprenticeships and published with the support of HEFCE - can be summarised as follows:

- Policy - Apprenticeships are a priority for government, which wants to move them from a predominately level 2 and 3 programme to a level 3 and higher-level programme, and to tackle technician and associate professional level skills gaps and shortages. The government also wants to explore the potential of alternatives to traditional full-time bachelor's degree programmes that currently dominate higher education provision.

- Demographics - The number of 18 to 20-year-olds reached a peak in 2011. The decline in 18to 20-year-olds will continue to decline until 2021 (rising thereafter) by when the population of this age group will be 14 per cent lower than in 2011. To maintain the current level of 18 – 20 age group recruitment to higher education in the short to medium term, HEIs will have to recruit a higher proportion of the cohort. Higher and degree apprenticeships could represent a way for HEIs to recruit new cohorts of learners - those who, in future, want to choose an alternative (and more cost effective route) to a 'traditional' full-time bachelor's degree. Involvement in apprenticeships could also represent a useful approach to widening participation and access both in delivering programmes at level 4 and above and supporting progression to higher education from advanced apprenticeships.

- International students - While international (non-EU) full-time undergraduate entrants increased in both 2012-13 and 2013-14, the level of growth was substantially lower than that experienced prior to 2010-11 and compared with competitor countries [136]. Future immigration controls and changes to student visas could have an impact on international entrants to higher education.

- The dominance of full-time bachelor's degree provision - English higher education provision is characterised by the dominance of full-time bachelor's degree provision. Full-time UK and other EU undergraduate entrants to universities and colleges in England in 2013/14 were 8 per cent higher than in 2012/13, following a dip from 2011/12. UK and other EU full-time undergraduate entrants to universities and colleges in England were 384,000 in 2010/11, 398,000 in 2011/12, 351,000 in 2012/13 and 378,000 in 2013/14[137]. In comparison:

▪ UK and EU part time undergraduate entrants fell from 259,000 in 2010/11 to 139,000 in 2013/14.
▪ Entrants to foundation degree programmes at English HEIs have fallen substantially between 2010/11 and 2012/13 with full time entrants falling from 21,000 to under 13,000.
▪ International comparisons on short cycle professional education and training qualifications make interesting reading. The percentage of adults aged 20 – 45 who have short cycle professional education and training as their highest qualification (OECD 2014 survey of adult skills) was lower in England than in France, Japan, the USA and Germany

- The employer as purchaser of higher education – The expansion of higher education provision in recent decades has been funded by the state and, latterly, by the state and individuals with access to state-backed loans. The reforms to apprenticeship focus on the employer as purchaser of the learning and accreditation required

---

[136] HEFCE. (2015) *Higher Education in England 2014: Key Facts*: www.hefce.ac.uk.

[137] HEFCE. (2015) *Higher Education in England 2014: Key Facts:* www.hefce.ac.uk.

for the apprenticeship. Apprenticeships will represent a new and important source of employer income for HEIs.

- Learners, loans and debt – As the apprenticeship funding model is based on the employer and government funding the cost of the training and accreditation provision, the apprentice will not pay tuition fees or therefore need to access a student loan if an HE programme is used to deliver the requirements of the Apprenticeship Standard. Essentially, when used to deliver the requirements of the Apprenticeship Standard the apprentice/student will have their tuition fees paid by a government funding contribution and their employer [138]. The apprentice/student will also be employed and earn a salary. The prospect of 'earning while you learn', no student debt and a job from day one will potentially be highly attractive proposition to many prospective HE students and their parents.

Conclusions

Higher education clearly has a fundamental role to play in supporting the delivery of the government's ambition for three million apprenticeship starts by the end of this parliament in 2020. We would make a plea that in focusing on this target there was a race to the top, a focus on quality and the delivery of apprenticeships including higher and degree apprenticeships that have the maximum impact on raising productivity. Placing apprenticeship development and purchasing in the hands of employers is the first and essential step in this process.

A second step is to introduce a target for higher and degree apprenticeships. We have suggested a target of a quarter of the three million should be higher and degree apprenticeships. As a nation we can't use apprenticeship as the key vehicle to raise productivity if it continues as predominately a level 2 and level 3 skills programme.

Oversight of the apprenticeship system is also fundamental. Perhaps the Skills Funding Agency should be remodelled to become not just a further education agency, but an apprenticeship agency focused on achieving the

---

[138] Subject to the caps and rules outlined in Section 7 of this guide.

government's apprenticeship target; making the system work for the employer as outlined in the apprenticeship reforms.

If, as government intends, we ensure that employers are in control of apprenticeships and we get the system right, then universities will respond and we will be several steps closer to having a world-class apprenticeship system delivering the skills employers, learners and the economy require.

Gone will be the days of a dominance of level 2 apprenticeships for questionable job roles – apprenticeships will be the equal of degrees/higher education not least because many will include the acquisition of bachelor's or master's degrees and professional qualifications. England will then be well on the way to a world-class apprenticeship system.

37.
DEVELOPING ASSESSMENT:
CHALLENGES AND BEST PRACTICE IN THE UK AND
INTERNATIONALLY

Judith Norrington
Skills Commission and Education and Skills Consultant

———————

Now, at a time when the ambition for UK apprenticeships is growing, both in terms of numbers and achievement, it is increasingly important to ensure that our assessment system is of the highest quality. Effective assessment plays a central role in confirming outcomes, increasing confidence, and assuring excellence.

This chapter will consider some of the major recommendations of the Richard Review of 2012, their efficacy for bringing about lasting improvement, and the challenges that may arise for those designing assessments. It will also discuss what good practice in assessment looks like, and what we can learn from the experience of those offering apprenticeships internationally to influence our future practice, or to provide greater confidence in our own assessment arrangements. Some suggestions to support the development of good practice have been included throughout.

After consulting widely with business, professional bodies, awarding organisations, colleges, training providers and government, Richard's conclusions were that assessment should be focused at the end of an apprenticeship, should extend over a number of days, should 'demonstrate that the apprentice can take the knowledge and expertise they have gained and apply it in a real world context to a new, novel problem' (2012, p. 67) and should test that the apprentice can meet the standards established by industry for that occupation.

Therefore, the Review proposes, it would be advantageous to include 'testers […] drawn from the ranks of employers as well as educators' (2012, p. 67) in a more central role within the assessment process. The assessment should also be conducted by a neutral party.

Richard appreciated that assessment would need to take a number of forms, including 'practical assessments, projects and assignments, or written exam papers, to test different aspects of the standard'. He also, perhaps more controversially, proposed that there should be a form of grading that was more than just pass or fail, calling for the development of 'clear and unambiguous criteria against which the attainment of apprentices can be accurately and consistently differentiated', and that these criteria should be 'understood by assessors and accurately and consistently applied'. He referred to the need to promote, 'consistency in measuring the levels of attainment of learners over time'. (2012, p. 73)

The Review makes it clear that a great deal rests on the shoulders of the assessment regime: 'testing is particularly important in the model of apprenticeships that I am proposing, where we are flexible about the content and curricula but rigorous on the outcome: the test is the mechanism by which we incentivise employers and training organisations to invest in the right type of training' (2012, p. 54).

Richard acknowledged that this would be an intricate task, stating 'In most cases, the skills required to do a job well will be multifaceted, and the test will need to be diverse and wide ranging in order to reflect this' (2012, p. 54). While it is relatively straightforward to list the concepts to be included within the assessment, in practice it is rather more complex to design a regime that will fulfil the original brief. Reviewing some of the building blocks of assessment demonstrates why this is.

Assessment is rarely an exact science. It relies fundamentally upon evaluation and judgement, and in vocational learning is dependent on properly constructed criteria. This must be balanced with a range of other factors including 'validity', ensuring that the methodology is appropriate for what is being assessed. This is not always straightforward in the case of apprenticeships, where so many different capabilities are looked for. Validity also includes concepts such as 'authenticity', that the work is only that of the candidate, 'currency', evidence that the attainment is up to date and relevant and 'fairness', that all candidates have an equal opportunity to attain the standard, for example by setting a time frame for assessment exercises.

Additionally, outcomes must also be reliable, consistent and repeatable. These features are to ensure fairness and comparability so that equally matched apprentices in the same occupation would be judged in the same way and receive the same result.

In the context of the real world with resource constraints, it is also prudent to ensure that any assessment approach is both manageable and cost effective. This means that, whenever you are designing assessment, difficult choices always have to be made, and to add to the pressure it is not always understood that these very choices have wider consequences: 'there is little appreciation for the impact of assessment method on performance and outcomes although a variety of studies demonstrate that this can be enormous' (McNamara, 2006, p. 38).

A high quality assessment regime is, in many ways, dependent on a high quality learning experience for the apprentice. Fuller and Unwin's work on the content of apprenticeships presents a useful image of what a 'good' apprenticeship looks like as part of their 'restrictive-expansive continuum' (2011, p. 36).

Apprenticeships, they suggest, should contain vocational skills and knowledge, a working environment where expertise can be developed through practise with others, and access to knowledge and skills that will enable apprentices to grow beyond, as well as within, their current roles and sectors. They comment, 'For young people, it is [...] the platform for progression – which is often missing from the British system and which is hard to get right' (2011, p. 37). Internationally, many apprenticeships appear already to be built into the fabric of the wider education and training system. For example in Denmark 'the system is highly flexible, offering learners the possibility of [...] returning to the VET system at a later point in time to add to their VET qualifications in order to access further and higher education' (Dibbern and Kruse, 2014, p. 2).

While UK apprenticeships have clear links with further education, there are comparatively fewer with higher education, though with the design of higher-level apprenticeships and the increasing number of degrees that are classified as vocational this may well change in the future. However, it is important to maintain the practical element of assessment which is so vital in judging competence.

In delivering a new ambition for apprenticeships, the assessment challenge is to design a regime that fully but efficiently captures all of the required elements.

The previous apprenticeship framework provided certainty in terms of rules and regulations that set out hours of training, both on and off the job, and details about maths, English and transferable skills requirements. The content that needed to be assessed was largely contained in qualifications. The changes proposed in the Review place a far greater responsibility on those delivering and assessing to plan their programme of delivery. In order to be successful it is essential to have well-written and comprehensive standards against which to assess. These must clearly state each separate outcome required, and be detailed enough for everyone involved to understand 'what is being assessed and what should be achieved' (Wolf, 2001, p. 3).

As of August 2015, the system is still in transition with 140 employer-led Trailblazer groups already having delivered, or presently working on, over 350 standards, of which around thirty-five have been approved and each of which has an assessment plan (BIS, 2015e). It may be too early to tell for certain how effective these standards will be, but there are encouraging examples of good practice.

All now include opportunities for apprentices to become members of relevant professional bodies, supporting their progression in the industry. The assessment plans closely reflect the needs of the industries and in many cases include greater flexibility. For example, the Butchery apprenticeship offers multiple assessment pathways (BIS, 2015a, p. 10). A strong feature of the Golf Greenkeeper assessment plan is the requirement for an apprentice to provide a personal statement that reflects upon their learning and development, their team and the business as a whole (BIS,

2015c, p. 2), providing the apprentice with analytical skills and a strong platform from which to be an independent learner in the future.

All the plans set out the clearly defined steps required to progress through the stages of assessment and the roles of the different participants involved. The vast majority now have built-in tasks which allow a demonstration of excellence rather than just the meeting of a minimum standard.

As time goes on, I believe that it would be advantageous to increase the breadth of standards and move towards a broader definition of occupation rather than focusing on individual job roles. This would allow individuals to move easily between related trades and provide them with a wider base of expertise and knowledge along with helping to minimise the number of different standards. At the same time, it is essential to ensure that everyone in a single occupation is judged the same way and using the same standards.

However, no matter how well constructed standards are they will only ever be as effective as their application in practice. Much of the success of the system relies on different parties working together. Although we have a different range of organisational partners in the UK, the more we can emulate the very cooperative approach adopted by social partners internationally the easier it will be to deliver high quality apprenticeships. In the case of Germany, 'In the practice of vocational training, all cooperation is based on consensus; no regulations concerning initial or further vocational training may be issued against the declared will of either of the social partners' (Hensen-Reifgens and Hippach-Schneider, 2014, p. 11).

There has also been a move towards greater differentiation in grading in the new Trailblazers' standards, spurred on in part by the Richard Review. The Review proposed that 'Differences in ability and accomplishments should be acknowledged [...] Also, similar to a University degree, I believe that the test at the end of an apprenticeship should be graded. Prospective employers should be able to use the grade in the test as evidence of the apprentice's ability and potential' (2012, p. 56). Given the scale of the change it is important to consider how straightforward this will be to achieve and how likely it will be to improve the quality of apprenticeships both for employers and the apprentices themselves.

Until recently it has been usual in the UK to regard competence-based achievement as either passed or failed: 'Competence is seen as finite – you can either do a task effectively or you can't' (SQA, 2014, p. 4). To pass a competence-based qualification a candidate has to demonstrate mastery or, put another way, must satisfy every one of the assessment criteria that appear in the standards and assessment plan to a level that would be deemed acceptable for someone operating effectively in that industry, a high hurdle to begin with. When knowledge and understanding are being assessed through written tests 'The pass mark for assessments is therefore usually set quite high (typically 70% or above),' and longer tests covering a wide range of skills are often broken down 'and require a pass on each section' (AlphaPlus Consultancy Ltd, 2014, p. 23).

This approach is often much more rigorous, though less publicly understood, than that which operates for general qualifications. In these, pass marks tend to be much lower and the question range is deliberately designed to allow individuals of different abilities to demonstrate their level of knowledge. Candidates can also compensate for a lack of knowledge in one area with a 'good' answer in another, and many topics are sampled rather than assessed completely.

Currently, some of the employer groups view grading as an opportunity to motivate their apprentices while others see the practicalities as more challenging. We may all believe that it is possible to distinguish someone who does an excellent piece of joinery from one who does not or who serves a meal in a way that makes you want to return to the restaurant or the reverse but it is not always straightforward to describe what actually makes the distinction.

In practical terms, additional grading 'must go beyond the level of effectiveness needed for a pass' (SQA, 2014, p. 4) and the criteria on which the grades will be judged must be distinctive enough to allow proper differentiation. Being able to distinguish between apprentices is useful but only when you can be sure the outcome is accurate.

One approach would be to think about what 'beyond' looks like in a particular industry. Different occupations will have different expectations of what constitutes an outstanding performance that could relate to attitudes and behaviours, or the way an individual conducts their role. The jury is still out as to whether grading will bring the advantages cited by Richard though it is notable that there is currently little consensus among employers over the scales and number of gradations to be used in assessing apprenticeships. Some are retaining a binary 'pass-fail' for competence-based elements.

Communicating the reasons for these differences will be important to avoid mixed messages. It would be unfortunate if, for example, grading the knowledge component and not the technical competence were to give the impression that one element of an apprenticeship was more valuable than another, particularly as we know that 'What learners learn is, in large part, influenced by their perceptions of how they will be assessed' (Tanggaard, and Elmholdt, 2008).

Additional levels of grading have the potential to provide useful information so long as they really are used to differentiate between the things that matter rather than merely those things that are easy to measure. The best quality outcomes will come from a system that is flexible enough to reflect the needs of each industry and which grades different elements appropriately, showing where additional strengths lie. Ideally information on achievements should be displayed in a profile where the apprentice's performance in different activities can be seen and understood by all parties.

To date most competence-based qualifications are taken when the apprentice is ready. The assumption behind this approach has always been that if you recruit with integrity, and take on individuals who will be capable of completing the training, everyone could pass. The Review suggests that there will be some apprentices who do not pass first time (2012, p. 56).

While assessment should be rigorous, it would be unfortunate if the change to a wider range of grades encouraged an assumption of a particular grade distribution. There should be no predetermined number of individuals receiving any particular grade. Above all the expectations and needs of the current and future industry should drive the approach used for grading.

A further fundamental change proposed in the Richard Review was to move the focus of significant assessment activity to a summative approach carried out predominantly at the end of the apprenticeship and covering all the relevant competencies. Richard states 'I do not believe that it is in the interest of the apprentice to have on-going tests and exams throughout, with accreditation of small bite-sized chunks' (2012, p. 55).

As one of the critical goals of an apprenticeship is to ensure that an apprentice has accumulated a comprehensive range of expertise and skills by its conclusion, there is a strong rationale for a significant proportion of the assessment to be concentrated towards the end and to be synoptic in character. This approach may require less time for assessment and leave more for learning, can enable more effective planning and can avoid unnecessary duplication of assessment requirements.

However, in practice the needs of individual industries must be taken into account. For various occupations, such as some of those in the energy industry and in social care, earlier summative assessments and certification would be needed to ensure that an individual can work safely before, in these examples, they are allowed on site or have the knowledge and understanding to work with a patient with Alzheimer's. The result would be an assessment regime consistent within an occupation but not uniform across different industries. Such prior certification would not preclude a professional discussion about skills, learned and exhibited, taking place at the end of the apprenticeship as part of the remaining summative assessment.

The Life and Industrial Sciences Laboratory Technician apprenticeship and the Financial Services Customer Delivery apprenticeship both include knowledge and practice tests during the apprenticeship (BIS, 2015d, 2015b) that, along with other ingredients, act as a 'Gateway' and demonstrate the apprentice's readiness for synoptic assessment at the end of their training.

In addition, some behavioural criteria, such as adapting working practice as conditions or circumstances change, lend themselves to assessment over a longer period and would be hard to identify accurately only at the end.

This summative approach could also place additional stress on the apprentice that might result in an uncharacteristically poor outcome for some. Good practice to mitigate these challenges would include effectively preparing the apprentice for their summative assessment, which many new assessment plans are already emphasising. To enhance quality in the future, techniques such as assessors seeking feedback on their own performance and initially working in pairs or groups in order to compare and moderate judgements made by different individuals would be helpful.

It is also important that the emphasis on a final assessment should not remove the opportunity for the apprentice to receive feedback along the way from those providing training and instruction or for that matter for those training to gain feedback about the programme they are offering.

This formative assessment is a critical ingredient of high quality apprenticeships. It ensures that a major gap in skills or knowledge can be discovered before it is too late to rectify it. More positively, many will have witnessed the added motivation that comes from receiving constructive feedback. This form of assessment also encourages the apprentice to be more actively involved in their own learning, helping them to develop important skills such as self-assessment. The more sophisticated understanding a person has of themselves as a learner, the more they can progress and improve their skills. Given that most industry requirements and skills change over time, some at a very fast pace, inculcating a habit of continued learning adds significant value.

There are a range of ways of utilising formative assessment to support further learning. Good practice examples that could be developed further include peer assessment, both with other apprentices and with other employees in the company, keeping a learning diary of achievements, undertaking individual and group projects, reflecting on formative assessment outcomes in a formal discussion and topping up knowledge or testing understanding from relevant e-learning modules. There is plenty of opportunity for trying out innovative forms of assessment during the formative phase of the apprenticeship and when it has been tried and tested, using the same approach in summative assessment.

Internationally formative assessment is also valued but there is mixed practice on the timing of summative assessment with some countries having, or moving towards, a model based on success in relevant qualifications. For example in the 'Accelerated Australian Apprenticeship Program' 'progression through an apprenticeship [...] is dependent on the satisfactory demonstration of occupational competencies [...] and is not solely tied to a specific duration' (OECD and ILO, 2011, p. 13).

Finland places a high value on formative assessment during the training, with a formal assessment discussion held around three times a year. Then 'At the end of the apprenticeship training, the employer and the workplace trainer provide the student with the final assessment of his or her vocational competence. Each module is assessed on the scale of 1-3 [...] Apprenticeship training usually aims at the attainment of a competence-based qualification' (Koukko and Kyrö, 2014, p. 5).

Other countries such as Germany and Switzerland have a regime that is at least two years in duration and combines written tests, performance and discussion at the end of the apprenticeship, which has worked successfully for those countries.

While there is no single approach to assessment internationally, the most effective examples include feedback during training, combine a range of different assessment methods, and have significant involvement from employers and social partners. 'The engagement of employers is a crucial element for the success of an apprenticeship system. Apprenticeships cannot expand and become a recognised pathway from school to work without the strong involvement of employers' (OECD, 2014, p. 3).

In the UK too, many employers are extending their central role in apprenticeships and becoming increasingly involved in assessment activities, a direction strongly supported in the Richard Review. Alongside industry experts and other educational and training specialists, employers are best placed to judge whether an apprentice meets the needs of their industry and are 'suitable for the needs of a real workplace' (2012, p. 56).

437

High-quality assessment has many common features but no single approach to delivery. Some occupations, particularly where small firms predominate, are collaborating extensively with education providers and awarding bodies who will offer an independent assessment service. There is already some experience of this approach, for example in horticulture where peripatetic examiners observe and ultimately certificate safe practice in using a chain saw, and this approach is being scaled up to meet the needs of different industries. This approach has the benefit of being time efficient for employers, although it will still require significant industry expertise from assessment providers.

Other industries are looking for a shared approach where employers will be involved in panels both assessing and moderating outcomes alongside education and training experts, providing a balance of expertise in assessment and moderation. However, the time that will need to be taken to establish a shared understanding of appropriate outcome standards should not be underestimated, as this will represent a significant commitment.

This approach has international parallels. In Finland, 'competence tests normally take place in authentic working life settings. [They] are assessed by the representatives of the employer, employees and the organiser of the competence-based qualification. When candidates have demonstrated that they can meet the vocational skills requirements, they are awarded a qualification certificate by the relevant Qualification Committee' (Koukko and Kyrö, 2014, p. 5). Once the UK apprenticeships based on the new standards have become established, benefit could be gained from involving past apprentices in the assessment process both to share their valuable expertise and to continue to refine the system.

In more than one case, industries including the Energy and Efficiency Independent Assessment Service (EEIAS) have established their own independent assessment service (Brooks and Smith, 2015, p. 6). In this example, employers work collectively not only to decide the standards but also to lead the assessment and quality assurance process. In such a model employers from one company can be part of a team assessing apprentices in another, helping the cross-fertilisation of good practice and training techniques across the whole industry. It also allows effective future planning for the workforce and provides a responsive service able to support new skill requirements in a fast moving industry.

An apprenticeship is all about learning while growing more productive in a job. Although apprentices will inevitably have slightly different experiences because of the focus of activity in their own company and role, it is important that they are able to learn about and carry out the full range of skills in the appropriate occupational standard.

In some cases some smaller companies cannot offer this alone. One of the ways of accommodating this is to form further group training organisations, a practice already followed internationally. In Hamburg, for instance, 'In these circumstances the Chambers will explore the feasibility of bringing together a number of companies to deliver the training contract.' Assessment is shared between the teaching institutes and panels of employers and employees, supported by the chamber of commerce (OfQual et al., 2013, p. 16). This approach would provide a model that could be followed with the chamber role being replaced by a professional body, an awarding body or provider.

An apprenticeship is likely to have an instrumental role in an individual's future career and it is therefore vital that its quality is properly assured. Such a system should safeguard the outcomes for both the apprentice and the employer and over time increase transparency and trust both within the industry and with external providers, as well as providing valuable evidence that can be used to drive quality improvement.

UK industries are considering a number of approaches, including working with awarding organisations already regulated by national bodies such as OfQual and the SQA and, in the case of England, with providers approved by the Skills Funding Agency to train apprentices. Others are exploring the possibility of establishing their own additional process to oversee and select providers and independent assessors to work with the industry. The challenge is to find the most efficient and effective way of quality assuring apprenticeships in a context where many of the parties, colleges, training providers and awarding bodies are already subject to other forms of regulation and quality assurance, as are employers within their own industries.

Good practice would be to add a further layer of quality assurance only when such an approach is deemed necessary to give these groups and employers in particular, greater confidence in the outcome.

Internationally the focus of quality assurance extends to the companies offering apprenticeships. In Germany companies go through a process to be authorised to offer training (Hensen-Reifgens and Hippach-Schneider, 2014, p. 3) while in the Netherlands 'Once companies are accredited, accreditation has to be renewed every four years' (Westerhuis and Smulders, 2014, p. 16). Generally in other European countries quality assurance is maintained through activities led by members of the social partnership. In the case of France the government, the regions, the academic authorities, the chambers of commerce and the institutions delivering some of the training all take a role in quality oversight alongside the employers and the trades unions (Pigeaud, 2014, p. 11-12).

Some industries in the UK, with the help of the UK Commission for Employment and Skills (UKCES), have established industrial partnerships. These could provide one locus for further industry-based activity, although the conditions are not completely comparable as we do not have the same highly regulated training market found commonly in other parts of Europe where, in many cases, gaining an apprenticeship is required prior to permanent employment.

In most mainland European countries, the employer enters into a contract with the apprentice. There are some positive examples of similar approaches to codification in the new standards, although on a less formal basis. The Butchery apprentices use a log book to document their time engaged in both on and off the job training and the different outcomes that need to be achieved (BIS, 2015a, p. 2). Whilst not advocating a formal requirement, creating a plan for each apprentice provides clarity for all parties and, so long as it involves a discussion and is not just a paper exercise, it can be used to support the apprentice's learning and assessment. An agreed plan is also an important ingredient when working with apprentices who are able to work, have some form of disability and need reasonable adjustments in their workplace and in their off-the-job training.

Evidence from across the early work of the Trailblazers provides some reasons to be optimistic about a number of the changes to the apprenticeship assessment system currently being implemented in the UK. However, with much of the development still going on it is too soon to fully determine how these will be carried out in practice.

To inform future progress it would be valuable to increase our understanding of apprenticeships as the new system evolves, to determine what has worked and where there is still work to do. One of the ingredients that would help is to increase the amount of research focused on this topic.

I have been struck by how important a place research is given in some of the mainland European countries where some of the most successful models of apprenticeships operate, and wonder whether there are any lessons that can be learned from this more central positioning of research, linked closely as it is to employers and social partners. For example, the Federal Institute for Vocational Education and Training (BIBB) in Germany. Their remit includes systematically analysing vocational education and training – including apprenticeships – learning from good practice across the EU, disseminating the results of successful projects and programmes and meeting peers to share what works and what does not.

The independent research community in the UK, including a number of universities and centres in each of the nations, has shown interest in this area. For example, the Centre for Learning and Life Chances in Knowledge Economies and Societies (LLAKES) at University College London and the London School of Economics and Political Science Centre for Vocational Education Research have provided valuable insight into a number of aspects of apprenticeships. Alongside the work of the universities there would be scope for practitioners in the system to undertake action research on assessment and other topics and for longitudinal studies of apprentices and their progression to be carried out in order to help increase our understanding of their experience and inform future development.

This time of change also provides a significant opportunity for innovation in the assessment system, although innovation will mean different things

441

in different sectors. In some cases this could involve making greater use of technology, such as implementing online knowledge testing or utilising online log books to record an apprentice's progress which their supervisor could view at any time. Workplace activity can be filmed when undertaken and used as evidence for formative or summative assessment. Innovation could also come from extending the membership of assessment and moderation panels to include not only additional employers but also individuals from other nations in relevant industries. This would provide opportunities to compare the outcome standards of different countries, share experience of different forms of assessment and exchange proven examples of good practice.

With the shift in focus towards synoptic assessment and a greater emphasis on sector specific knowledge, it is increasingly important for the apprenticeship system that individuals with current industry experience take part in assessment activities. There are already positive references in some of the Trailblazers' standards to providing training for this purpose. In occupations where there are fewer apprentices, particularly in niche and specialist areas, there have historically been a very small number of experts to call on to support assessment activities and it would be helpful to avoid this in the future by passing on assessment expertise to a wider group. Taking on an assessment or moderation role could be encouraged by rating such activity positively in performance management reviews or included as a part of continuous professional development.

In many of the countries where apprenticeships have a more central role, and have been embedded into the structure of the economy over a longer period, the governance of the system is different from our own. Government is one of the participants but employers with their social partners, educationalists and researchers are predominantly relied upon to drive the system forward. These models appear to have greater stability and have the advantage that changes are made with the involvement of all participants and on a more evolutionary basis. We might strengthen our own system by learning from this approach, with government setting out policy parameters for change but then stepping back and enabling the experts in the different fields to take forward developments and to increase their ownership of the process.

There is still work to be done to improve collaboration across the apprenticeship system, building a consensus around industry needs, creating a more seamless training experience for apprentices, and recognising the contribution of all parties. Developing and extending longer-term relationships between employers, providers, assessors, awarding organisations and professional bodies is key. To be effective these need to be both consensual and robust enough to support long lasting, sustainable development, particularly as the approach taken to delivering skills and carrying out assessment will need to continuously adapt as industry requirements evolve.

The creation, where they do not exist already, of local forums involving representatives of all these parties, could be beneficial, paralleling the training committees seen internationally. These groups could also usefully explore the likely future needs of industry and society, facilitating planning for future content and assessment.

Apprenticeship training should be dynamic, making extensive use of appropriate technology, benefitting from a growing understanding of appropriate pedagogy and utilising research-based information to ensure that the skills and attributes taught and assessed are at the forefront of industry needs. There should be some increased emphasis on characteristics that are complex to assess but important in the workplace, such as the qualities outlined in *Remaking Apprenticeships* which include resourcefulness, maintaining a professional attitude, and carrying out a role in a way that any external or internal customer would perceive to be of high quality (Lucas and Spencer, 2015, p. 66-67).

It is also essential that all those within the apprenticeship system are not only unequivocal about the importance of quality but are prepared to take action to support continuous improvement.

Ultimately, if we are to fulfil our ambition for apprenticeships and get the best out of the new system we need clear, effective standards; skilled, well-trained individuals with a mixture of intimate industry knowledge and assessment expertise; and an environment that supports innovation and cooperation in which to operate. The success of this incoming system will be dependent on how well these elements come together in practice.

References

AlphaPlus Consultancy Ltd. (2014) *Validation of vocational qualifications: Final Report.* (www.gov.uk).

BIBB. *Vocational Education and Training – An Overview:* www.bibb.de..

BIS. (2015) *Assessment Plan – Butcher.* (www.gov.uk).

BIS. (2015) *Assessment Plan: Financial Services Customer Adviser Standard.* (www.gov.uk).

BIS. (2015) *Assessment Plan: Golf Greenkeeper.* (www.gov.uk).

BIS. (2015) *Life and Industrial Sciences Trailblazer Apprenticeship Assessment Plan.* (www.gov.uk).

BIS. (2015) *PM unveils plans to boost apprenticeships and transform training.* (www.gov.uk) .

Brooks, B. and Smith, B. (2015) '*The Energy and Efficiency Independent Assessment Service (EEIAS): An Update*', a PowerPoint presentation..

Dibbern, O. and Kruse, K. (2014) *Apprenticeship-type schemes and structured work-based learning programmes: Denmark.* (cumulus.cedefop.europa.eu).

Fuller, A. and Unwin, L. (2011) The content of apprenticeships. In Dolphin, T. & Lanning, T (eds.) *Rethinking Apprenticeship.* London: Institute for Public Policy Research.

Hensen-Reifgens, K. and Hippach-Schneider, U. (2014*) Apprenticeship-type schemes and structured work-based learning programmes: Germany.* (cumulus.cedefop.europa.eu).

Hoeckel, K., Field, S. and Norton Grubb, W. (2009) *Learning for Jobs: OECSD Reviews of Vocational Education and Training: Switzerland.* (www.oecd.org).

Koukko, A. and Kyrö, M. (2014) *Apprenticeship-type schemes and structured work-based learning programmes: Finland.* (cumulus.cedefop.europa.eu).

Lucas, B and Spencer, E. (2015) *Remaking Apprenticeships: Powerful Learning for Work and Life.* (www.winchester.ac.uk).

McNamara, T. (2006). 'Validity in Language Testing: The Challenge of Sam Messick's Legacy.' *Language Assessment Quarterly*, 3(1), 31-51.

OECD (2014) G20-OECD-EC '*Conference on Quality Apprenticeships for Giving Youth a Better Start in the Labour Market*', background paper prepared by the OECD. (www.oecd.org).

OECD and ILO (2011). *'Giving Youth a Better Start'*, a policy note for the G20 Meeting of Labour Ministers. (www.oecd.org).

OfQual, CCEA, CollegesWales/ColegauCymru and SCQF (2013) 'Apprenticeships: Report on a study to visit Germany', a report for the UK European Coordination Group for VET Initiatives. (www.rewardinglearning.org.uk).

Pigeaud, R. (2014) *Apprenticeship-type schemes and structured work-based learning programmes: France.* (cumulus.cedefop.europa.eu).

Richard, D. (2012) *The Richard Review of Apprenticeships.* (www.gov.uk).

SQA. (2014) *Assessment in Apprenticeships: white paper 1.* (www.sqa.org.uk).

Tanggaard, L. and Elmholdt, C. (2008) Assessment in Practice: An inspiration from apprenticeship. *Scandinavian Journal of Educational Research.* 52 (1). p. 7-116.

Westerhuis, A. and Smulders, H. (2014) *Apprenticeship-type schemes and structured work-based learning programmes: The Netherlands.* (cumulus.cedefop.europa.eu).

Wolf, A. (2001) Competence-Based Assessment. In Raven, J & Stephenson, J. (eds.) *Competence in the Learning Society.* New York: Peter Lang.

38.
REDESIGNING AND DELIVERING
'FIT FOR PURPOSE' APPRENTICESHIPS:
A CASE STUDY FROM THE WATER INDUSTRY

Phil White
Technical and Health and Safety Training Manager, United Utilities

Summary
Trailblazer apprenticeships will have replaced traditional frameworks completely by 2017. United Utilities (UU), with a workforce of 5,500, and with an ambition to be one of the best water and wastewater companies in the UK, felt it was important to get ahead of this trajectory. We therefore decided to play an active role in the development of two trailblazer apprenticeships as part of an evolving internal technical training strategy developed in the business since 2012.

This chapter describes how shared ambition, collaboration between water companies, and sheer persistence under pressure can achieve successful standards design extremely rapidly when industry takes the lead.

It describes how we successfully designed the brand new Utilities Engineering Technician and Water Process Technician Standards with corresponding and innovative assessment plans, seven underpinning frameworks, and 146 units in a highly compressed timeframe between November 2014 and August 2015. All of this was completed with the active support of twenty-one subject matter experts deployed from six water companies and signed off by government one week before a pioneering cohort of thirty-five apprentices started their 3- or 4-year Trailblazer apprenticeship journey in September 2015.

By sharing the way in which we approached this process, the context, our rationale, the drivers and the financial and non-financial benefits that can accrue, we hope our experience will inspire other employers to grasp the opportunity presented by trailblazers and to invest in the future competence and quality of apprentices coming into the workplace.

What was the problem we needed to solve?
UU had always recruited apprentices in the past. Many of its directors, senior and middle managers were apprentices themselves and continued to be advocates. However, between 2005 and 2010, as the business evolved, organisational changes had the unintended consequence of fragmentation.

The decision to recruit apprentices became a function of different individuals across distinct business units. There was no centralised approach to apprenticeship recruitment and, as a result, only twenty apprentices joined the company. Moreover, the approach to recruitment was inconsistent and the training of those apprentices was outsourced.

It was felt by the business that the programme had lost its way, so in 2010 the Apprenticeship Programme was formally re-established. This was at the request of the Managing Director whose sponsorship drove change at a strategic level. Activity was overseen by a dedicated Apprentice Steering Group, made up of eight senior leaders in the business, and implemented by an Apprentice Working Group.

This renaissance led to a seven-fold increase in apprentices being recruited to the business between 2011 and 2015 (133 apprentices) compared to the previous five-year period. Moreover, the type of apprentices being recruited had diversified from engineering, process and network roles to other support areas such as human resources, information technology and finance. This had never been tried before at UU.

While this new approach was emerging and some success was being achieved, a number of different problems were identified particularly around 2012 when, as newly appointed Technical Health and Safety Training Manager, I was asked to review the business's entire approach to technical training, including apprenticeships.

I spent three months across the business and spoke with our supply chain and quickly discovered a range of concerns about the quality of outcomes

447

from our training which at that time was still outsourced. Completion rates were good. Certificates were issued. Apprenticeship retention rates averaged 88 per cent and most people would be forgiven for not looking past those kinds of seemingly successful indicators. However, it was clear to us that there were a number of areas we needed to improve. Our core capabilities needed strengthening and our review found that:

- Some providers were not as responsive to our skills development needs as we needed them to be, and we were not being demanding enough nor articulating our expectations clearly

- Often, the college provision was regarded as too generic

- Managers and mentors in the business were reporting that engineering apprentices completing their first year joined the business and effectively had to be re-trained because what they were learning did not match the day-to-day realities of their role. The learning outcomes were not sufficiently focused and so they were not business ready. This put additional pressure on the mentors and meant duplication of effort

- There was not enough emphasis on true skills development

- UU was being driven by the providers, not the business need, and lacked the in-house technical learning and development expertise to change that dynamic

- The funding regime for apprenticeships appeared to be driving the qualifications being delivered and taught, rather than the skills requirement driving the funding

- The assessment approach did not encourage the testing of genuine competence

Serendipitously, the Richard Review came at the right time for UU. It confirmed our thinking, namely that apprenticeships were not fit for our industry (they were out-dated) and that the emphasis had been on completion rather than quality. Having looked at all of our technical and health and safety training arrangements across topics such as lifting and slinging, manual handling, electrical testing and disinfection training, we knew we needed a better model that would raise the bar in terms of outcomes for all our training. It felt like the government, through trailblazers, was passing over the baton to industry and giving us the power to change the system. We had identified the problems; now we needed to develop the solutions.

What solutions did we develop?

We developed a business case for an internal technical training strategy as we had concluded that United Utilities needed its own dedicated technical training team, and somewhere to deliver that training.

With significant MD sponsorship, the strategic thinking and plenty of determination to succeed in a fluctuating policy environment, we used the internal structures we had in place to sign off our proposed approach and did something unusual in the water and wastewater sector by curtailing our outsourcing model.

We built a full time in-house team of technical trainers and invested in a dedicated technical training centre in Bolton. We had sixty employees from within United Utilities apply for the seven technical trainer roles, and each of the successful applicants, mostly recruited in September 2013, had significant operational experience. They were then trained up and qualified to be technical trainers, and in the space of twelve months had gone from turning spanners to helping shape our future workforce.

This now means all our technical apprentices spend the first year being trained by our own staff at our own centre and are then returned to the business in year two with much more positive feedback about their job competence and ability to hit the ground running.

As well as developing our internal arrangements to overcome the problems identified with our own technical training at large, we pursued a proactive strategy of working with other water companies to specifically develop two trailblazer apprenticeships.

The Energy and Efficiency Industrial Partnership in the utilities sector was instrumental in initially gathering support within our sector and then forming an action-oriented group (the Water Development Group) to start our trailblazer journey. We collaborated with Northumbrian Water, South West Water, Yorkshire Water, Thames Water and Southern Water to shape the new Apprenticeship Standards for the roles of Utilities Engineering Technician or Water Process Technician.

We were convinced that by taking over the design we would make tangible differences not only to what would be taught to apprentices of the future, but to how they would be assessed. Ultimately therefore we would be effecting a step-change in apprentice competence levels to the benefit of the water sector.

This approach was right for United Utilities because:

- We needed business-ready apprentices and a future-proof apprenticeship strategy
- Our prime driver was increasing the quality and core capability of apprentices, not cost savings, though we knew efficiencies would ultimately manifest themselves in productivity rates
- It was a great opportunity to shape more rigorous, relevant, specific, contemporary and industry focused standards for our sector
- It complemented the wider evolution of our internal technical training delivery model
- It gave us a voice in the landscape of sector qualifications development and encouraged more openness to collaborate with other like-minded companies in our sector to the benefit of all our future talent and attraction strategies
- We knew we could make positive change and had the right people with the right mind-set with the right support in place to make the difference

Despite encountering pressures on time, costs and the support that would be needed, not only from people in our own business but subject matter experts from the other water companies, we knew the value of what would be achieved far outweighed these pressures. Our view was clear that pursuing this route would lead to improved competence of apprentices in our sector and the reduction or complete elimination of repeat training to cover items that had previously been missed.

How did we develop the trailblazer apprenticeships?
Through a series of technical working groups convened across different subject matter experts (typically three from the six main water companies involved) for each of the seven frameworks across the two trailblazer apprenticeships, content was redefined; innovative assessment plans with an increased emphasis on end-assessment processes were devised; and flexibility was built in to the approach to be relevant to the different contexts in which the apprentices would find themselves working in future. In the old frameworks we found references to assets that were no longer used in the industry, but taught because they would feature in the assessment. Any outdated content was immediately removed.

As a process, the collaboration by subject matter experts was rapid - the 146 units were written in only two months. Every aspect of the previous curriculum for the apprenticeships in focus was reviewed and enhanced. This is an important point. While the task in hand was to develop trailblazer apprenticeships, our approach was actually curriculum development from a technical perspective rather than just standards development per se.

The process has therefore led to a number of tangible improvements and differentiators for trailblazers compared to the predecessor apprenticeship frameworks:

- There is a greater emphasis on relevant technical knowledge that equips the apprentices to make confident decisions in real situations e.g., when things are going wrong in a treatment works at two in the morning!
- Every bit of content in each unit has been developed with three aspects in mind: why we should do things in a certain way; how we do things; and the consequences of not doing things correctly. When we reviewed the approach of our outsourced providers back

451

in 2012 we found a lot of emphasis on the 'why' but not enough emphasis, content-wise, on the 'how', or on the consequences

- The end point assessment is much more innovative. Previously the approach was just a portfolio, ticking things off as the learning was delivered. The problem with that is an apprentice could complete some learning in year 2, but by year 4 what they learned may have been forgotten or not be current. The end point assessment for the trailblazers is much more rigorous – including trade tests, scenario tests and activities that take individuals out of their comfort zone. The assessment forces the apprentice to drag all the knowledge from their entire journey into one place

- We've added more performance criteria than ever before, but importantly we have made them specific to roles, and captured the contemporary landscape of the water industry in 2015

- The trailblazers are more specific to the water industry e.g., the content is specific about the types of pumps and gearboxes that an apprentice will actually need to be able to understand and use in their job not just any generic pump or gearbox

- Flexibility is achieved in the frameworks through the completion of mandatory units complemented by optional units that are appropriate to the processes and assets that the apprentice covers. One such example for water treatment is a mandatory filtration unit (however there are four filtration units available to choose from to reflect the diversity of contexts and assets used in different water companies)

- Following this intense, but rigorous technical process and negotiations with BIS about funding caps for the trailblazers, the new standards were finally signed off in August 2015. We had very valuable help and support along the way from EU Skills and the Energy and Efficiency Industrial Partnership. This was vital in helping manage relations with BIS and overcoming potentially bureaucratic burdens and challenges that would have distracted technical experts from the already rapid process required.

The current situation and looking to the future

Thirty-five trailblazer apprentices started in September 2015 across United Utilities (25) and Northumbrian Water (10), pioneers who will complete their apprenticeships in 2018 and 2019. Within UU, much has been done to achieve buy-in from senior leaders for the internal training strategy, technical training model and benefits of recruiting apprentices. The momentum and trust has been created.

The people in the business who need to know about trailblazers - and why we have moved away from the Specification of Apprenticeship Standards Frameworks they previously knew and understood - are now advocates of the fresh arrangements. They are convinced it will lead to more competent, business ready apprentices year on year. At optimum, we believe we have the capability to train over 140 apprentices per year at our training centre in the future.

Summary points
- Employers in the water sector have now demonstrated that they can manage the apprentice journey from standards development and curriculum design to funding arrangements and delivery. This is a huge shift from the previous model where out-dated standards were provided by awarding bodies and changes to these standards were often difficult to implement
- Employers have been responsible for creating new standards but more importantly are working collaboratively to ensure that these remain current and can come together quickly to change them if not through the new awarding body that we have created in our sector[139]
- These new standards are fit for purpose and have been created by working groups of subject matter experts for the water industry throughout the sector. Collaboration on the skills agenda has been the key to successful standards development and this openness in approach for trailblazers has extended into other key initiatives such as the Licenced Operator curriculum review.

Key messages to employers
- Be willing to challenge the status quo in your business and the sector. You cannot afford to sit on the sidelines and say apprenticeships are not fit for purpose and not do anything about

---

[139] The Energy Efficiency Independent Assessment Service.

it. Get involved. Work with others in your sector. Collaborate with clear purpose and intention. Take that leap of faith. Put time aside, put energy and resources in to changing the landscape around us

- If quality and increasing core workforce capability is your primary driver, and if you are becoming increasingly sophisticated with your workforce planning forecasting, taking control of this part of your talent or attraction strategy makes perfect business sense

- Focus on curriculum enhancement, innovate where it is needed with your delivery model and think about the profound effect you can make for others over the next couple of decades by acting now

- Through investing in the apprenticeship levy, seek to maximise your return using standards that are fit for purpose

Key messages to government
- For industry to be responsive, government needs to do likewise. Timeframes for trailblazer development were very tight, negotiation on funding caps were intense, sign off was only just in time for real apprentices to start their journey.

- In terms of the target for three million more apprentices by 2020, our view is that apprentices are vital and their value needs to be further appreciated by industry and society at large. Having an ambition is helpful, but having a number – three million – raises concerns. Industry must shape the requirement, the need, the volume required and at all times quality must usurp quantity. We would never want to see the quality of apprenticeships watered down to achieve a numeric target.

Key messages to future apprentices and parents
- A common misconception regarding apprenticeships is that they are purely a path towards a recognised technical trade. Apprenticeships however are much more wide-ranging than this,

454

available across many sectors in industry and therefore they are a viable alternative to further education routes

- The academic achievements of apprenticeships are often not particularly well publicised. However it is not uncommon for apprenticeships to offer routes to qualifications that match those achieved through university education

- Apprenticeships offer a real opportunity to develop a career, earning while you learn and gaining vital work skills in the process. An added bonus of apprenticeships is the lack of student debt, as qualifications are gained through the sponsorship of an employer

39.
THE IMPORTANT ROLE OF GROUP TRAINING ASSOCIATIONS IN GROWING STEM APPRENTICESHIPS: CASE STUDIES

Craig Crowther
Network Development Manager, GTA England

Summary
This chapter highlights the background and value proposition of Group Training Associations (GTAs) that have been delivering apprenticeships for over fifty years. It examines how the GTA model is ideally placed to help deliver an increasing number of employer-led apprenticeships to the standards demanded by employers and deserved by learners. The employer case studies highlight how existing and new GTAs across the country are delivering a greater number of high quality apprenticeships in STEM employment areas as the 'training partner of choice' for employers, and in particular for SMEs.

Background
Since the 1960s, Group Training Associations (GTAs) have provided solutions to the workforce development needs of employers large and small around the country, delivering high quality technical training and achieving success rates consistently above the national average. They have created real opportunities for many thousands of young people and adults to pursue worthwhile careers across a range of sectors vitally important to the UK's economic growth.

In 2009, a report was commissioned to look at the Group Training Association model and its potential to expand. The report made a number of key observations as highlighted in the extract below:

'In summary, GTAs are valued by the government and its agencies for the high quality of their work and their solidity. Rooted in engineering apprenticeships, many have been established for a half-century and they consistently deliver successful completion rates among trainees which are among the very best in the sector. However, their national strategic role – and above all their strategic potential – is under-appreciated, and their diversification into other sectors is largely unacknowledged and capable of further development/expansion. Extending the GTA model in the light of experience in, for example, Australia is seen as an important opportunity, particularly in the light of the current economic downturn.

As well-established charitable entities providing an employer-led public service, the GTAs closely match the government's desired focus for investment. They are one of the few wholly credible models for employer-led training which meet all the public interest tests. Whilst they are not uniform, they are generally entrepreneurial and enterprising in character and business-like in structure and staffing. They are at the heart of the British economy.' (AELP and Beyond Standards, 2009)

One of the recommendations from the report was the establishment of a membership body for GTAs and this led to the creation of GTA England in 2009.

The original findings of the 2009 report were further reinforced in 2012, when a 'Commission of Inquiry into the Role of Group Training Associations' was established, and the Commission report was authored by Professor Lorna Unwin of the Institute of Education. It concluded that:

'GTAs should be central to the Government's plans for economic growth, rebalancing the economy, increasing the stocks of technician and higher-level skills, and the expansion and improvement of apprenticeships.' (Unwin, 2012)

Despite a change of government and the significant policy implications of this for skills, and the harshest economic climate for over fifty years, much has been achieved by GTA England and member GTAs since its establishment in 2009. Member organisations have continued to demonstrate their resilience and flexibility to respond to the needs of employers and the changing skills landscape. GTA England has taken significant steps in working with government and in 2013 succeeded in opening up access to capital funding for GTAs. Despite all of these efforts, it is recognised that there is still much to do to ensure GTAs play a full and recognised role in addressing the very real skills challenges facing the UK economy.

The distinctiveness and success of the GTA model means that they are ideally placed to help deliver the ambitions outlined in *'Fixing the Foundations: Creating a More Prosperous Future'* (BIS, 2015) As organisations that were born out of the previous levy system in direct response to the needs of employers, the network is willing and ready to help BIS and the government achieve the ambitious targets they have set.

In *'Fixing the Foundations; Creating a more prosperous future',* the Chancellor and Business Secretary articulated their desire for apprenticeships to be 'well-funded, high quality, and meet employers' real needs' - such apprenticeships are already being delivered across the GTA network, and GTA England's members are keen and committed to playing their part in helping deliver the three million high quality apprenticeships over the next five years.

GTA England members currently engage with over 24,000 employers collectively, over 90 per cent of which are SMEs, and are training over 12,000 apprentices employed with these companies. STEM apprenticeship starts across the network have increased from the baseline of 3,900 in 2013/14 to 5,206 in 2014/15 – an increase of over 1,300 (34 per cent) and current reporting from members projects a further increase for 2015/16 STEM apprenticeship starts to 5,798 – a further 11 per cent increase in the year and a total increase of just short of 1,900 (49 per cent).

The GTA value proposition

Group Training Associations deliver the robust governance and public good benefits of a college, with the best of employer responsiveness associated with private training providers. The GTA value proposition is built on the following principles:

- GTAs provide good or outstanding learning in an environment that reflects the modern workplace in terms of standards, behaviour and equipment
- GTAs provide the right balance between hands-on experience and high quality teaching that benefits the learner and their employer
- GTAs individually and collectively are committed to achieving learner outcomes significantly above the national average and share best practice to help achieve this objective

GTA England members deliver the high quality provision demanded by employers, and have consistently out-performed other FE and skills providers. All current GTA England members were graded as 'Good or Outstanding' by Ofsted at their most recent inspection. The recognition of the work done by GTA England to improve quality was recently publicly recognised by Ofsted in the 'Future of Apprenticeships' Report.

'Group Training Associations (GTAs) provide a very successful model for collaboration between employers and providers. This has become a model of industry/provider partnership resilient to policy changes and has responded very effectively to the training demands of industry. Training companies that are members of GTA England generally provide high-quality training.' (Ofsted, 2015)

Working with STEM employers
There are some excellent examples of innovative practice taking place across the GTA network that have led to the significant increases in STEM apprenticeships being delivered, bucking the national trend. In the remainder of this chapter we will examine some of these characteristics and approaches that have allowed GTAs to continue to grow apprenticeship provision and some specific examples of these characteristics in action with employers.

GTAs were created through a levy system to address a market failing identified by employers. They identified a lack of quality provision providing hands-on practical learning alongside core knowledge right across the country. A network of 150 GTAs was created to address this market failing that was both geographical and pedagogical. The way in which GTAs were created has had a significant impact on how they have subsequently developed and led to their almost unique relationship with employers.

GTA staff came from industry and worked with industry, conducting genuine training needs analysis (TNA) at all levels across the workforce. They then worked with the employer to devise programmes that addressed the needs identified during the TNA or in many cases Organisational Needs Analysis (ONA). This approach remains today and continues to ensure that programmes are fit-for-purpose and relevant to the needs of employers and their employees, more often referred to elsewhere in the sector as learners.

This highlights a further differentiator in the way that GTAs approach apprenticeships – in the vast majority of cases, learners on apprenticeship programmes are first and foremost an employee of a company, and their learning, while giving them recognised qualifications, is also delivered in a way that is in line with that employer's identified skills needs.

GTAs offer support not just to SMEs but also to many large companies who are outsourcing training and seeking partnership arrangements with apprenticeship experts.

One of the largest GTAs in the network, Training 2000, has been established for some fifty years and provides high quality training from its seven UK training centres. As well as its work with SMEs across Lancashire, it has a number of innovative and flexible arrangements to support apprenticeship delivery in large employers.

For example, it has a significant partnership with a major aerospace company whereby the majority of the apprenticeship delivery staff operating in their Apprenticeship Academy in Derby are employed by Training 2000. These staff are badged in company work-wear and embrace the culture, ethos and standards of the host while benefitting from the extensive support and resources of Training 2000. This partnership approach means that learners feel an integral part of a leading international employer from day one while also receiving high quality training and assessment in their own workplace.

West Anglia Training (WATA) is an excellent example of a GTA that is working closely to support employers and the regional growth agenda. WATA was established in 1976 as a charitable organisation and has operated as a Group Training Association for nearly forty years. Since 2013, they have embarked on an ambitious growth plan to ensure that more learners can benefit from the outstanding learning opportunities provided in direct response to skills gaps and shortages.

One aspect of this growth plan involved the design and building of a new skills centre for scaffolding and engineering in direct response to increased demand from employers for these areas of training. Building work commenced on the new centre in February 2014 and the first apprentices started their programmes only three months later.

The speed of response, characteristic of GTAs, is further demonstrated with the more recently created Highways Academy in direct response to an identified shortage of skilled workers to construct major infrastructure projects in the region. Working collaboratively with the local LEP, Highways England and the major contractors, WATA identified the skills gaps that needed addressing.

They then secured funding and sponsorship for the development of a new major external highway learning environment, and developed the curriculum for new programmes to meet the new trailblazer standard. This is in direct response to the identification of a need to train thousands of skilled highways operatives and engineers needed to complete the road infrastructure projects being undertaken that are so essential to the region's future prosperity.

Gen2 in Cumbria currently train over 1,250 apprentices. They have also experienced significant growth in their apprenticeship numbers over the last two years with 2014/15 seeing an intake of over 500 apprentices, 400 of which were in STEM subjects and 22 per cent of those were females – way in excess of the national average.

Their numbers of employed apprentices in learning peaked at 1273 during 2014/15, and with the recent opening of the Energy Coast UTC, Gen2 continue to excel at meeting the needs of the nuclear and associated industries.

An important element of the Gen2 offer has been the development of its higher education provision, and they now have over 300 part-time learners on programmes ranging from HNCs to BEng (Hons), many of whom have progressed through their advanced apprenticeship route. Through innovative programme development and partnerships, Gen2 are able to offer apprentices progression on one site through to degree level while being employed – an offer proving increasingly attractive to a growing number of learners as demonstrated in the recruitment numbers above.

Rochdale Training is another GTA that has experienced exponential growth in apprenticeship numbers over the last three years through enhanced employer engagement. In 2012/13 they had 385 apprenticeship starts. In 2013/14 this increased to 508 and in 2014/15 this reached 720. This is an 87 per cent increase on 2012/13. They have also grown their 16-18 Apprenticeships by 77 per cent during the same time in key priority advanced manufacturing areas, significantly bucking the national trend.

Alongside significant growth in the numbers of apprentices being trained in existing GTAs, GTA England has seen a recent increase in the number of employers looking to create new Group Training Associations to address their skills needs through increased support for and investment in apprenticeships.

One such example is amongst employers working on the River Thames. GTA England are in discussions with representatives from the Port of London Authority and Thames Tideway Tunnel Project to facilitate the creation of a new GTA, the Thames Training Alliance, to address the urgent training requirement for a significant increase in the number of engineers, boatmasters and mariners skilled and qualified to operate on this river.

There will not be enough suitably qualified employees to maintain current levels of safe passenger and freight operations, never mind meeting the future collective aspirations to see passenger numbers grow in line with the Mayor's River Action Plan or indeed the demands of the Thames Tideway Tunnel (TTT) and other infrastructure projects.

The employers recognise the value proposition of the GTA model and are aiming to create a new GTA early in 2016 to address the issues outlined by recruiting a minimum of 30 to 40 apprentices per year for the next ten years. It is such initiatives, led by employers, with 90 per cent SMEs, that will be vital if the government is to achieve its ambition of three million high quality apprenticeships by 2020.

Summary

GTAs have played an important role in delivering high quality apprenticeships in key sectors for over fifty years, even when such activity did not have a high political profile. Their collective learning over fifty years, and the inherent qualities of the GTA model, mean that GTAs remain as significant and important today as they have been at any point in their history.

The Commission of Inquiry in 2012 was clear in concluding that 'GTAs should be central to the Government's plans for economic growth... increasing the stocks of technician and higher-level skills, and the expansion and improvement of apprenticeships.'

Ofsted were clear in their October 2015 apprenticeships report that 'Group Training Associations (GTAs) provide a very successful model for collaboration between employers and providers' and that the GTA approach 'has become a model of industry/provider partnership resilient to policy changes and has responded very effectively to the training demands of industry.'

There is wide agreement that GTAs have a vital role to play in continuing to help employers achieve their growth and productivity ambitions and government their apprenticeship growth ambitions, now and for many years to come. So let's use them.

References:
AELP and Beyond Standards. (2009) *Investigating the GTA Model.* Bristol
BIS. (2015) *Fixing the Foundations; Creating a more prosperous future.*
    Department for Business, Innovation and Skills. London.
Ofsted. (2015) *Apprenticeships: developing skills for future prosperity.*

Unwin, L. (2012) *Report of the Commission of Inquiry into the Role of Group Training Associations.* London.

CONCLUDING MESSAGES AND PRIORITIES

David Way CBE

———

This final chapter reflects my own views, having considered the thoughts and experiences of all who have contributed to the book.

While I set out five themes that I believed were important in considering apprenticeships, I did not seek to constrain the views of experts about the state of apprenticeships today and what more needs doing. It is therefore especially interesting to see the common messages that have come through.

None of the following should be taken as criticism of those involved in apprenticeships past and present. Apprenticeships are in a much better place today than ten years ago and everyone I have met and who is influential on apprenticeships wants the best for their future.

This book's conclusions are intended to be an honest but fair and constructive account of what is important if we are to produce apprenticeships of which everyone is justly proud.

KEY MESSAGES:
We start in a good place...
There is much appreciation of the priority that has been given by three successive governments to revive apprenticeships and to give them much greater attention and profile than for many decades.

The ambition to achieve three million more apprenticeships has created a positive and collaborative culture.

The active engagement of key employer intermediary bodies and trades unions is unprecedented, has helped get the attention of more employers and created the right conditions for growth.

....and the general direction of travel looks good.

The ambition of three million more apprenticeships and the government's readiness to be held account for its achievement is a clear signal of the enduring importance of apprenticeships. This is very important to business who value stability and the ability to see clear longer-term planning horizons if they are to effect change through partnership working with government.

The continuing rhetoric and action concerning employers being in the driving seat is similarly welcome. The expectation is that this will carry through into setting more standards for apprenticeships and in influencing key issues such as quality and the apprenticeship levy.

Support for current reforms if generally strong, including for trailblazers and the expansion of traineeships.

Expectations for the Institute for Apprenticeships and for the Careers and Enterprise Company are high and getting higher.

There are potential opportunities for further growth that look very promising

The government's ambition to ensure apprenticeships grow across the public sector is important not only in opening up huge areas of employment to those entering via a vocational route but as a clear demonstration of its faith in its own policies. While current ambitions are well below the 5 per cent benchmark adopted by many other large employers of apprentices, this will be a major change programme for most. Many health authorities and local authorities have set an excellent lead for others to follow.

Large infrastructure projects such as Crossrail have shown what can be achieved when there is a determination to include apprenticeships as a core part of the skills strategy for the work. With so many large projects now in hand, the clear ambition to repeat this approach across the projects in the National Infrastructure Plan is very welcome.

A clear lead on how to use procurement practices to maximise apprenticeship opportunities will be very welcome also.

466

Other successful reforms need scaling up...especially higher apprenticeships if the UK is to secure the higher technical skills needed by the economy

Considerable progress has been made in introducing higher apprenticeships and, more recently, degree apprenticeships. The UK has a good understanding of how to design and implement these important programmes and there is both growing demand from employers and keen interest from many in higher education.

Completing the pathways in many more occupations is vital if young people and their families are to be convinced that taking the vocational route rather than going to university at eighteen leaves open the HE route at a later stage of their working lives.

This process of creating the top tier of higher-level apprenticeships and making it easier for people to find their way up the vocational ladder needs to be prioritised if the much-needed link between apprenticeships and increased productivity is to be exploited.

This will also help draw many more higher education institutions into providing apprenticeships and dilute the sense that talented young people choose university purely to pursue academic subjects. The higher apprenticeship or degree apprenticeship option, with a job attached, has the power to transform the attitude of young people, parents and advisers to the vocational pathway.

This should also create the right conditions for the reforms being considered that would link apprenticeship applications to the UCAS-style clearing-house arrangements.

Higher-education institutions should also be drawn into all aspects of apprenticeship governance.

Some long-standing critical issues still need addressing to transform apprenticeships.....especially careers information, advice and guidance
There are so many advantages that flow from an effective service for young people offering high quality career information, advice and

guidance both for young people and for the economy. Too many young people are ruling out career choices too early on the basis of inadequate knowledge. There is insufficient connectivity between the careers that young people are choosing and the high quality jobs of the future.

There is much to do to convey the exciting potential of employment opportunities facing young people in this high tech age and where apprenticeships are the best way to develop the range of skills and attitudes that are needed. The UK requires more informed and passionate advisers who are strong advocates of apprenticeships and can inspire young people.

We all know this but the problem needs fixing. Employers are keen to help. This message comes through loud and clear from contributors and needs to be addressed as a matter of urgency. This is not new and the government understands the issues. The message here is primarily one of urgency and prioritisation of solving a problem that has long been critical for employers, young people and apprenticeships.

The new Careers and Enterprise Company is welcome, especially if they can give shape and coherence to the complex and disparate arrangements currently in place in England.

Employers and others are seeking a strategy that will transform the attitudes of all young people, their advisers and parents towards apprenticeships. Many employers are already contributing actively to this through their own commitment of time, but the overall experience is of a disparate and less effective approach than apprenticeships need if there is to be real traction on the many schools that recognise they need help to fulfil their responsibilities to all of their pupils.

The UK has a critical need for tech skills. Ill-informed advice that means capable young people, and especially women, do not take up the opportunities that exist will not only adversely affect many future career prospects but deny the economy many of the critical skills needed to meet the productivity ambitions for the UK.

Improved information is also vital if we are to maximise the prospects for greater social mobility and prevent young women in particular from making career choices that reinforce previous over-representation in low-paid sectors.

There is an overwhelming call for quality......
While there is very strong support for the government's ambition for three million more apprenticeships, this should not be at the expense of quality.

The belief expressed repeatedly by contributors to the book is that if the UK is to achieve true transformation in the way in which apprenticeships are regarded by employers, individuals and society at large then it is quality even more than quantity that will drive this greater prize.

If apprenticeships are seen to be of such obvious quality that many more high achievers start to choose them then that will begin to have the effect of all young people considering the full range of options available to them rather than simply staying on the academic conveyor belt.

…..and for great teaching.
Employers and those within the skills system wish to see a systematic approach to quality that begins with designing quality into apprenticeships, strengthened teaching and improved assessment. Of these, contributors see least attention being given to improving teaching and training. A serious programme of support to those who teach and train apprentices is called for. Great apprenticeships need great teaching. We know a lot about teaching and learning styles that best suit apprenticeships. We need to be sure that this knowledge is deployed to best effect.

The leadership role of the Education and Training Foundation will be critical to address this, and government should prioritise this for action as part of achieving three million quality apprenticeships.

The UK systems for apprenticeships need simplification, stability, clear responsibilities and good management.

This is another message that has been given before by others but remains of the greatest importance. The changes made to the skills organisational structure and programmes are excessive. It is hard to find any public sector part of the skills system that was in the same form even ten years ago. It makes engagement by employers off-putting. Driving the system everyone says they want is made so much harder for employers if they cannot readily understand how it works.

Government should take advantage of the strong cross-party consensus on skills to discuss and agree a clear and sustainable skills landscape. Ideally this would extend to clear and sustained ways in which the four nations of the UK, with their devolved responsibilities, work together.

If the UK is to agree a skills landscape that would reach beyond the next election, then it makes sense to ensure that the full range of partners and politicians are involved in these discussions and decisions.

Stability in structures would help employers feel confident that relationships they establish have a reasonable prospect of enduring. It should also enable accountabilities to be much clearer. Clarity about the role of government and those of employers and other skills bodies underpins the most successful international apprenticeship systems.

There is a strong belief that a simple route to an apprenticeship is the key to many more employers 'buying' an apprenticeship. As Jason Holt makes clear, the benchmark for investing or buying a new service is set by online services such as Amazon.

A number of businesses have made the point that simplicity does not mean a 'one size fits all' approach. The UK benefits from the choice of routes to apprenticeships that include university technical colleges and Group Training Associations. If we are to retain the strength of this diversity, we need to invest more heavily in good and clear signposting to these services. Again, stability helps acquaint employers with the choice available.

The creation of institutions that are properly resourced and have good governance will give better prospect of future change being managed more effectively. They would not be judged by how many new initiatives they had launched but by the progress that was being achieved, by outcomes, and the initiatives brought to fruition.

Settling on a lasting infrastructure should include a clear commitment to Industrial Partnerships and Sector Skills Councils. Sectors do matter. There is a keen appetite in many employers to collaborate on apprenticeships but many need an accessible vehicle to channel their efforts. Industrial Partnerships have shown signs of giving greater leadership than many employers have experienced before and need time to mature. Their time horizons for turning round the supply of skills within a sector extends beyond a term of government and they too would benefit from cross-party endorsement and support.

Overall management of the skills system should include maximising the deployment of new approaches that work e.g., degree apprenticeships and traineeships; as well as ensuring key research is undertaken and promulgated; that lessons from all parts of the UK, including all four countries and LEPs or City Regions, are systematically considered for potential wider dissemination and application in the UK; and that there are similar exchanges of best practice with other apprenticeship nations.

The work of LEPs, mayors and city regions in particular offer huge potential from fresh creative approaches and local coordination. However they also risk not learning from other approaches and repeating mistakes that could be avoided. Best practice and lessons from pilots need to be shared.

One further issue affects companies that operate across the countries of the UK. How does the UK both benefit from the devolved arrangements and ensure that businesses do not face excessive bureaucracy or complexity? Where will this be managed so that the benefits of devolved arrangements are retained but the interests of cross-UK employers and the needs of a mobile workforce are appreciated and reflected in decision taking?

Employer ownership has turned into leadership......
The book has illustrated the impressive role that many employers are taking to lead the development and expansion of apprenticeships. It has sampled individual and collaborative employer leadership of

471

apprenticeships in action. Employer organisations have also taken up the cause of apprenticeships to good effect. Trailblazers have provided a great vehicle for employers to come together and set new standards. It is vital that this leadership momentum continues.

I have already made reference to the leadership role being provided by Industrial Partnerships and by increasing numbers of LEPs and city regions. Industrial Partnerships have brought CEOs together and made unprecedented progress in tackling deep-rooted skills issues.

The Apprenticeship Ambassador Network is also a real strength of current employer leadership and should be developed further. The development and integration of more Local Apprenticeship Ambassador Networks is welcomed. They have an important role in not only harnessing the efforts of LEPs to apprenticeships but also in bringing knowledge of what already works elsewhere to these discussions.

...but they need ready access to best practice
Employers value ready access to best practice and to networks of businesses who themselves are seeking to maximise the benefits of apprenticeships for themselves and their apprentices. Increasingly companies are looking for international best practice and there appears to be an absence of a ready accessible host for this intelligence and interpretation.

We are fortunate in the UK to have such a strong group of researchers into workplace learning. Employers would benefit from easier access to research findings about successful apprenticeships and to learning about introducing new initiatives.

Within the apprenticeship reforms, it should be made easier for employers to obtain advice from their peers and experts in the field on best apprenticeship practice.

SMEs need financial help but they need so much more.
While employers appreciate the additional financial help available to SMEs, this in itself is not sufficient. They need extensive practical support to find their way through the initial complexity of apprenticeships to a

point where they feel confident in taking on the commitment and personal responsibility associated with employing an apprenticeship. They need help and this is a problem that still needs fixing.

Jason Holt's excellent report on the additional help needed by SMEs should be updated under Jason's guidance. His chapter in the book makes it clear that he has great confidence in current reforms but that they need to 'keep shining a light' on what will work for SMEs. Simple navigation tools are a continuing critical aid for SMEs.

There are many recent developments that build on the initial report's recommendations. These include supply chain initiatives, Talent Bank in the energy sector, and the clearing-house ideas developed by a range of companies to circulate apprenticeship applicants around those SMEs in the sector.

There are a number of ideas within the book that emphasize the importance of keeping arrangements as simple and local as possible. One of the most compelling proposes greater collaboration at a local level between all training providers and large employers to make it easier for employers to find a 'go to' place for help. These are approaches that need testing out and developing with large companies through the Industrial Partnerships, city regions or LEPs.

SMEs need tailored messages about how to access the skills system and messages to which they can relate, including return on investment calculations that reflect the reality of running an SME rather than a large business.

The potential of the apprenticeship levy as a game changer for skills needs employer ownership.
The introduction of the levy is the biggest change in apprenticeships for many years and has the potential to ramp up the private sector investment in apprenticeships. Much depends on ensuring employers buy into the new arrangements and make changes in training that will boost skills across the country rather than write it off as a further tax.

It will also be vital to ensure that the new levy arrangements match the rhetoric of being truly employer led. This feels like an opportunity for simplicity rather than a series of eligibility rules. This is employers' own money being reinvested and needs to work for their benefit through skills.

For example, if employers pay their levy and commit to growing apprenticeships then there needn't be restrictions on prior qualification levels. So many employers report the importance of second chance learning. These are often people who find their best direction in life after a false start. They should be helped rather than penalised and if an employer believes they are helping with their skills needs and want to spend their levy money on their training through an apprenticeship then why would the state intervene?

Similarly, if traineeships are being used successfully to prepare young people for apprenticeships then why can't employers use the levy funds for this purpose? Government needs to not just listen to employers but let them take the lead in maximising the return to skills on reinvesting their levy funds.

If the levy is used to breathe fresh life into some of the successes emerging from the Employer Ownership pilots, such as the upskilling of supply chains, then this would help instil the attitude that this is indeed intended as a boost to skills and not a tax. There are valuable messages from the CITB's experience about how to achieve this.

The management of this change by government is critical to the success of the levy which needs quickly to slip into the background as a process so that we can all concentrate on the positive effects of increasing skills investment. As Toby Peyton-Jones said earlier in the book, the levy funds apprenticeships, it doesn't ensure they are delivered or are of high quality. Those other debates must continue and not be drowned out by noise over the levy. This is felt especially acutely by SMEs who are mostly unaffected by the levy.

The Institute for Apprenticeships will have a key role and must be given the authority it needs.

The idea of The Institute for Apprenticeships has excited many people who see huge potential for it. Like many good ideas though, people seem to be putting more and more on the Institution's shoulders and it is not yet clear if they all accord with the government's plans.

The attraction of the Institute is based on the belief that it will have 'end to end' responsibility for the apprenticeship system and be able to ensure that quality is maintained throughout. It will be independent of government and involve the full range of partners necessary to deliver the transformational change in apprenticeships that everyone wants to see. The best model for this is the Low Pay Commission, though the Bank of England Monetary Commission has been used to illustrate the supportive but independent nature of the Institute.

If the Bank of England Monetary Committee is a good parallel, then the equivalent to its inflation target responsibility should arguably be the closing of critical skills gaps and shortages.

My personal experience of running an organisation, the NAS, with so-called 'end to end' responsibility, was that it was a great concept and gave you a mandate to look into everything and to recommend change. However, when it came to actions, there were usually other bodies that had the legal responsibility for decision-taking. It will therefore be interesting to see how this well-intentioned and necessary ambition for the Institute is followed through.

While some of the Institute's responsibilities have already been determined and welcomed, for example for Apprenticeship Standards; clarity and reassurances are awaited at this time on other important issues. These include research and sharing best practice from the UK and abroad; ensuring the capacity of the sector to handle change; driving a simpler, less bureaucratic approach for employers; collective action to address more acute skills shortages; and its interface with other skills bodies across the UK.

In an employer-led system, don't forget the apprentices
Every apprenticeship needs an apprentice and we need to ensure they are attracted and achieve great things through this experience. They can be a really powerful driver of apprenticeship growth because, quite simply, so

many of them are absolute stars. They provide both a stimulus and a reality check for apprenticeships, reminding us that we should be preparing them for a career and not simply their first job.

When I ran the National Apprenticeship Service, there were attempts to activate a network of Apprentice Ambassadors to help give direction to our work with young people. This baton has been picked up by other organisations, such as Youth Employment UK, who are clear that apprentice views need to be central to plans for apprenticeships. It will be important that the skills bodies overseeing the growth of high quality apprenticeships respond positively to this message.

The voices of young people need to be heard in order to help design experiences that they will really appreciate and compare well with friends who remain on the route to university. We often hear the voices of successful apprentices but we need to capture the thoughts of those who only see the barriers to apprenticeships and make other choices. How many of them end up as NEETs or unemployed graduates?

This will also help to drive quality. Many large employers have begun to recruit apprentices alongside their traditional graduate in-takes. They want the best of those young people entering the labour market whatever their age and preferred learning route. The talent of these apprentices and their proximity to graduates means that the standards in their apprenticeships are rising quickly to match the aspirations of these new apprenticeship cohorts.

Once in an apprenticeship, apprentices are able to promote what makes their success more likely including the support of mentors and buddies, networking with other apprentices and the enhanced pride that comes from public celebration events that mirror graduation events.

Apprentices are the best advocates of apprenticeships to other young people and to employers. As employers frequently said to me 'you mean I would get people like them to be my apprentice?'

They also force the system to be honest about its weaknesses e.g. how many people on vocational courses would prefer to be on an apprenticeship; and how easy is it to move from a level 2 apprenticeship all the way to degree level apprenticeships?

An approach to promoting apprenticeships is needed that transforms behaviours
While millions of pounds have been spent by government on marketing campaigns that have made a difference, responses to these campaigns have inevitably ebbed and flowed. The professional advice in this book recommends that an approach is needed that goes beyond temporary windows of intensive broadcasting, and the high intensity National Apprenticeship Week, that looks at a sustained approach influencing all parts of the apprenticeship system and, crucially, changes behaviours.

The world of apprenticeships is as exciting as the modern world of work and yet there is little appreciation of the opportunities to pursue careers such as accountancy and law through apprenticeships. Advice in this book makes it clear that a skilled approach to apprenticeships presents them not as an alternative to going to university but as a smart option in their own right.

There also needs to be a much lesser tolerance of those practices that give apprenticeships a bad name. Legislation to define an apprenticeship has the potential to help. More resources to ensure that employers and training providers who exploit the apprenticeship system are deterred, including on wages and on training that is not recognised by the individual, would stop so much of the good and hard work undertaken by everyone else from being undone.

Training providers will remain an important part of the solution under an employer led approach.
The growth of apprenticeships over the past decade has depended largely on the work of colleges and training providers. Their job has not been made any easier by many of the barriers to expansion outlined above, especially the complexity and changes surrounding funding and programme priorities.

The important role of training providers will undoubtedly continue. They are often the ones to introduce employers to apprenticeships and to helpfully encourage and nurture new employers as they navigate the apprenticeship system for the first time. For many SMEs, they are their HR department. I have no doubt that they will adapt and innovate as they always have in response to the latest set of changes. The increased interest from employers in skills as a result of the levy is a great opportunity.

Training providers have already largely adapted to a world in which apprenticeships are the main source of public funding. If the articles in this book are a guide, employers are looking for training providers who are able to go on the apprenticeship journey with them and value long term relationships. This potential for stable, maturing relationships can only be a good thing for quality delivery.

It is also clear that many employers are taking up the opportunities presented by trailblazers and the message of employer ownership to reconfigure their programmes, often at a higher level of skill. This commitment by employers will be welcomed by most training providers who will see it as an opportunity to engage, innovate and expand.

The movement towards higher apprenticeships will be challenging for some but the opportunity to engage with higher education institutions offers an exciting prospect. Indeed the next few years is likely to see the map of training providers changing considerably, not least with many more HEIs becoming serious players in apprenticeships. This will be an interesting challenge for funding agencies.

Conclusion
This book has demonstrated the importance of looking at apprenticeships in the round. In taking stock of Apprenticeships today and as the UK continues on its expansion plans, there is a lot of well-intended advice from those who are at the heart of apprenticeships.

They are clear that government, employers and all those with responsibility for apprenticeships in the UK must keep their main eye on transformational change and not just on the three million additional apprenticeships (for England anyway.)

Three million more apprenticeships is a very decent place to start. To use an aeronautical analogy, they are a decent altitude measure but we don't want just to go up. We need to go in the right direction. We therefore need to be studying the full instrument panel.

I believe that in order to assess progress towards transformational change in apprenticeships by 2020, the following are the most important:

1. The effects of the introduction of the levy on employer investment in skills – does it achieve its first goal of boosting investment?

2. More importantly, the outcomes. The impact that the reforms have on overcoming skills shortages, especially for those sectors that are critical for productivity plans and national infrastructure projects. Within the three million apprenticeships growth, these critical technical and construction skills shortages need to be addressed.

3. Evidence that young people of all abilities are genuinely able to consider the full range of career choices, with real intelligence about career prospects and vocational routes; and that this leads to significant growth in under-25s . Has this perennial problem been solved at last?

4. Stability in the institutions that govern the skills system or are critical to its successful operation, and clarity about their accountabilities. Will they survive beyond 2020?

5. Strong management of the skills system, including the discipline to only change that which is essential and to do this really well in the interests of employers and individuals; a determination to tackle anything that undermines the brand (e.g., not paying apprentice wages); and the ability to link the creative energies offered by local/regional initiatives to aid implementation and learning from others.

6. Simplification of the system, and improvements to the ways in which employers are able to engage, especially SMEs.

7. A programme to improve the overall quality of apprenticeships with enhanced teaching standards and effective learning methods at its heart.

8. Progression routes in place that enable people who aspire to do so and have the ability to move up the apprenticeship ladder as far as they can go,

including to higher and degree apprenticeships. This is vital if parents and high achievers are to be attracted to apprenticeships and employers are to get the higher skills needed to power their businesses.

9. Promotion of apprenticeships that will harness the creative strengths of professionals and led by employer and apprentice ambassadors to change attitudes and behaviours towards apprenticeships as they are today.

10. Capacity and capability of the sector, FE, HE and schools to work together in the interests of the employer and the apprentice.

Others will undoubtedly add to this list but this will be the basis on which I, and many others, will be judging the Race to the Top and looking to contribute to the solutions.

In monitoring progress, it will be vital to keep an eye on this wider set of outcomes that the government reforms are mostly seeking to address. These reforms are backed by most partners in the skills world, but need to go further in their impacts, not only achieving the ambition for three million more apprenticeships, but providing clear evidence of a change in the status of apprentices that makes each one rightly proud of their achievement, and employers confident that they have indeed found the best way of powering their future business prospects. We can all then be equally proud that we have a skills system that enables individual talents to be fully realised.

John Cridland describes a desirable future virtuous circle: one in which skills and productivity boost one another higher and higher as each succeeds and propels the other. An exciting and intoxicating journey.

A Race to the Top that we can all agree was well run.

INDEX

481

Index

Index

Lightning Source UK Ltd.
Milton Keynes UK
UKOW07n0759300616

277327UK00001B/1/P